Green Society, Environmental Strategies and Sustainable Development

Aziza Chakir • Rohit Bansal
Johanes Fernandes Andry
Fazla Rabby • Muhammad Hussain
Editors

Green Society, Environmental Strategies and Sustainable Development

Springer

Editors
Aziza Chakir
Faculty of Law, Economics and Social Sciences
Hassan II University
Casablanca, Morocco

Johanes Fernandes Andry
Universitas Bunda Mulia
Jakarta, Indonesia

Muhammad Hussain
School of Mechanical, Materials, Mechatronic and Biomedical Engineering
University of Wollongong
Wollongong, NSW, Australia

Rohit Bansal
Vaish College of Engineering
Rohtak, Haryana, India

Fazla Rabby
Stanford Institute of Management and Technology
Glenfield, NSW, Australia

ISBN 978-3-031-94085-9 ISBN 978-3-031-94086-6 (eBook)
https://doi.org/10.1007/978-3-031-94086-6

This Springer imprint is published by the registered company Springer Nature Switzerland AG
The registered company address is: Gewerbestrasse 11, 6330 Cham, Switzerland

Preface

In this age of unprecedented environmental crises, the time is ripe for a radical rethinking of our human–nature relationship. The consequences of climate change, resource depletion, and biodiversity loss not only endanger the natural world but also threaten the very fabric of our societies. Research and practice at the nexus of society, environment, development, and sustainability is perhaps the most important frontier and the key opportunity for re-mapping the future of humanity. This book is written to light up that junction, offering insights and solutions that could help us navigate the road to a more resilient and sustainable world.

In an era where the pace of change can feel overwhelming, the intersection of emerging technologies and sustainable practices presents a glimmer of hope. Leveraging data up to October 2023 and a commitment to innovation, we strive to overcome key environmental challenges and promote prudent energy consumption and eco-friendly production. By addressing both aspects, we can not only combat climate change but also foster social equity and boost economic development, ensuring individuals are progressively able to co-exist with their environment.

The following pages aim to achieve the following several goals. We initially wanted to raise public awareness of building a sustainable and green society. Equipping them with skills in a myriad of environmental practices, we can take the first steps toward instilling new habits that honor, replenish, and revive our earth. By doing this, we will offer our readers a holistic perspective of challenges faced today while also offering them actionable solutions that can enable sustainable development.

Additionally, this book highlights the benefits of implementing green policies and practices—not only to the environment but also for people and society. With one human being taking another life, we start to recognize how integrated economic, political, and environmental problems are: it is an interconnected web, not a linear cause and effect straight line. We hope to share successful environmental policies and inspiring sustainable development projects around the globe to encourage collaboration and innovation between all types of stakeholders: governments, corporations, NGOs, and individuals.

These chapters, like a reader, are not meant for passive consumption, you are bell hooks reading the content. You learn to give practical help to transform habits and lifestyle, reinforcing the principle that all small gestures contribute to a bigger social phenomenon.

We are in a moment that requires urgency, creativity, and cooperation, that we will be measured by in history. By doing so, we can measure a more sustainable future—remembering that thin threads connecting society to the environment. I hope this is the benefit that all the news, experience, and knowledge together can achieve: a better world.

Casablanca, Morocco — Aziza Chakir
Rohtak, Haryana, India — Rohit Bansal
Jakarta, Indonesia — Johanes Fernandes Andry
Glenfield, NSW, Australia — Fazla Rabby
Wollongong, NSW, Australia — Muhammad Hussain

Introduction

Addressing the urgency of the situation, we aim to provide a bridge toward the vast potential of a sustainable future. The urgent need to address environmental challenges has become a defining concern of the twenty-first century. As societies worldwide grapple with climate change and resource depletion, the pursuit of a sustainable future has never been more crucial. This book thus arises as a needed response to the call for a new plan on how to leverage novel technologies to solve some of the greatest challenges in history and bring coherence to the crises we face, by offering new tools to combat climate change, for sustainable energy generation, energy efficiency, sustainable mobility, and preservation of our ecosystems. In a post-industrial, consumer-driven world, we explore the edges of many worlds—lead the ecosensing—as we come to understand the carnaged result of our forms and our systems of life on the planet.

It is crucial to understand the societal and environmental interface in the context of energy and society. These activities are having a powerful impact on our climate, which goes beyond just the natural world. This leads to their interlinked nature and the importance of recognizing this relationship as we work toward a green society founded on sustainability in the environment and economic development.

This book has several central functions. The first is to increase the public awareness level of the importance of creating a sustainable and green society by applying environmental practices. We aim to educate readers on the current environmental challenges we face and the solutions we must pursue in order to achieve sustainable development. Green policies create many benefits for people, communities, and the planet.

In addition to that, *Green Society, Environmental Strategies and Sustainable Development* displays worldwide instances of positive environmental policies and sustainable growth strategies, providing valuable insights. At a time where the news is often disheartening and discouraging, these case studies are a ray of hope representing steps, not just the consideration of steps, for progress in the right direction. Our hope is that everyone—governments, corporations, NGOs, individuals—will work together to create a more sustainable, environmentally friendly planet.

This book is not just a series of theories, so to speak, but it is a call to action. This book includes helpful strategies and resources to motivate readers to practice and incorporate green habits and lifestyles into their daily lives, eventually cultivating a sustainable lifestyle. We can take steps now to be better prepared for the future—a future that is resilient and sustainable—and one that acknowledges the need for balance between humans and the world around them. Ultimately, this book invites readers to contribute to building a greener, healthier planet for future generations. This book is intended for policymakers, environmentalists, researchers, educators, and students seeking practical strategies for promoting sustainable development and fostering a greener society. It also serves as a valuable resource for industry leaders and community advocates dedicated to environmental stewardship.

Contents

Emerging Green Technologies and Renewable Energy Sources for Sustainable Eco-Innovation: Understanding the Concept of a Green Society

Ankur Kumar, Mansi Rawat, Priya Bhatnagar, and Zahid Hussain

1 Introduction

In today's era, people are much more focused on economic growth which is driven by their production activities that have undeniably brought prosperity and development in countries. But they have brought several issues related to environmental changes such as global warming, water pollution, and many more**.** Related to this, green economic transformation is emerging as a pivotal solution to obtain sustainable development strategies. While considering these transformations, GI allows the product or processes that lead the company to get a higher level of environmental sustainability while managing the economic viability (Qin et al., 2024). GI has gained worldwide popularity as it includes innovation in the technology, which aims to balance and save energy (Kumar et al., 2024b, 2024d, 2024e), pollution reduction, apply the process of recycling and reuse, and also balance economic growth and environmental protection to become an essential strategic choice for any organization. Several research were emphasized that profitability in economic and environmental suitability showed a great significance. This green innovation not only enhances the performance of the environment but also attains sustainable competitive benefits. These advantages showed a positive impact on the organization's

A. Kumar (✉)
SCEE, IIT, Mandi, India

M. Rawat
Graphic Era Deemed to Be University, Dehradun, India

P. Bhatnagar
Department of Psychology, Aryabhata College, Delhi, India

Z. Hussain
Department of Business Administration, Khadim Ali Shah Institute of Technology (KASBIT), Karachi, Pakistan

A. Chakir et al. (eds.), *Green Society, Environmental Strategies and Sustainable Development*, https://doi.org/10.1007/978-3-031-94086-6_1

growth rate and are also relevant with communities to integrate GI into their core strategies (Kumar et al., 2025). Higher economic growth has now been sustained in India and is also expected to continue in the future. People are now adopting green innovation that can open opportunities for sustainable growth (Lema & Lema, 2012). A country like India now accounts for a substantial amount of the production of the world's goods and services. Several studies were conducted in the past few years, and they demonstrated that GI can significantly increase business and research performance, which showed a positive impact on economic growth (Rahman et al., 2023). It helps to enhance the market position, attract eco-friendly customers, offer green services, and gain competitive benefits because these advantages were focusing on businesses, several organizations, and researchers (Schiederig et al., 2012). However, the growing rate of environmental contamination caused by industries can create complexity and significantly affect the business operation. Thus, total factor productivity (TFP) (Bansal et al., 2024a, 2024b), which considers environmental and energy resources, can positively work on sustainable development as it is managing environmental practices effectively (Wang et al., 2021). GI addresses these challenges by applying good technical advancement, including material and energy efficiency, recycling of product, reduction in pollution, green product development (GPD), and managing environmental practices effectively.

Still, the implementation of green technology is not easy. Companies and organizations are required to work extensively with external stakeholders, suppliers, and customers to conceive the important information, resources, and other expertise required. This collaborative work is important for product development and processing that meet the desired sustainability goals (Bansal et al., 2024a, 2024b). Moreover, under rising environmental concern, GI has emerged as a crucial strategic tool to achieve organizational and economic success in financial and ecological terms. Studies related to GI technology not only consist of a conceptual framework for sustainable strategies but are also highly impactful in balancing financial considerations and include environmentally sustainable innovation. An organization can address the climatic challenge and assure long-term growth by working on a green technology, support, and environmentally sustainable project. This research highlights the significance of minimizing waste and using resources more effectively (Raja et al., 2025). This study combines data analysis with a review of existing research to give a complete picture of how GI can support sustainability. It also raises important questions for future research, encouraging the use of GI to promote circularity and sustainability globally. This makes it a valuable guide for researchers, policymakers, business leaders, and practitioners looking to address climate challenges and promote sustainable development (Liu, 2022).

2 Key Areas of Green Technology Innovation

2.1 Renewable Energy Sources

Efficient ways to reduce emissions in the energy sector include improving the efficiency of coal-fired power plants, expanding the use of gas and nuclear energy, increasing the adoption of renewable energy sources, and encouraging consumers to use energy more rationally (Shan et al., 2021). Beyond environmental benefits, renewables enhance energy independence, save foreign currency, and create jobs. In Slovakia, renewables (excluding large hydroelectric power plants) currently play a minor role in the energy mix, accounting for only 3.5% of total energy consumption and contributing about 16% to electricity generation (Satar et al., 2023). While nuclear and thermal power plants will remain dominant, the global shift toward clean energy highlights the need for increased renewable use (Matuszewska-Janica et al., 2021). Achieving this requires supportive legislative and economic measures, such as incentive pricing, subsidies, tax relief, and investment in research and domestic manufacturing (Painoli et al., 2021). According to the research, India is implementing one of the largest programs in renewable energy. India ranks second in biogas utilization and fifth in wind power energy that is derived from natural source. Renewable sources contribute to about 5% of the total power generating capacity in India (Kothari et al., 2021). India's renewable energy expansion drives economic growth, enhances energy security, and mitigates climate change while ensuring affordable and sustainable energy access. Strong government policies and favorable market conditions have positioned India as a global leader in renewable energy. This sector not only boosts electricity generation but also creates significant employment opportunities (Bansal et al., 2022).

Table 1 shows that various key technologies and solutions significantly contribute to building a green society by promoting sustainability across multiple sectors. Renewable energy sources such as solar, wind, and geothermal reduce reliance on fossil fuels and lower carbon footprints. Advanced energy storage solutions, including lithium-ion and hydrogen storage (Bansal et al., 2023), improve energy efficiency and grid stability. Smart grids enhance energy distribution through IoT-based management, minimizing waste. Green infrastructure innovations, such as smart cities and carbon-negative cement, support sustainable urbanization. Sustainable transport technologies like EVs and hydrogen fuel cells reduce emissions and encourage cleaner mobility. AI-powered waste management and biodegradable materials help minimize landfill waste and promote a circular economy. In agriculture, precision farming and AI-driven irrigation conserve resources and enhance food security. Water conservation technologies, including smart irrigation and wastewater recycling, ensure a sustainable water supply. Carbon capture methods like DAC and BECCS help combat climate change by removing CO_2 from the atmosphere. Lastly, eco-innovation through AI-driven sustainability and green R&D fosters new environmental solutions and sustainable industries (Majid, 2020).

Table 1 Emerging green technologies and renewable energy sources

Category	Key technologies and solutions	Impact on green society
Renewable energy	Solar, wind, hydro, geothermal, biomass energy	Reduces dependence on fossil fuels, lowers carbon footprint
Energy storage	Lithium-ion, solid-state batteries, hydrogen storage	Enhances energy efficiency and grid stability
Smart grids	IoT-based energy management, decentralized power systems	Optimizes energy distribution and minimizes waste
Green infrastructure	Smart cities, self-healing concrete, carbon-negative cement	Promotes sustainable urbanization and eco-friendly buildings
Sustainable transport	Electric vehicles (EVs), hydrogen fuel cells, hyperloop	Reduces emissions and promotes cleaner mobility
Waste management	AI-powered recycling, waste-to-energy, biodegradable materials	Minimizes landfill waste and promotes circular economy
Agriculture	Precision farming, hydroponics, AI-driven irrigation	Conserves resources and ensures food security
Water conservation	Smart irrigation, desalination, wastewater recycling	Reduces water wastage and ensures sustainable supply
Carbon capture	Direct air capture (DAC), carbon sequestration, BECCS	Helps mitigate climate change by removing CO_2
Eco-innovation	AI-driven sustainability, biodegradable products, green R&D	Encourages sustainable industries and new environmental solutions

2.2 *Energy Conservation or Energy Efficiency*

Energy conservation and efficiency play a crucial role across all sectors of the economy. It is often regarded as the most cost-effective way to acquire additional energy, as it focuses on minimizing energy wastage without compromising productivity or human comfort (Ao et al., 2023). Significant reductions in energy consumption can be achieved by optimizing lighting levels in residential, commercial, and industrial settings. Enhancing efficiency not only lowers costs but also reduces resource consumption and environmental impact (Raberto et al., 2019).

Studies emphasize that gaining a deeper understanding of a plant's thermodynamic behavior can help identify key areas for efficiency improvement. Research findings indicate that the overall energy efficiency, measured as the ratio of net electrical energy output to coal energy input, stands at approximately 37%, while the exergy efficiency is around 36%.

Energy intensity plays a key intermediary role, particularly in middle- and high-income nations (Mittal & Bansal, 2023). These insights provide policy guidance for integrating digital economy growth with climate strategies, accelerating low-carbon transitions, and supporting China's carbon peaking goals. According to the research findings, energy efficiency has emerged as a key policy strategy to address the twin challenges of energy security and climate change. Energy efficiency reflects the inclusive development and consumption of energy to a large extent (Yasmeen et al., 2023). Other researchers reported that digital economy growth shows a powerful impact on carbon emission reduction which is crucial for global climate governance and sustainable economic development (Kumar et al., 2024a). Their research shows that digital economy expansion significantly curbs carbon emissions, with the effect varying by income level and influenced by energy structure and efficiency. Energy intensity plays a key intermediary role, particularly in middle- and high-income nations (Mei et al., 2023).

2.3 Sustainable Transportation

The sustainability of transportation systems is a key concern in urbanization, with increasing air pollution in major cities highlighting the issue (Bickman, 2020). As population growth and migration drive the use of public and personal vehicles, the environment suffers (Dubey & Gunasekaran, 2015). Transportation plays a crucial role in economic, social, and environmental sustainability, especially in urban areas. Green transportation (GT) minimizes negative impacts on health and the environment by incorporating renewable fuels, electric vehicle technology, biogas-powered buses, and efficient public transit systems. Effective GT systems can reduce risks, traffic congestion, and pollution while enhancing safety, energy sustainability, and traffic flow (Shah et al., 2021). Some studies also mention that the transport sector reflects the largest increase in the CO_2 emissions among all. However, it is also a major contributor to global CO_2 emissions, accounting for around 15% of emissions today and projected to rise to 30–50% by 2050 (Sharma et al., 2024a, 2024b). Technological innovation offers a promising pathway to mitigate these impacts by enabling zero-carbon technologies, renewable energy adoption, and efficient transportation systems (Awan et al., 2022). Moreover, the sustainability of transport in the cities is linked to high-quality implementation of innovative systems and to the need to gain public trust. Other study reported presents sustainable transport solution by enhancing EV charging efficiency through the integration of photovoltaic (PV) systems with AC–DC power factor correction (PFC).This approach not only optimizes energy utilization but also improves power quality by reducing harmonic distortion and maintaining a unity power factor (Gurung et al., 2022). By increasing efficiency and reliability, this system advances sustainable EV charging infrastructure, contributing to cleaner and more energy-efficient transportation (Kumar et al., 2024b, 2024d, 2024e).

2.4 Circular Economy Practices

The circular economy (CE) operates as a closed-loop supply chain system that prioritizes regeneration and restoration (Sharma et al., 2024a, 2024b). In contrast, the sharing economy, a branch of the digital economy, utilizes data as a core production factor to grant users temporary access to physical and digital resources, catering to their specific needs efficiently (Lim, 2020; Chen & Wang, 2019).

This study focuses on the CE framework, which seeks to shift the traditional linear model of production, consumption, and disposal toward a more resource-efficient circular system (Kumar et al., 2024b, 2024d, 2024e). The primary goal of CE practices is to support the three pillars of sustainability—environmental, social, and economic (Khan et al., 2022). Research suggests that the effective implementation of CE strategies enhances financial and environmental outcomes (Dawoodbhoy et al., 2021). Specifically, these strategies emphasize optimizing resource utilization and streamlining product delivery, ultimately improving product quality, and boosting firms' operational efficiency.

By adopting CE principles, businesses can improve environmental performance and encourage sustainable consumption. The integration of circular practices leads to a reduction in hazardous chemicals and pollutants while promoting resource conservation through recycling and remanufacturing. Additionally, CE strategies contribute to cost reduction, waste minimization, productivity enhancement, and environmental preservation. Research highlights CE as an effective tool for sustainable business growth, as it reduces reliance on resource-intensive consumption patterns (Ghisellini et al., 2016). The collaboration between industries in adopting CE practices fosters economic, environmental, and operational advantages, significantly influencing organizational performance (Sarfraz et al., 2021).

This study focuses on the CE concept, which seeks to transform the traditional linear manufacturing approach of "produce, use, and dispose" into a circular model that maximizes resource utilization. At its core, CE practices aim to achieve the triple bottom line of sustainability, encompassing environmental, social, and economic objectives (Khan et al., 2022). Research demonstrates that adopting CE practices can significantly improve both financial and environmental outcomes. Specifically, these practices prioritize the efficient delivery of products and optimal resource utilization, leading to enhanced product quality and improved operational performance for firms. Moreover, CE adoption bolsters environmental performance by promoting sustainable consumption. It achieves this by minimizing the use of hazardous chemicals, reducing pollutants, and conserving resources through recycling and remanufacturing. By cutting production costs, lowering system waste, and increasing productivity, CE practices contribute not only to environmental preservation but also to economic gains. One study highlights CE as a powerful tool for sustainable organizational growth, as it mitigates (Palta & Kumar, 2024).

Resource-intensive consumption (Ghisellini et al., 2016): This suggests that the integration of CE practices, particularly through collaboration among industries, drives substantial economic, environmental, and operational benefits, thereby enhancing overall organizational performance.

Figure 1 shows the interconnection between green technology, green innovation, and green culture in promoting sustainability. The diagram categorizes key elements under each domain. Green technology includes renewable energy sources, sustainable transport, waste management, and energy efficiency, all aimed at reducing environmental impact. Green innovation highlights advancements such as green

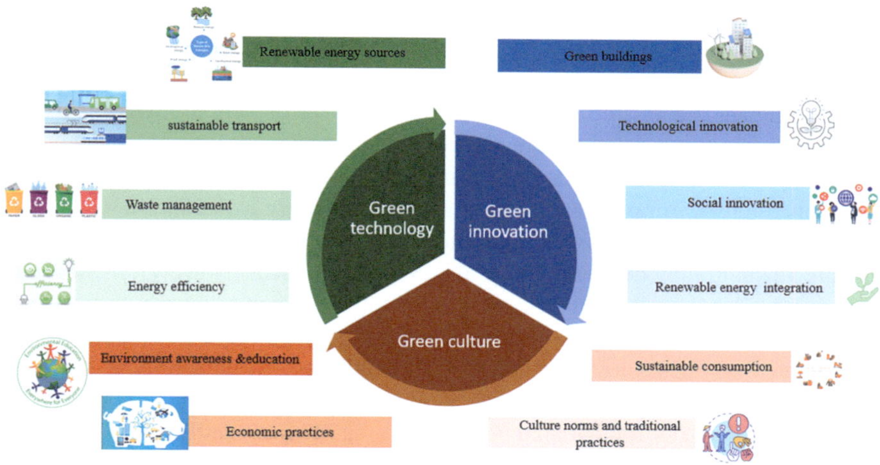

Fig. 1 Green technology, green innovation, and green culture

buildings, technological and social innovation, and renewable energy integration, emphasizing the role of innovation in sustainability. Green culture focuses on environmental awareness and education, economic practices, sustainable consumption, and traditional cultural norms that support eco-friendly living. The figure illustrates how these three pillars collectively drive a greener and more sustainable society.

2.5 *Green Building and Infrastructure*

Policies like Leadership in Energy and Environmental Design (LEED) offer a framework that integrates green building certification with renewable energy solutions. This approach promotes sustainability by enhancing energy efficiency, reducing operational costs for both property owners and occupants, and encouraging the adoption of renewable energy. Collaboration with policymakers and regulatory bodies plays a crucial role in advancing green building initiatives and supporting the widespread implementation of sustainable construction practices (Cole et al., 2017). According to the study, researcher concluded that green building and sustainability require integrating various aspects of a project from the start to ensure success. Key principles include staying updated on sustainable technologies, designing for both human and environmental benefits, choosing eco-friendly materials, reducing material use, minimizing community impacts, and maintaining existing resources. These practices promote environmentally conscious and long-lasting construction solution. Researcher studies have shown that accessible green spaces improve mental and physical well-being by offering opportunities for relaxation, exercise, and social connection, even amidst restrictions. This highlights the need for urban planning to integrate GI, such as pocket parks and green corridors, into urban landscapes. Just like free-flowing, sustainable buildings, and infrastructure, accessible green spaces foster adaptability and improve the quality of urban life (Pamukcu-Albers et al., 2021). Sustainable building materials are essential for creating eco-friendly and energy-efficient green buildings. By reducing resource depletion, lowering energy use, and minimizing waste, these materials help protect the environment while enhancing building performance and occupant well-being. Embracing sustainable materials in green infrastructure paves the way for a more resilient and sustainable future (Gore et al., 2023). Green infrastructure (GI) offers a vital solution for urban climate adaptation by delivering diverse ecosystem services. However, its full integration into city planning faces challenges, including design standards, regulations, socio-economic factors, financing, and innovation. Addressing these requires a people-centered, context-specific approach, supported by strong political will and stakeholder engagement. These elements are crucial for securing public support, funding, and long-term GI maintenance, ensuring resilient and sustainable urban development (Zuniga-Teran et al., 2020).

Figure 2 visualizes the adoption and impact of emerging green technologies by 2030 across key sustainability sectors. The bar graph highlights the expected growth in various sectors contributing to environmental sustainability. Renewable energy is projected to see the highest adoption rate, reaching approximately 85%, as solar, wind, and other renewable sources continue to replace fossil fuels. Electric vehicles (EVs) follow with a 75% growth expectation, driven by advancements in battery technology and supportive policies. Smart grids are anticipated to expand by 70%, optimizing energy distribution and improving efficiency through IoT and AI integration. Waste recycling is expected to grow by 65%, reflecting increased efforts in circular economy practices and AI-powered waste management. Water conservation technologies, such as smart irrigation and desalination, are projected to grow by 60%, ensuring sustainable water use. Lastly, carbon capture utilization is expected to increase by 55%, playing a crucial role in mitigating climate change by capturing

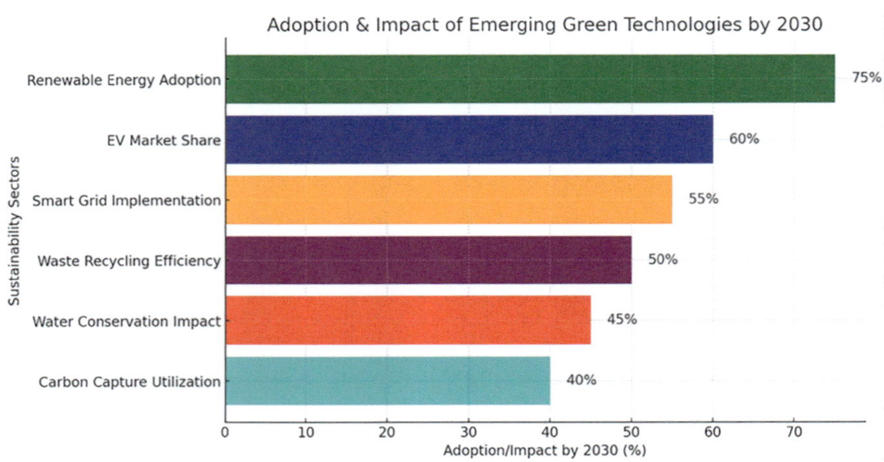

Fig. 2 Adoption and impact of emerging green technologies by 2030

and storing CO_2 emissions. This figure underscores the significant advancements in green technologies and their vital role in fostering a sustainable future (Palta & Kumar, 2025).

3 Emerging Technologies and Trends

3.1 Artificial Intelligence and IoT for Energy Management

In smart environments, managing energy efficiently is a big challenge, especially for businesses with remote branches. These systems need to scale easily and work on their own by learning patterns from energy data and applying the best practices automatically (Singh et al., 2024). They should also adapt to changes in the system without human involvement to ensure optimal energy use with promising results. The blend of IoT and machine learning is not novel per se, as it is a natural way of considering the need of a system to be autonomous for energy saving (Tomazzoli et al., 2023). The study indicates that the utilization of AI and IoT enhances domestic energy efficiency through the development of cutting-edge, evidence-based, AI-driven energy efficiency recommendation systems (RSs) (Sayed et al., 2022). IoT devices are essential for real-time digital twin (DT) applications and energy management improvements. However, using IoT in real-world smart and connected systems is still limited by challenges like short device battery life, limited network capacity, and high maintenance costs (Agostinelli et al., 2021). The benefit of IoT devices is that they enable automation; however, a significant amount of energy is required for billions of connected devices to communicate with each other. IoT-enabled sensors and devices provide real-time data on energy consumption, facilitating predictive analytics and adaptive control mechanisms. In smart cities and villages, IoT-driven solutions contribute to sustainability by optimizing energy distribution and reducing waste, aligning with SDGs and promoting inclusive socio-economic growth (Chui et al., 2018). Furthermore, IoT enhances ICT-driven policymaking by enabling data-driven decisions, ensuring efficient energy supply and demand management. At the intersection of these domains, IoT-powered systems enable smarter resource consumption through intelligent automation, algorithms, and seamless data integration (Khan et al., 2024).

Figure 3 illustrates the block diagram of an IoT (Internet of Things) system, highlighting its key components and their interactions. The IoT system consists of four main elements:

Device Layer: This includes various IoT-enabled devices such as sensors, actuators, and computing devices that collect and transmit data.

Network Layer: It ensures seamless communication between IoT devices using wired or wireless networks like Wi-Fi, Bluetooth, or cellular networks.

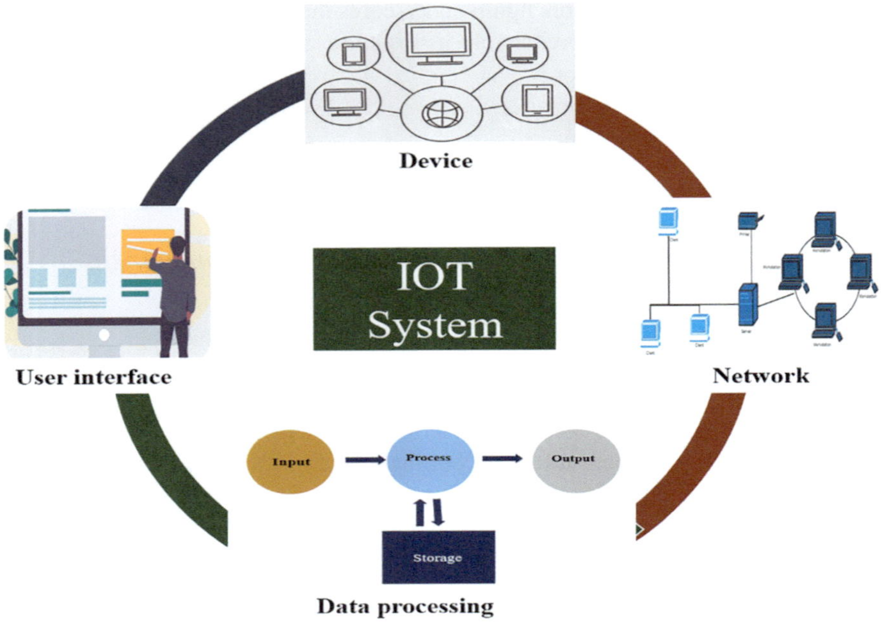

Fig. 3 Block diagram of IOT system

Data Processing: The collected data undergoes processing, which includes input, processing, storage, and output. This step is crucial for analyzing and deriving insights from the data.

User Interface: The processed information is presented to users through interactive dashboards, applications, or automated systems, enabling decision-making and control.

3.2 Green Nanotechnology

Green nanotechnology (GNT) plays a crucial role in achieving sustainable development by balancing social progress, environmental protection, and economic growth (Singh et al., 2024). It contributes to eco-friendly nanomaterials and nanodevices that reduce pollution, conserve resources, and enhance energy efficiency. GNT supports the circular economy by minimizing waste and promoting responsible production and consumption. Additionally, it aids in environmental cleanup, efficient water management, and sustainable agriculture through advanced nanomaterials. By integrating informatics and decision-making systems, GNT enhances the assessment and synthesis of nanomaterials, aligning with the 2030 SDG targets and fostering a resilient, low-carbon future (Khan, 2020a, 2020b). Recent theoretical and experimental reviews on nanotechnology and nanomaterials highlight their potential for future technologies, but their production and application face significant challenges. One major issue is that defects can compromise the fundamental properties of nanomaterials, affecting their performance (Tran et al., 2013). For instance, carbon nanotubes, known for their exceptional strength, can suffer from reduced tensile strength due to impurities, discontinuous tube lengths, and random orientation. Another critical challenge is the high cost of synthesizing nanomaterials, as their production often requires specialized equipment and controlled environmental conditions, limiting large-scale manufacturing and widespread adoption (Srivastava & Bhargava, 2022). Green nanotechnology ensures safety by mitigating environmental threats and promoting sustainable practices. Predictive analysis suggests that it is an effective and secure technology, contributing to 13 out of the 17 UN Sustainable Development Goals (SDGs) within the 2030 timeframe (Aithal & Aithal, 2021). With the growing global population, resource scarcity and environmental degradation have become critical challenges, necessitating pollution-free technologies for sustainability. Green nanotechnology emerges as a promising solution, integrating nanoscience with eco-friendly practices to develop cleaner technologies for environmental remediation and clean energy (Khan, 2020a, 2020b).

3.3 Carbon Capture and Storage (CCS)

Carbon capture and storage (CCS) is a crucial three-stage process involving carbon capture, transport, and storage to mitigate CO_2 emissions. In India, around 64 districts contribute to nearly 60% of total emissions, posing challenges for sustainable socio-economic growth (Shaw & Mukherjee, 2022). While India has significant geological sequestration potential in basaltic rocks, coal seams, and saline aquifers, no large-scale CCS projects are currently operational. International collaborations are helping India advance its CCS technology, making the next 10–15 years critical for developing and deploying large-scale carbon capture initiatives (Bui et al., 2018). Carbon capture and storage (CCS) is a crucial mitigation option for reducing greenhouse gas emissions in hard-to-abate sectors like cement, steel, and chemicals. Using the MIT EPPA model, we analyze global industrial CCS deployment under a 2 °C climate target. Without CCS, mitigation costs rise significantly, highlighting its role in enabling energy-intensive industries to cut emissions while sustaining growth (Paltsev et al., 2021). Carbon capture from flue gases occurs through three pathways: pre-combustion, oxy-combustion, and post-combustion. In pre-combustion capture, technologies like IGCC and IRCC convert fuel into syngas, separate CO_2, and produce a cleaner energy source (Allangawi et al., 2023). To make CCS technologies viable for large-scale applications, technological and economic improvements are needed, particularly in gas–liquid absorption, a widely used CO_2 capture method. IGCC (integrated gasification combined cycle) with CCS offers a promising low-cost, high-efficiency solution by enabling poly-generation and utilizing diverse fuel sources, thereby enhancing both carbon capture potential and energy security (Cormos, 2012). Carbon capture and storage (CCS) and biological carbon capture and utilization (bio-CCU) are two urgent mitigation options that researchers have found to be necessary in response to the increasing global temperatures caused by CO_2 emissions. While CCS is crucial for large-scale emission reduction in industries, microalgal bio-CCU in a biorefinery model presents an eco-friendly alternative, turning waste CO_2 into biofuels and valuable products. Advancements in reactor engineering and CO_2 delivery mechanisms aim to enhance capture efficiency, making bio-CCU a complementary solution alongside CCS for achieving sustainable carbon management (Sen & Mukherjee, 2024).

Fig. 4 illustrates the projected growth of key green society sectors from 2025 to 2045, highlighting the increasing adoption of renewable energy, sustainable transport, waste management, water conservation, and carbon reduction. The graph demonstrates a steady rise in these sectors as technological advancements, policy initiatives, and environmental awareness drive their expansion. Renewable energy is expected to see the highest adoption, reflecting the global shift toward clean energy sources like solar and wind. Sustainable transport, including electric vehicles and hydrogen fuel cells, shows consistent growth due to improved infrastructure and regulations. Waste management and water conservation also exhibit a rising trend, emphasizing the importance of recycling, circular economy models, and efficient water usage. Carbon reduction strategies steadily increase, driven by carbon capture technologies and green building solutions. Overall, the graph underscores a clear trajectory toward a greener future, demonstrating the progressive impact of sustainability-focused initiatives over the next two decades.

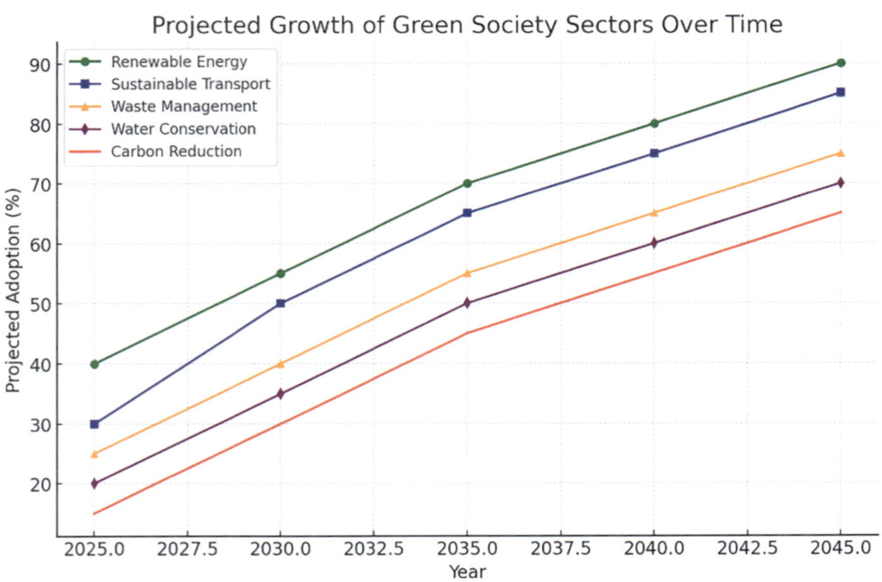

Fig. 4 Projected growth of key green society sectors (2025–2045)

3.4 Bioenergy and Algae-Based Fuels

The alarming rise in atmospheric CO_2 necessitates effective mitigation strategies, with algae-based bioenergy emerging as a sustainable solution. Algal cultivation not only captures CO_2 but also serves as a feedstock for biofuels, biogas, and other value-added products, reducing dependence on fossil fuels. Advanced bioelectrochemical systems (BESs) further enhance CO_2 sequestration while enabling bioenergy production and wastewater treatment. Optimizing these processes can make large-scale algae-based biofuels a viable alternative for a low-carbon future (Li & Yao 2024).

Algae is a promising raw material for biofuel production due to its rapid growth, high lipid content, and versatility in producing biodiesel, biogas, and bioethanol. While challenges like high costs and large-scale harvesting remain, advancements in cultivation and processing can enhance biofuel production, promoting environmental sustainability and resource conservation (Kumar & Ghangrekar, 2024). Algae is an emerging renewable energy source with high potential in biofuel production and biorefineries. Its high biomass yield, low land requirement, and valuable co-products make it a sustainable alternative, while its role in wastewater treatment further enhances its viability in biofuel and other industries (Saran et al., 2024). Algae-based biofuels, a third-generation renewable energy source, offer a sustainable alternative to fossil fuels by addressing environmental pollution and reducing the food versus fuel debate. Cultivated in closed, open, or hybrid systems, algae provide economic, social, and environmental benefits, making them a promising solution for future energy needs (Mahmood et al., 2023).

Fig. 5 graph illustrates the expected distribution of renewable energy sources in the energy mix by 2030. The data highlights the dominance of solar and wind energy, which are anticipated to be the primary contributors due to advancements in technology, decreasing costs, and supportive policies. Hydropower remains a significant contributor, leveraging existing infrastructure, while geothermal and biomass energy play smaller but essential roles in diversifying the renewable energy landscape. This projection underscores the global transition toward cleaner energy sources and the increasing reliance on sustainable alternatives to reduce carbon emissions.

4 Economic and Social Impacts

Innovation in green technology and biodiversity conservation plays a crucial role in driving economic growth and social well-being. Advancements in renewable energy, sustainable agriculture, and eco-friendly manufacturing create new job opportunities, enhance resource efficiency, and reduce environmental degradation. Biodiversity preservation supports ecosystem services, improves food security, and promotes ecotourism, contributing to economic resilience. Socially, these

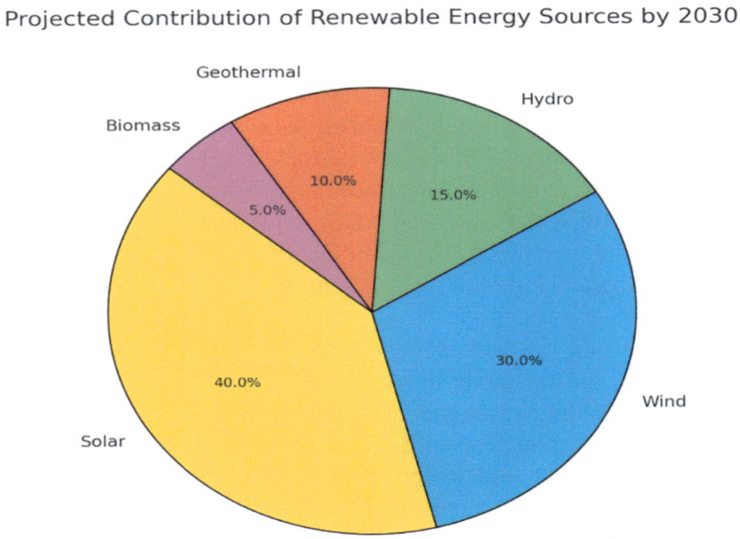

Fig. 5 Projected contribution of renewable energy sources by 2030

innovations improve public health, ensure sustainable livelihoods, and foster a greener, more inclusive future (Qamruzzaman & Karim, 2024). The transition to cleaner energy production is crucial for resource-efficient manufacturing, with significant economic and social benefits. Green energy sector holds vast potential for growth, yet it remains underutilized due to limited government and private sector involvement. Expanding green manufacturing can drive economic development, create jobs, and improve public health by reducing environmental damage, fostering a more sustainable and prosperous society (Rana et al., 2021). Social innovation in biodiversity conservation has significant economic and social impacts by promoting sustainable land use, responsible resource management, and climate resilience. Civic-driven initiatives, policy innovations, and consumer–producer collaborations create new economic opportunities, support green jobs, and enhance ecosystem services that benefit communities. Additionally, participatory governance and nature-based solutions improve social well-being by fostering inclusive decision-making and environmental stewardship, ensuring a sustainable and prosperous future (Ziegler et al., 2022). The transition to renewable energy has profound economic and social impacts, driving sustainable development, job creation, and energy security while reducing environmental harm. Challenges like infrastructure costs, policy barriers, and public acceptance must be addressed through government and industry collaboration. Investments in biorefinery technologies, wind, solar, and hydropower not only stimulate economic growth but also enhance social well-being by providing cleaner energy, improving public health, and ensuring equitable access to resources. A shift toward renewable energy fosters economic resilience and a more sustainable, inclusive society (Jaiswal et al., 2022). The integration of digitalization and green technology innovation plays a crucial role in driving economic and social growth. Digital advancements enhance green economic growth by promoting green technology innovation, optimizing industrial structures, and improving resource efficiency. The spatial spillover effect further amplifies regional economic benefits, fostering sustainable development and environmental resilience. Investments in digital infrastructure and policies supporting green energy consumption can accelerate innovation, create jobs, and enhance social well-being, ultimately leading to a more sustainable and inclusive economy (Hao et al., 2023).

Figure 6 bar graph illustrating the anticipated reduction in environmental impact by 2030, driven by emerging green technologies and sustainable eco-innovation. The data showcases improvements in key areas, including CO_2 emissions, fossil fuel reliance, water conservation, air quality, deforestation rates, and plastic waste reduction.

Table 2 outlines the major challenges hindering the adoption of green technologies and renewable energy sources, along with their causes, innovative solutions, and impacts on sustainability. High initial costs, caused by expensive R&D and infrastructure, can be mitigated through government incentives and green financing, encouraging wider adoption. Energy storage limitations due to renewable intermittency can be addressed with advanced battery technologies and hydrogen storage, ensuring a reliable energy supply. Traditional grids struggle with renewable integration, but smart grids and AI-driven energy distribution optimize energy use and minimize power loss (Kumar et al. 2024c). The slow adoption of green transport, primarily due to a lack of EV charging infrastructure, can be improved by expanding charging networks and battery swapping, boosting EV usage and reducing carbon emissions. Waste management inefficiency, driven by poor recycling and public awareness, can be countered with AI-driven sorting and circular economy policies, significantly reducing environmental pollution (Beres et al. 2021). Water scarcity, resulting from overuse and pollution, can be tackled through smart irrigation, desalination, and wastewater recycling, ensuring sustainable water use. Rising carbon

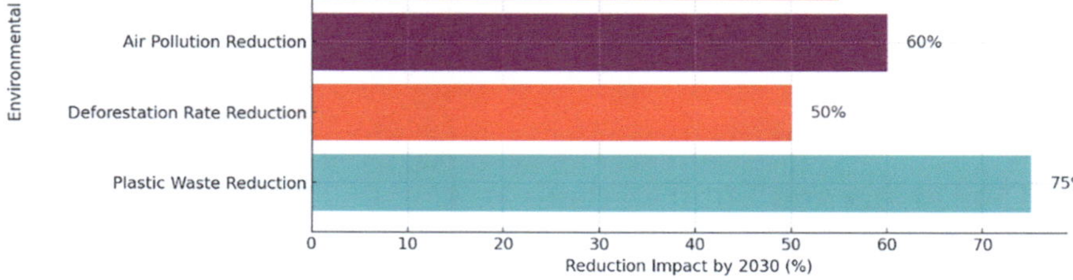

Fig. 6 Projected environmental impact reduction by 2030

Table 2 Challenges and corresponding eco-innovative solutions for emerging green technologies and renewable energy sources

Challenge	Cause	Solution/innovation	Impact on green society
High initial costs	Expensive R&D and infrastructure investment	Government incentives, subsidies, green financing	Encourages adoption of sustainable technologies
Energy storage limitations	Intermittency of renewable sources	Advanced battery technologies, hydrogen storage	Ensures reliable renewable energy supply
Grid integration issues	Traditional grids not designed for renewables	Smart grids, AI-driven energy distribution	Optimizes energy use and reduces power loss
Slow adoption of green transport	Lack of EV charging infrastructure	Expansion of charging networks, battery swapping	Increases EV usage and reduces carbon emissions
Waste management inefficiency	Poor recycling and lack of awareness	AI-driven sorting, circular economy policies	Reduces landfill waste and environmental pollution
Water scarcity	Overuse and pollution of freshwater sources	Smart irrigation, desalination, wastewater recycling	Ensures sustainable water use and conservation
Carbon emissions and climate change	Industrial pollution and fossil fuel dependency	Carbon capture, renewable energy adoption	Mitigates global warming and environmental degradation
Lack of public awareness	Limited education on sustainable practices	Eco-education, awareness campaigns, sustainability policies	Encourages responsible environmental behavior
Land use conflicts	Competing needs for urbanization, farming, and energy	Vertical farming, offshore wind farms	Balances land use and promotes eco-friendly solutions

emissions and climate change, caused by industrial pollution and fossil fuel dependency, necessitate carbon capture and renewable energy adoption to curb environmental degradation. Public unawareness of sustainability issues, a key barrier, can be resolved through eco-education, awareness campaigns, and sustainability policies, fostering responsible environmental behavior. Finally, land use conflicts between urbanization, agriculture, and renewable energy can be managed with vertical farming and offshore wind farms, optimizing land use while promoting eco-friendly solutions. Together, these innovations drive a greener, more sustainable future.

5 Conclusion and Future Trends

Green technology innovation and biodiversity conservation are pivotal for driving sustainable development, mitigating climate change, and ensuring long-term economic and social well-being. Advancements in renewable energy, energy efficiency, sustainable transportation, and circular economy practices contribute to resource conservation, emissions reduction, and economic resilience. The integration of emerging technologies such as artificial intelligence, IoT, green nanotechnology, and carbon capture further enhances pollution control, energy management, and sustainable resource utilization. Additionally, bioenergy and algae-based fuels present promising alternatives to fossil fuels, reinforcing the shift toward low-carbon solutions. However, the large-scale adoption of these innovations requires strong policy frameworks, increased investment in research and development, and collaboration across industries. Biodiversity conservation plays an equally significant role in sustaining ecosystem services, promoting food security, and fostering ecotourism, all of which contribute to economic stability and climate resilience. Social innovation in conservation efforts, including participatory governance and nature-based solutions, fosters inclusive decision-making and environmental stewardship. Moreover, digitalization enhances green economic growth by optimizing industrial structures, improving resource efficiency, and accelerating innovation. Looking ahead, future trends in green society will focus on expanding renewable energy, strengthening circular economy models, integrating digital green technology, and investing in sustainable infrastructure. The transition to a green economy is essential for balancing economic development with environmental protection, ensuring cleaner energy sources, reducing waste, and enhancing social well-being. Achieving a sustainable future demands collective efforts from governments, industries, and individuals to embrace and promote green solutions. By prioritizing sustainability and innovation, societies can ensure economic resilience, environmental protection, and a healthier planet for future generations.

Acknowledgements We wish to extend our heartfelt thanks to all those whose valuable contributions have enhanced this manuscript.

References

Agostinelli, S., Cumo, F., Guidi, G., & Tomazzoli, C. (2021). Cyber-physical systems improving building energy management: Digital twin and artificial intelligence. *Energies, 14*(8), 2338.

Aithal, S., & Aithal, P. S. (2021). Green nanotechnology innovations to realize UN sustainable development goals 2030. *International Journal of Applied Engineering and Management Letters (IJAEML), 5*(2), 96–105.

Allangawi, A., Alzaimoor, E. F., Shanaah, H. H., Mohammed, H. A., Saqer, H., El-Fattah, A. A., & Kamel, A. H. (2023). Carbon capture materials in post-combustion: Adsorption and absorption-based processes. *C, 9*(1), 17.

Ao, L., Bansal, R., Pruthi, N., & Khaskheli, M. B. (2023). Impact of social media influencers on customer engagement and purchase intention: A meta-analysis. *Sustainability, 15*(3), 2744.

Awan, A., Alnour, M., Jahanger, A., & Onwe, J. C. (2022). Do technological innovation and urbanization mitigate carbon dioxide emissions from the transport sector? *Technology in Society, 71*, 102128.

Bansal, R., Pruthi, N., & Singh, R. (2022). Developing customer engagement through artificial intelligence tools: Roles and challenges. In *Developing relationships, personalization, and data herald in marketing 5.0* (pp. 130–145). IGI Global.
Bansal, R., Shukla, G., Gupta, A., Singh, A., & Pruthi, N. (2023). Optimizing augmented reality and virtual reality for customer engagement. In *Promoting consumer engagement through emotional branding and sensory marketing* (pp. 24–35). IGI Global.
Bansal, R., Saini, S., Ngah, A. H., & Durga Prasad, T. (2024a). Proselytizing the potential of influencer marketing via artificial intelligence: Mapping the research trends through bibliometric analysis. *Cogent Business & Management, 11*(1), 2372889.
Bansal, R., Martinho, C., Pruthi, N., & Aggarwal, D. (2024b). From virtual observations to business insights: A bibliometric review of netnography in business research. *Heliyon, 10*(1), 1.
Beres, L. K., et al. (2021). Mobile ecological momentary assessment and intervention and health behavior change among adults in Rakai, Uganda: Pilot randomized controlled trial. *JMIR Formative Research, 5*(7), 1–12. https://doi.org/10.2196/22693
Bickman, L. (2020). Improving mental health services: A 50-year journey from randomized experiments to artificial intelligence and precision mental health. *Administration and Policy in Mental Health, 47*(5), 795. https://doi.org/10.1007/s10488-020-01065-8
Bui, M., Adjiman, C. S., Bardow, A., Anthony, E. J., Boston, A., Brown, S., & Mac Dowell, N. (2018). Carbon capture and storage (CCS): The way forward. *Energy & Environmental Science, 11*(5), 1062–1176.
Chen, Y., & Wang, L. (2019). Commentary: Marketing and the sharing economy: Digital economy and emerging market challenges. *Journal of Marketing, 83*, 28–31.
Chui, K. T., Lytras, M. D., & Visvizi, A. (2018). Energy sustainability in smart cities: Artificial intelligence, smart monitoring, and optimization of energy consumption. *Energies, 11*(11), 2869.
Cole, L. B., McPhearson, T., Herzog, C. P., & Russ, A. (2017). Green infrastructure. In *Urban environmental education review* (pp. 261–270).
Cormos, C. C. (2012). Integrated assessment of IGCC power generation technology with carbon capture and storage (CCS). *Energy, 42*(1), 434–445.
Dawoodbhoy, F. M., et al. (2021). AI in patient flow: Applications of artificial intelligence to improve patient flow in NHS acute mental health inpatient units. *Heliyon, 7*(5), e06993. https://doi.org/10.1016/j.heliyon.2021.e06993
Dubey, R., & Gunasekaran, A. (2015). Sustainable transportation: An overview, framework and further research directions. *International Journal of Shipping and Transport Logistics, 7*(6), 695–718.
Ghisellini, P., Cialani, C., & Ulgiati, S. (2016). A review on circular economy: The expected transition to a balanced interplay of environmental and economic systems. *Journal of Cleaner Production, 114*, 11–32.
Gore, S., Swarnkar, P., Singh, G. K., & Singh, S. (2023). *Sustainable solutions for global infrastructure growth through green building practices*. SAAR Publications.
Gurung, D., et al. (2022). Stigma against mental health disorders in Nepal conceptualised with a 'what matters most' framework: A scoping review. *Epidemiology and Psychiatric Sciences, 31*, e11. https://doi.org/10.1017/S2045796021000809
Hao, X., Li, Y., Ren, S., Wu, H., & Hao, Y. (2023). The role of digitalization on green economic growth: Does industrial structure optimization and green innovation matter? *Journal of Environmental Management, 325*, 116504.
Jaiswal, K. K., Chowdhury, C. R., Yadav, D., Verma, R., Dutta, S., Jaiswal, K. S., & Karuppasamy, K. S. K. (2022). Renewable and sustainable clean energy development and impact on social, economic, and environmental health. *Energy Nexus, 7*, 100118.
Khan, S. H. (2020a). Green nanotechnology for the environment and sustainable development. In *Green materials wastewater treatment* (pp. 13–46). Springer.
Khan, S. H. (2020b). Green nanotechnology for the environment and sustainable development. In *Green materials for wastewater treatment* (pp. 13–46). Springer.

Khan, S. A. R., Umar, M., Asadov, A., Tanveer, M., & Yu, Z. (2022). Technological revolution and circular economy practices: A mechanism of green economy. *Sustainability, 14*(8), 4524.
Khan, S. B., Kumar, A., Mashat, A., Pruthviraja, D., Imam Rahmani, M. K., & Mathew, J. (2024). Artificial intelligence in next-generation networking: Energy efficiency optimization in IoT networks using hybrid LEACH protocol. *SN Computer Science, 5*(5), 546.
Kothari, D. P., Ranjan, R., & Singal, K. C. (2021). *Renewable energy sources and emerging technologies*. PHI Learning Pvt. Ltd.
Kumar, S., & Ghangrekar, M. M. (2024). Algae-based bioenergy production as a carbon mitigation technology. In *Algae as a natural solution for challenges in water-food-energy nexus: Toward carbon neutrality* (pp. 833–850). Singapore.
Kumar, A., Khan, A. A., & Singh, J. (2024a). Enhancing the diagnosis of cardiovascular disease: A comparative examination of support vector machine and artificial neural network models utilizing extensive data preprocessing techniques. *WSEAS Transactions on Computers, 23*, 1.
Kumar, A., Dhanka, S., Bansal, R., Sharma, A., Singh, J., Khan, A. A., & Maini, S. (2024b). Agriculture in the era of industry 5.0. Industry 5.0 and emerging technologies. *Transformation Through Technology and Innovations, 565*, 111.
Kumar, A., Singh, J., & Khan, A. A. (2024c). A comprehensive machine learning framework with particle swarm optimization for improved polycystic ovary syndrome (PCOS) diagnosis. *Engineering Research Express, 6*, 035233. https://doi.org/10.1088/2631-8695/ad76f9
Kumar, A., Tyagi, P. K., Tyagi, S., & Ghorbanpour, M. (2024d). Integrating green nanotechnology with sustainable development goals: A pathway to sustainable innovation. *Discover Sustainability, 5*(1), 364.
Kumar, A., et al. (2024e). *Smart Crop Selection: Harnessing Machine Learning for Sustainable Agriculture in the Era of Industry 5.0* (pp. 111–134). https://doi.org/10.1007/978-3-031-70996-8_6
Kumar, A., Sharma, A., Dhanka, S., & Maini, S. (2025). Machine learning for combating mental health stigma. In *Transforming neuropsychology and cognitive psychology with AI and machine learning* (pp. 333–366). IGI Global Scientific Publishing.
Lema, R., & Lema, A. (2012). Technology transfer? The rise of China and India in green technology sectors. *Innovation and Development, 2*(1), 23–44.
Li, G., & Yao, J. (2024). A review of algae-based carbon capture, utilization, and storage (algae-based CCUS). *Gases, 4*(4), 468–503.
Lim, W. M. (2020). The sharing economy: A marketing perspective. *Australasian Marketing Journal, 28*, 4–13.
Liu, H. (2022). Applications of artificial intelligence to popularize legal knowledge and publicize the impact on adolescents' mental health status. *Frontiers in Psychiatry, 13*(May), 902456. https://doi.org/10.3389/fpsyt.2022.902456
Mahmood, T., Hussain, N., Shahbaz, A., Mulla, S. I., Iqbal, H. M., & Bilal, M. (2023). Sustainable production of biofuels from the algae-derived biomass. *Bioprocess and Biosystems Engineering, 46*(8), 1077–1097.
Majid, M. (2020). Renewable energy for sustainable development in India: Current status, future prospects, challenges, employment, and investment opportunities. *Energy, Sustainability and Society, 10*(1), 1–36.
Matuszewska-Janica, A., Żebrowska-Suchodolska, D., Ala-Karvia, U., & Hozer-Koćmiel, M. (2021). Changes in electricity production from renewable energy sources in the European Union countries in 2005–2019. *Energies, 14*(19), 6276.
Mei, B., Khan, A. A., Khan, S. U., Ali, M. A. S., & Luo, J. (2023). Variation of digital economy's effect on carbon emissions: Improving energy efficiency and structure for energy conservation and emission reduction. *Environmental Science and Pollution Research, 30*(37), 87300–87313.
Mittal, G., & Bansal, R. (2023). Driving force behind consumer brand engagement: The Metaverse. In *Cultural marketing and Metaverse for consumer engagement* (pp. 164–181). IGI Global.
Painoli, A. K., Bansal, R., Singh, R., & Kukreti, A. (2021). Impact of digital marketing on the buying behavior of youth with special reference to Uttarakhand state. In *Big data analytics*

for improved accuracy, efficiency, and decision making in digital marketing (pp. 162–182). IGI Global.

Palta, P., & Kumar, A. (2024). Effect of vermicompost additive on physical, chemical and dielectric properties of soil and its modeling. *The Journal of Microwave Power and Electromagnetic Energy, 58*(3), 186–206. https://doi.org/10.1080/08327823.2024.2378668

Palta, P., & Kumar, A. (2025). Soil dielectric response to chemical fertilizers in Northern India's key agricultural areas. *Physics and Chemistry of the Earth, 137*, 103806. https://doi.org/10.1016/j.pce.2024.103806

Paltsev, S., Morris, J., Kheshgi, H., & Herzog, H. (2021). Hard-to-abate sectors: The role of industrial carbon capture and storage (CCS) in emission mitigation. *Applied Energy, 300*, 117322.

Pamukcu-Albers, P., Ugolini, F., La Rosa, D., Grădinaru, S. R., Azevedo, J. C., & Wu, J. (2021). Building green infrastructure to enhance urban resilience to climate change and pandemics. *Landscape Ecology, 36*(3), 665–673.

Qamruzzaman, M., & Karim, S. (2024). Green energy, green innovation, and political stability led to green growth in OECD nations. *Energy Strategy Reviews, 55*, 101519.

Qin, Y., Xu, Z., Wang, X., & Skare, M. (2024). Artificial intelligence and economic development: An evolutionary investigation and systematic review. *Journal of the Knowledge Economy, 15*(1), 1736–1770.

Raberto, M., Ozel, B., Ponta, L., Teglio, A., & Cincotti, S. (2019). From financial instability to green finance: The role of banking and credit market regulation in the Eurace model. *Journal of Evolutionary Economics, 29*, 429–465.

Rahman, K. T., Bansal, R., & Pruthi, N. (2023). Consumer adoption of technologies: A mindset-oriented approach. In *Contemporary studies of risks in emerging technology, part B* (pp. 243–255). Emerald Publishing Limited.

Raja, M., Swamy, T. N. V. R. L., Das, S., Bansal, R., Fahlevi, M., & Aziz, A. L. (2025). Does human resource information system influence performance management? *Cogent Business & Management, 12*(1), 2438862.

Rana, M. W., Sufang, Z., Jamil, K., Jaffri, N. R., & Hamid, I. (2021, October). Eco-innovation and its impact on environmental performance in the context of green technology. In *2021 International conference on engineering and emerging technologies (ICEET)* (pp. 1–5). IEEE.

Saran, C., Devi, A., Saratale, G. D., Ferreira, L. F. R., & Bharagava, R. N. (2024). Bioenergy generation and cogeneration from algae-based biorefineries. In *Value added products from bioalgae based biorefineries: Opportunities and challenges* (pp. 189–211). Singapore.

Sarfraz, M., Ivascu, L., Belu, R., & Artene, A. (2021). Accentuating the interconnection between business sustainability and organizational performance in the context of the circular economy: The moderating role of organizational competitiveness. *Business Strategy and the Environment, 30*, 2108–2118.

Satar, M. S., Aggarwal, D., Bansal, R., & Alarifi, G. (2023). Mapping the knowledge structure and unveiling the research trends in social entrepreneurship and inclusive development: A bibliometric analysis. *Sustainability, 15*(7), 5626.

Sayed, A., Himeur, Y., Bensaali, F., & Amira, A. (2022). Artificial intelligence with IoT for energy efficiency in buildings. In *Emerging real-world applications of internet of things* (pp. 233–252). CRC Press.

Schiederig, T., Tietze, F., & Herstatt, C. (2012). Green innovation in technology and innovation management–an exploratory literature review. *R&D Management, 42*(2), 180–192.

Sen, R., & Mukherjee, S. (2024). Recent advances in microalgal carbon capture and utilization (bio-CCU) process vis-à-vis conventional carbon capture and storage (CCS) technologies. *Critical Reviews in Environmental Science and Technology, 54*(24), 1777–1802.

Shah, K. J., Pan, S. Y., Lee, I., Kim, H., You, Z., Zheng, J. M., & Chiang, P. C. (2021). Green transportation for sustainability: Review of current barriers, strategies, and innovative technologies. *Journal of Cleaner Production, 326*, 129392.

Shan, S., Genç, S. Y., Kamran, H. W., & Dinca, G. (2021). Role of green technology innovation and renewable energy in carbon neutrality: A sustainable investigation from Turkey. *Journal of Environmental Management, 294*, 113004.

Sharma, A., Dhanka, S., Bansal, R., Kumar, A., & Maini, S. (2024a). *Industry 5.0 with sustainable development* (pp. 347–370). Springer. https://doi.org/10.1007/978-3-031-70996-8_17

Sharma, A., Dhanka, S., Kumar, A., & Maini, S. (2024b). A comparative study of heterogeneous machine learning algorithms for arrhythmia classification using feature selection technique and multi-dimensional datasets. *Engineering Research Express, 6*(3), 035209. https://doi.org/10.1088/2631-8695/ad5d51

Shaw, R., & Mukherjee, S. (2022). The development of carbon capture and storage (CCS) in India: A critical review. *Carbon Capture Science & Technology, 2*, 100036.

Singh, J., Pandey, B., Karna, S., Arora, A. S., & Kumar, A. (2024). Enhancing the thermographic diagnosis of maxillary sinusitis using deep learning approach. *Quantitative InfraRed Thermography Journal, 00*(00), 1–15. https://doi.org/10.1080/17686733.2024.2349976

Srivastava, S., & Bhargava, A. (2022). *Green nanoparticles: The future of nanobiotechnology* (pp. 1–13). Springer.

Tomazzoli, C., Scannapieco, S., & Cristani, M. (2023). Internet of things and artificial intelligence enable energy efficiency. *Journal of Ambient Intelligence and Humanized Computing, 14*(5), 4933–4954.

Tran, Q. H., Van Quy, N., & Le, A. T. (2013). Silver nanoparticles: Synthesis, properties, toxicology, applications and perspectives. *Advances in Natural Science, 4*, 1–21.

Wang, H., Cui, H., & Zhao, Q. (2021). Effect of green technology innovation on green total factor productivity in China: Evidence from spatial durbin model analysis. *Journal of Cleaner Production, 288*, 125624.

Yasmeen, R., Zhang, X., Tao, R., & Shah, W. U. H. (2023). The impact of green technology, environmental tax and natural resources on energy efficiency and productivity: Perspective of OECD rule of law. *Energy Reports, 9*, 1308–1319.

Ziegler, R., Balzac-Arroyo, J., Hölsgens, R., Holzgreve, S., Lyon, F., Spangenberg, J. H., & Thapa, P. P. (2022). Social innovation for biodiversity: A literature review and research challenges. *Ecological Economics, 193*, 107336.

Zuniga-Teran, A. A., Staddon, C., De Vito, L., Gerlak, A. K., Ward, S., Schoeman, Y., & Booth, G. (2020). Challenges of mainstreaming green infrastructure in built environment professions. *Journal of Environmental Planning and Management, 63*(4), 710–732.

Sustainable World: Concepts and Frameworks

Aziza Chakir, Aziz Atmani, and Ilham Boughaba

1 Introduction

Sustainability is an increasingly important concept as the world faces climate change, resource depletion, social inequality, and other global challenges. Sustainability, plain and simple, is the ability to provide for today's needs without destroying the ability for tomorrow's (Brundtland Commission, 1987). This definition has evolved over time, encapsulating a multidimensional approach that covers environmental, economic, and social dimensions. Specifically, if you see sustainability as the harmony of ecological health, economic viability, and social equity, integrated in a manner that serves as a guiding principle in policy and practice to bring the human activities in balance with natural systems, then this world is sustainable.

Future Sustainability Masters: The history of sustainability dates back to early environmental movements and the realization of humanity's limited resources. An important milestone came when the 1972 United Nations Conference (United Nations, 1972) on the Human Environment in Stockholm recognized ecological limits and the need for concerted international action (United Nations, 1972). Later milestones, such as the Earth Summit in Rio de Janeiro in 1992 and the adoption of the Sustainable Development Goals (SDGs) in 2015, have entrenched sustainability as a key principle of global development (United Nations, 2015). These frameworks resonate with a growing recognition of the complex interconnections between social, economic, and ecological systems, highlighting the fact that sustainable development is as much about the quality of people's lives as it is about the natural environment.

In times of challenges like these, the complexity of need can be overwhelming and the significance of sustainability cannot be overvalued. Since the industrial

A. Chakir (✉) · A. Atmani · I. Boughaba
Faculty of Law, Economics and Social Sciences, Hassan II University, Casablanca, Morocco

A. Chakir et al. (eds.), *Green Society, Environmental Strategies and Sustainable Development*, https://doi.org/10.1007/978-3-031-94086-6_2

revolution, unsustainable practices have led to biodiversity loss, climate change, and social injustice, which have become existential risks to human life (IPCC, 2021; World Economic Forum, 2023). Not only to prepare for these challenges but also to overcome them we should embrace sustainable practices which bring about resilience and adaptability for the unpredictable future. Since countries battle with the repercussions of unsustainable expansion, crafting and enforcing durability frameworks are critical to a sustainable long run. This chapter examines various concepts and frameworks of sustainability to guide the integration of sustainable principles into policies and practices aimed at achieving a sustainable world.

2 Core Concepts of Sustainability

Sustainability has undergone a considerable evolution over the last few decades, becoming a key principle in navigating the multifaceted problems facing modern societies. When we talk about sustainability, it mainly involves four interconnected aspects which are known as the four pillars of sustainability: Environmental sustainability, economic sustainability, social sustainability, and cultural sustainability. To develop frameworks that can lead the path to a sustainable world, it is important to know what these dimensions are. This chapter discusses these central ideas in greater detail.

2.1 Environmental Sustainability

Environmental sustainability is the ability of the natural environment to support human life and economic activity in the long term while also maintaining ecosystem integrity and biodiversity. It stresses the need for responsible use of natural resources so that (respectively) future generations can also benefit from them (Altieri et al., 2021).

2.1.1 Key Principles

Integrity of Ecosystem: An important factor in environmental sustainability is a healthy ecosystem. This means protecting biodiversity, soil health, water quality, and clean air (Bennett et al., 2020). These are processes that an ecosystem carries out, including carbon sequestration, nutrient cycling, and water filtration, which sustain humans.

Resource Management: Sustainable resource management seeks to maximize the consumption of both renewable and non-renewable resources. These strategies involve waste minimization, incentivizing recycling, and harnessing renewables such as wind and solar (IPCC, 2022). Reuse and recycling are important aspects of this transformation; the circular economy is rapidly becoming a core strategy for achieving environmental sustainability (Geissdoerfer et al., 2018).

Climate Change Mitigation and Adaptation: Climate change mitigation and adaptation is an essential aspect of environmental sustainability. Mitigation efforts reduce greenhouse gas emissions, and adaptation strategies help improve resilience to climate impacts. Newer frameworks like the Paris Agreement show that there is truly a need for global cooperation to achieve climate goals UNESCO, 2018, 2005.

Sustainable Agriculture: As the global population continues to grow, agricultural practices need to adapt to feed everyone while reducing environmental impacts. Sustainable agriculture encourages agroecological, organic, and permaculture methods to prioritize soil health and ecological balance (Altieri et al., 2021).

2.2 Economic Sustainability

To achieve this sustainability, you need to have a system where the economy is resilient enough to care for what you need now without denying future generations their needs. It is focused less on quarterly returns than on a longer-term view of economic health, social equity, and environmental stewardship.

2.2.1 Key Principles

Sustainable growth: Sustainable growth means investing in green technologies and industries that create jobs without damaging the environment (OECD, 2020; Chakir et al., 2021).

Equitable Distribution of Resources: Sales must be spent on a vast house without idle resources. By implementing policies that promote fair wages, access to education, and social welfare programs, we can ensure that the benefits of economic growth are distributed more equitably among all members of society (Piketty, 2014).

Adaptability and Resilience: Economies should have the ability to adapt to shocks (financial crises, pandemics, and climate). Long-term sustainability will require promoting adaptive capacities through diversification of economic activities and social safety nets (World Bank, 2021a).

Promote Sustainable Investment Practices to Reach Economic Sustainability: To drive capital toward sustainable projects and initiatives, more and more investors are adopting environmental, social, and governance (ESG) criteria in their investment decision-making (Friede et al., 2015).

2.3 *Social Sustainability*

Social sustainability is another part of this board and involves ensuring the welfare of humans and communities, in addition to social equity, cultural identity, and community engagement. It covers the social aspects of sustainability, such as health, education, and social justice.

2.3.1 Key Principles

Social Equity and Justice: An important aspect of social sustainability is the focus on social equity, which seeks to address systemic inequalities and ensure that all members of society have access to the basic rights and resources needed to thrive, such as education, healthcare, and employment opportunities (Rawls, 1999). Discrimination will not be tolerated; inclusivity must be embraced, and policies must be introduced.

Social Sustainability: Community engagement and participation investing and giving a voice to the local populations while providing a platform to highlight concerns and solutions bring greater sense-of-ownership, kuarain (accountability), and responsibility (Pretty, 2020).

Economic Equity and Development: Fostering inclusive economic opportunities and reducing inequality leads to more stable societies. Highlighting and celebrating indigenous knowledge systems and practices can contribute to social cohesion and resilience (UNESCO, 2018).

Health and Well-Being: The social dimension of sustainability also means physical and mental health. The availability of healthcare services, clean drinking water, and nutritious food is essential to promote wellness and quality of life for individuals and communities (WHO, 2021). Social sustainability also involves public health initiatives which promote healthy lifestyles.

2.4 *Cultural Sustainability*

When we think of sustainability, seldom we think of cultural sustainability, which effectively enhances social cohesion and community identity. This refers to the ability of a society or culture to preserve its values, practices, and identity despite rapid globalization.

2.4.1 Key Principles

Traditional Knowledge: Embracing and sharing traditional knowledge, wisdom, and practices promotes resilience and adaptability. Rights of indigenous peoples and minority cultures are protected, so their traditions and knowledge are preserved and respected (UNESCO, 2005).

Preserving Cultural Heritage: Protecting cultural heritage such as historical landmarks, traditional arts, and languages is an important part of keeping up cultural identity. Cultural sustainability can be further promoted by supporting community-led initiatives that integrate archeological processes with local populations (Santagata, 2019).

Making Knowledge Accessible: Intergenerational dialogue and knowledge transfer are essential to sustaining cultural practices and values. Educational systems that integrate local knowledge and cultural histories tend to create a sense of belonging and identity (Miller, 2020).

Cultural Innovation: Encouraging cultural innovation and creativity can contribute to enhanced economic and social progress. Support for artists, cultural entrepreneurs, and creative industries could help enhance cultural sustainability while fostering economic growth (Elkington, 1997).

These four concepts mentioned above constitute sustainability. Frameworks are needed to bring these together and all key actors from policymakers, businesses, and communities should collaborate to create these pathways for a resilient, equal, future.

3 Theoretical Frameworks

Through this exploration, guiding structural theories have emerged to assist policymakers, businesses, and communities as they work to balance economic advancement, environmental cohesion, and social equity in service of a sustainable world. In this chapter, we will discuss four analytical frameworks: The triple bottom line, the Sustainable Development Goals (SDGs), the circular economy, and doughnut economics. These frameworks all highlight unique aspects and approaches that may contribute toward a more sustainable world.

3.1 The Triple Bottom Line

According to the triple bottom line (TBL) framework—which was first defined by John Elkington in 1994—the economic, environmental, and social factors all play a role in sustainability (Elkington, 1997). It argues that businesses and societies should measure their success not just in monetary profit (the so-called bottom line) but also their impact on people and the planet. This framework contains three pillars and encourages a more systemic view of sustainability, encouraging actions that create long-lasting benefits for stakeholders.

TBL has recently been considered as one of the important frameworks for sustainability reporting and decision-making in several studies. An increasing number of companies are integrating TBL metrics into the performance measurement of each of the three dimensions (Global Reporting Initiative (GRI), 2023). Moving to such an approach increases transparency as well as establishes a new culture of accountability and continuous improvement. Moreover, the social equity dimension of TBL relates to the really higher need for actions focused on labor rights, community, and diversity as part of corporate social responsibility (CSR) (Garriga & Melé, 2023).

However, the TBL framework has not been without criticism, notably with respect to those challenges to measuring and incorporating (that is, accounting for) social and environmental factors into economic performance (Hahn et al., 2023). Regardless of these challenges, TBL remains a prominent framework in the sustainable development dialogue and helps to encourage organizations to rethink their impact on people and the planet.

3.2 UN Sustainable Development Goals

The Sustainable Development Goals (SDGs) are a universal and integrated framework to tackle global challenges relating to poverty, inequality, climate change, environmental degradation, peace, and justice, adopted by the United Nations in 2015 (United Nations, 2015). The SDGs are a collection of 17 goals and 169 targets that guides governments, companies, and civil society toward addressing the world's greatest challenges by 2030.

Arguing in favor of such initiatives like SDGs, he adds, they have succeeded in dragging the world toward sustainable development: "SDGs have served to gain global action and to mobilize resources." Recent research highlights the significance of multi-stakeholder partnerships in achieving these aims. As an example, a recent work that obtained a nondimensionalized number of frequencies under different conditions that was published in *Nature Sustainability* illustrates how cooperative action among government, private sectors, and NGOs can permit an acceleration toward the goal in SDG 17 (Partnerships for the Goals) while advancing some results on the launch of goals (Sachs et al., 2023).

Also, the SDGs have been set to be adopted as to embed sustainability into national policy and corporate strategy. More than ever companies are aligning their sustainability efforts with the SDG (KPMG, 2023), understanding that tackling these global challenges can support innovative practices and improve competitiveness. This said, there are major obstacles to achieving the SDGs, such as funding gaps, data accessibility, and political will (Mazzucato, 2023). You have to keep both the dialogue real and the dialogue going and hold all the people who are involved in the dialogue accountable for those targets, those ambitious targets, and so on.

3.3 The Circular Economy

The circular economy (CE) paradigm is a well-cited transition from the traditional linear model of "take, make, dispose" toward a more regenerative approach that highlights resource efficiency, waste reduction, and continual reuse (Ellen

MacArthur Foundation, 2019). This concept is based on systems thinking that focuses on developing closed-loop systems that reduce the environmental impact and maximize the economic and social benefits.

Recent developments in CE literature show that, indeed, CE can contribute significantly to sustainability aims. It is against that backdrop that a recent study published in *Resources, Conservation and Recycling* outlines how new business models enable a shift toward a circular economy, emphasizing the role of digital technologies in optimizing resource recovery processes and managing the lifecycle of products (Linder & Williander, 2023). The European Union's Circular Economy Action Plan, though revised in 2020, specifies such targets, indicating the extent to which policymakers have been inclined to promote CE (European Commission, 2020).

However, such transition to circular economy does have some challenges. Addressing barriers including consumer behavior, regulatory frameworks, and infrastructural limitations is essential for enabling widespread adoption (Kirchgeorg et al., 2023). Until these challenges are addressed or solved, the CE framework can provide a viable approach for realizing a decoupled economy, where economic growth does not compromise the sustainability of the global economy.

3.4 Doughnut Economics

Kate Raworth first presented the concept of doughnut economics in her groundbreaking book, *Doughnut Economics: Seven Ways to Think Like a twenty-first-Century Economist* (2017) as a new way to think about the relationship between economy and ecology. The "doughnut" shape is the safe and just space for humanity, sandwiched between an ecological ceiling (on the right) and a social foundation (on the left) (described by Oxfam in the Infographic below) The aim is to work within this doughnut, where no one is left behind on life's essentials, and we don't slip into the other extreme of destroying the planet.

The ongoing conversations about doughnut economics find new life in our time of, simply, the new normal when it comes to en masse discussions around climate change and inequality are concerned. Reflections in 2023 on the model published in *Ecological Economics* reinforced that the doughnut model is relevant in directing investigation of urban planning and urban development, as the authors argue that cities can take up the model to achieve sustainability and social equity (Wilkinson & Pickett, 2023). Moreover, the doughnut model has already been adopted in several cities, such as Amsterdam and Copenhagen, as an overarching principle for formulating their urban policies, proving its applicability (City of Amsterdam, 2020).

Doughnut economics is a highly innovative approach to sustainability and socio-economic challenges, but it has not been without criticism, especially in terms of its feasibility or practical application in complex economic systems (Coyle, 2023). However, they contend that converting the vision outlined in the framework into effective policies and practices necessitates strong governance frameworks and cross-sector cooperation (Chakir et al., 2015, 2020).

The theoretical frameworks reviewed in this section provide different lenses and approaches within the sustainable development discourse, namely the triple bottom line, the Sustainable Development Goals (SDGs), the circular economy (CE), and doughnut economics (DE). At the same time, many define sustainability in similar terms but offer different tools to describe them in a sense of resources and we welcome different voices and aspired and can also have a layered application in themselves. Following these frameworks will be essential for the entire world as it evolves with daunting environmental and societal issues, as we will need to have them guide our decision-making process to promote sustainability in the near future and beyond.

4 Implementing Sustainability

Emerging environmental degradation, social inequality, and economic instability conditions are accelerating the need for a broader based approach to sustainability. In brief, this chapter examines: (1) the need and purpose of sustainability practices; (2) the efforts that entail for the formulation of policies, business models, and active engagement; and (3) the outcomes and feedback mechanisms that come from each of these three pillars. This nexus reinforces a solid framework for long-term development which enables societies to tackle modern issues and also set up a foundation for future generations.

4.1 Policy Frameworks

Shaping appropriate policy frameworks at the local, national, and global level has a big impact on sustainable development. Building blocks and net of actors provide legal and institutional frameworks to further more sustainable practices across different actors. Integrative approaches that consider economic, social, and environmental dimensions are now trending in sustainability policy.

4.1.1 International Agreements and Targets

A landmark international accord to combat climate change (Le Blanc, 2015). NDCs must be established by countries to mitigate greenhouse gas emissions and increase resilience to the adverse effects of climate change. In line with this, as part of the UN and following the 2021 UN Climate Change Conference (COP26) (California Air Resources Board, 2021), there has been a major debate over global temperature targets and the subsequent ambition needed from Nationally Determined Contributions (the updated NDCs) (UNFCCC, 2022).

Moreover, the Sustainable Development Goals (SDGs), which were adopted by the United Nations in 2015, pose a globally accepted platform for comprehensive achievements, covering 17 goals that at the same time target poverty, inequality, and climate action. According to Le Blanc (2015), the SDGs are interconnected, where progress in one area can directly impact progress in other areas. Improving SDG 13 (Climate Action) can help improve SDG 1 (No Poverty) and SDG 7 (Affordable and Clean Energy) and so on.

4.1.2 National and Local Policies

More and more nations have transitioned to assigning local from international sustainability policies. The European Union's Green Deal (European Commission, 2019) represents an ambitious and comprehensive framework toward a climate-neutral continent by 2050, with specific measures for diverse sectors (energy, transportation, and agriculture). Meanwhile, local governments are developing sustainability action plans that prioritize community needs. Frantzeskaki et al. (2019, 2020) examine how cities utilize sustainability indicators to influence local policies and shape citizen decision-making.

4.1.3 Governance and Incentive Structures

Even regulatory frameworks, like emissions trading systems and renewable energy portfolios, are considered as indispensable instruments of sustainability by implementation needs. Such mechanisms will provide market-based incentives for businesses to switch to cleaner technologies. California's Cap-and-Trade Program, for instance, is one of the success stories of the policy approach that has effectively decreased greenhouse gas emissions while still allowing the economy to flourish (California Air Resources Board, 2021). In addition, public procurement policies can require sustainable practices in government contracts, which can drive the demand for sustainable products and services.

4.2 Business Models for Sustainability

A sustainable world cannot be achieved unless policies are strong and business provides new models for environmental and social responsibility. Several papers recently published have described business approaches to making sustainability a foundational business tenet (Honeyman, 2020).

4.2.1 Circular Economy

This encourages recycling (as well as reuse) and decades of sustainable and green design and usage. (PDF) Ellen MacArthur Foundation 2021 Circular sata A Satellite Account for 213 Circular State Stuart/FUTURE The state of circles Circular statement (Ellen MacArthur Foundation, 2021). Here, companies like Unilever and IKEA are already paving the way, implementing circular practices in their supply chains to reduce waste and make sure resources are used as much as possible.

4.2.2 Social Entrepreneurs and B Corps

One of the trends in the business world that are more oriented toward a social and environmental goal in addition to profit registration are social enterprises and B corporations. They are certified to meet rigorous standards of social and environmental performance, accountability, and transparency by the non-profit B lab. Moreover, recent studies have shown that B corporations enjoy higher employee satisfaction and more loyal customers than traditional firms (Honeyman, 2020). The phenomenon of social enterprises proves that businesses can do well and be good at the same time.

4.2.3 Green Supply Chain Management

Sustainable supply chain management concentrates on integrating sustainability principles with supply chain processes. Businesses are swiftly integrating practices such as ethical sourcing, carbon footprinting, and lifecycle assessments. Research by Dubey et al. (Dubey et al., 2020) demonstrates that higher SSCM can contribute to superior operational performance and subsequently achieve competitive advantage. Finally, collaboration across supply chain actors can enhance sustainability initiatives that lead innovation and resilience.

4.3 Casting a Wider Net: Engaging Wider Communities

This shows a bottom-up approach can be utilized to boost sustainability. It is precisely this kind of localized approach that enables organizations to align their sustainability efforts with community values, culture, and context.

4.3.1 Participatory Approaches

Building local participation into sustainability planning can encourage a sense of stewardship and responsibility over local programs. Community workshops and collaborative planning sessions are participatory approaches that invite a range of "voices" and expertise into sustainability planning. A study by Kearns et al. (2021) emphasizes participatory budgeting as a strategy that allows community members to directly decide on the distribution of resources for sustainability projects.

4.3.2 By YNGI on October 19, 2023

Around the world, local movements have sprung up—urban agriculture, renewable energy cooperatives, and waste reduction programs. Consumer exposure to local food systems, for example, community-supported agriculture (CSA) programs connect consumers with local farmers, promoting sustainable food systems and supporting local economies (Lea & Worsley, 2022). Neighborhood solar cooperatives are another example, allowing local residents to invest in renewable energy, thereby lowering fossil fuel dependency and energy costs.

4.3.3 Teaching and Learning Campaigns

It is essential to raise awareness and educate the community on sustainability issues. Greening the Globe: As part of the existing policies, additional educational initiatives in schools, workshops, and public campaigns can bring an increased awareness to people about sustainable practices and, therefore, affect behavioral changes. According to research conducted by Rojas-Romero et al. (2021).

Incorporating sustainability takes place through a comprehensive ecosystem involving policy frameworks, creative business models, and the involvement of the people in their diversity. This holistic approach has the potential to tackle diverse challenges, ensuring a sustainable world that meets present needs while safeguarding future generations. From here onwards, cross-sectoral and cross-stakeholder collaboration would be indispensable for addressing the complex challenges of sustainability and ensuring that progress is fair and inclusive for all.

5 Challenges to Achieving Sustainability

A sustainable world cannot be achieved without understanding and conquering the many pitfalls and obstacles on the path toward sustainable development. While sustainability is recognized as one of the crucial development models for the future at the global level, many factors prevent its successful implementation. Identifying four major types of challenges—economic, political and institutional, social and cultural, and technological—this chapter explores the particularly complex question of barriers to the attainment of sustainability.

5.1 Economic Barriers

One of the biggest challenges in reaching sustainability is economic barriers. From a conventional perspective, economic development had to take precedence over ecological balance, which resulted in a depletion of resources and an increase in environmental degradation and destruction (Daly & Farley, 2019). The main economic problems are:

5.1.1 Short-Term Profit Orientation

The short-term profit-seeking of corporations can be at odds with environmental sustainability, resulting in decisions that harm the environment. Since 2010, annual CEO pay increases have averaged about 10 percent, and the business ethos of pursuing returns at any cost is increased also through shareholder pressure and the volatility of financial markets, both of which incentivize practices that maximize short-term returns at the expense of sustainable practices (Porter & Kramer, 2019). However, there are numerous cases where exceptions are made, and not always for a good reason, such as very high subsidies for the fossil fuel sector even as we are told that we need to shift to renewable energy sources (International Renewable Energy Agency (IRENA), 2022).

5.1.2 Financial Constraints

Another major barrier toward sustainability initiatives is funding access. A lot of sustainable projects demand upfront financing that are not viable options for developing economies or small businesses. Financing is indeed a major challenge, with an estimated $5 trillion per year required for achieving the Sustainable Development Goals (SDGs) (United Nations Environment Programme, 2021). Analytical tools and innovative financing mechanisms, such as green bonds and blended finance, are being developed but not widely used (World Bank, 2022).

5.1.3 Market Failures

Externalities are one type of market failure which can distort incentives for sustainability. Pollution, for instance, generates costs that are not reflected in the prices of goods and services, resulting in excessive consumption of harmful products (Stiglitz, 2020). Solutions such as carbon pricing and taxes to counter these failures have been suggested, but implementation remains uneven and controversial (World Bank, 2021b).

5.2 *The Resistance from Society and Cultural Oppositions*

Sustainability initiatives rely on social acceptance, which is often met with resistance due to deeply rooted cultural norms and values.

5.2.1 Knowledge and Awareness of the Public

Public perceptions surrounding sustainability issues can play a key role in public policy success or failure. Skepticism regarding the need or effectiveness of sustainable measures can be caused by disinformation and lack of acknowledgement (Verplanken & Roy, 2016). As one example of such misinformation, climate change denial remains an impediment to the introduction of effective climate policies in different areas, driven by misinformation (Cook et al., 2021).

5.2.2 Values and Cultural Styles

Cultural values and lifestyle choices can create barriers to sustainable practices, as well. Consumerism and a desire for material wealth are pervasive across many societies, complicating the transition to a more sustainable behavior state (Cook et al., 2021). In order to tackle these cultural barriers need, deep education and awareness campaigns that align with local values and emphasize the advantages of sustainability are needed.

5.2.3 Equity and Inclusion

Sustainability initiatives must also focus on social equity if they are to gain broader public support. Marginalized communities deal the most with environmental destruction but lack power in the decision-making process. The need to ensure the inclusivity and equity of sustainability efforts is vital for fostering social acceptance and mobilizing collective action (Sachs et al., 2021).

5.3 *Technological Limitations*

Technology can play a role in improving sustainable practices, but there are barriers that need to be overcome.

5.3.1 Consolidation: Accessibility and Affordability

Low-Income Populations and their Unequal Access to Sustainable Technologies: Many sustainable technologies, such as renewable energy systems and energy-efficient appliances, are still unreachable or unaffordable for low-income populations (International Renewable Energy Agency (IRENA), 2022). Such measures are needed to bridge the technology gap, especially in developing countries and should draw on targeted investments and policies to enable access to clean technologies.

5.3.2 Over-Reliance on Technology and the Innovation Matrix

Relying too heavily on technology may give way to complacency when it comes to tackling sustainability issues. Technological innovations can drive significant progress, but substantive progress requires changes in behavior, policy, and economic structures to complement the technology (Hoffman, 2021). Moreover, innovation gaps are found where emerging technologies might not be developed or deployed in a timely manner, to address urgent sustainability needs (Reinhardt, 2021).

5.3.3 Technologies Effect on Environment

Finally, any new technology needs to be responsibly analyzed in terms of its environmental implications. An example, although one that is often touted as a sustainable alternative, is the manufacturing of electric vehicles, where the extraction and processing of lithium and cobalt (for their batteries) introduce environmental and ethical concerns (García et al., 2022). True sustainability requires a more holistic approach to technology development—one that considers lifecycle impacts.

There are many challenges and interdependencies to overcome but there are many pillars to achieving a sustainable world that requires a multidimensional focus. Four types of innovation, namely economic, political-institutional, socio-cultural, and technological, as well as their barriers need to be tackled simultaneously to create an enabling environment for sustainability. Acknowledging and addressing these concerns can enable stakeholders to collaborate around a shared vision for the future of the river that recognizes the need for ecological integrity, social equity, and economic viability.

6 Case Studies

Case studies focus on three key areas of interest: Renewable energy initiatives, urban sustainability, and conservation and biodiversity efforts. Each case study explores activities that demonstrate sustainable practices, as well as the successes and challenges experienced while scaling sustainable solutions.

6.1 *Initiatives on Renewable Energy*

Implementing renewable energy is a cornerstone of sustainable development, as it promotes climate stability, energy security, and economic growth. There are several notable case studies from around the globe that highlight the successful implementation of renewable energy projects, such as Germany's solar energy initiatives and Denmark's wind energy programs.

6.2 *Urbanization and Sustainable Development*

Sustainable urbanization is a pressing challenge that demands innovative solutions in urban planning and development. Two case studies that illustrate sustainable urban development practices are the Eco-City Project in Tianjin, China, and the Minute City Initiative in Paris, France.

6.3 *Conservation and Biodiversity Projects*

Conservation is vital for preserving biodiversity and providing ecological services. Two case studies that reflect successful conservation and biodiversity projects are:

- Climate Change Adaptation: The Great Green Wall Initiative aims to combat desertification and promote sustainable land management in the Sahel region of Africa. Launched in 2007, the initiative seeks to restore 100 million hectares of degraded land by 2030 while enhancing food security and improving livelihoods for local communities (African Union, 2022).
- Fundo Amazônia in Brazil: Established in 2008, this fund helps protect the Amazon rainforest by financing projects focused on sustainable development and deforestation prevention. Heavily supported by international donations, the initiative sponsors projects related to sustainable agriculture and forestry, conservation, and the rights of indigenous peoples, among others.

These case studies featured in this chapter provided an overview of the different methods and frameworks used in renewable energy initiatives, sustainable urban development, and conservation projects. Although there has been significant progress at the same time, we must continue to innovate and collaborate and use sustainability as a driving force to meet ongoing challenges. This can help policymakers and practitioners develop a more sustainable future for all, through the knowledge of these case studies.

7 Future Directions for a Sustainable World

Amidst these ongoing struggles, recent initiatives have gained momentum to promote sustainability and resilience on planet Earth. Here, we will be considering the potential avenues of the future trends that could have the most impact on sustainability, such as developments in technology, education and awareness, and international cooperation and governance.

7.1 Advancements in Technology

Innovation in technology is central to the goal of sustainability. New developments in a variety of fields offer promising strategies to minimize environmental burdens and to make resource use more efficient.

7.1.1 Renewable Energy Technologies

Transforming fossil fuels to renewable energy sources is essential for reducing greenhouse gas emissions. Wind, solar photovoltaic (PV) systems, and storage technologies have allowed for this change. For example, the International Renewable Energy Agency (World Economic Forum, 2023) found that the price of solar PV has dropped by over 85% since 2010, cementing its reputation as one of the cheapest sources of energy on the planet. This is complemented by advances in battery technologies, specifically lithium-ion and emerging solid-state batteries, which are improving energy storage capacity and allowing for greater inclusion of renewables (Nykvist & Nilsson, 2015).

7.1.2 Circular Economy Technologies

The idea of a circular economy, which focuses on recycling, reuse of materials, and sustainable design for products, is taking hold. Materials science innovations, including biodegradable plastics, and novel recycling techniques will also be critical in alleviating waste and maximizing resource usage. Research by Kirchgeorg et al. (Kirchgeorg et al., 2021) shows that digital technologies (such as Internet of Things (IoT) and blockchain) can contribute to increasing supply chain transparency and help with recycling and, in turn, contributing to a circular economic model.

7.1.3 Smart Agriculture

Agriculture has a major toll on the environment, but technology may alleviate some of that. Precision agriculture technologies leverage sensors, drones, and data analytics, enabling farmers to improve resource efficiency, lower the use of chemicals, and increase crop yields (Al-Chalabi, 2015). Additionally, vertical farming and hydroponics would motion the slicing of land use and water consumption: Two important utilities in conventional agriculture but facilitated by the advances in technology (Al-Chalabi, 2015).

7.1.4 Green Transport

Another area for innovation is the transportation sector. EVs are quickly becoming more common due to improved battery technology and a larger charging infrastructure. As a sign of that transition to cleaner modes of transport, global EV sales amounted to ten million units in 2022 (International Energy Agency (IEA), 2023). Also, some consider hydrogen or biofuels to work in addition to electrification.

7.2 Awareness and Education

Education is a key pillar in creating a more sustainable world. It is important to raise awareness and promote sustainability literacy among individuals, communities, and organizations to drive behavioral changes and promote the implementation of sustainable practices.

7.2.1 Education for Sustainability

More and more educational institutions are including sustainability in their curricula. This cross- and interdisciplinary integration empowers students to tackle complex sustainability challenges, enabling them to become change agents. UNESCO (UNESCO, 2021) states that education for sustainable development (ESD) raises critical thinking and problem-solving skills which promote sustainable development. Topics include education, non-demographic technology, and contributions from sources such as the Global Action Programme on ESD, general education, crops, and soil, all essential in developing materials like the ESD initiative that promote quality education and sustainable practices.

7.2.2 Public Awareness Campaigns

More importantly, public awareness campaigns can be useful tools to educate citizens on the negative impacts of rapid development on the environment and encourage them to lead more an eco-friendly life. Information can be readily disseminated and responses can be initiated through social media, documentaries, community workshops, and various forms of community engagement. Research by Dinter et al. (Dinter et al., 2022) show that well-designed campaigns through a combination of persuasive messages and targeted outreach have been shown to have a significant impact on participant behaviors and public perceptions of sustainability issues which, consequently, may increase participation rates for recycling programs and energy-saving programs.

7.2.3 Continuing Education and Professional Development

As sustainability is such a dynamic field, you have to commit to ongoing learning and continuing professional development. This is where well-organized training programs by the organization can promote sustainability. Lifelong learning and workforce reskilling and upskilling are vital strategies for adapting to evolving economic and environmental contexts to promote sustainability through ongoing education and development (World Economic Forum, 2023).

7.3 Global Collaboration and Governance

To address the global sustainability challenges we face, we need strong cooperation and governance systems that span national borders. Effective policies and initiatives require collaboration between governments, businesses, and civil society.

7.3.1 Multi-Stakeholder Partnerships

Sustainable development and innovation require the cooperation of governments, businesses, and civil society in multi-stakeholder partnerships that will implement sustainable practices. So this is an obvious connection and initiatives like the World Business Council for Sustainable Development (WBCSD) aim to connect different sectors to sustainable development challenges. Research by Bocken et al. (Bocken et al., 2022) highlights the importance of partnerships in aligning interests and leveraging resources, which contributes to firm stability and sustainability goals.

7.3.2 Local Governance and Community Engagement

Global and national frameworks are important but local governance and grassroots actions are crucial to the deployment of sustainable projects. By enabling local communities to be part of decision-making processes, ownership and accountability are created. Case studies (e.g., ICLEI, 2023) provide convincing examples of how local governments implement sustainability programs appropriate for their own context that can successfully increase resilience and social equity.

A sustainable world is a long way off, but every effort in technology, education, and global cooperation will make a difference. Advanced technology has the potential to be part of the solutions to challenging environmental problems, and education and awareness can foster a culture of sustainability. Moreover, both sustainable policies and sustainable practices require good governance and cooperation among multiple stakeholders. These directions are essential for us as we prepare for the future and move toward a better, more sustainable, and equitable world for all generations to come.

8 Conclusion

We have touched upon the nuanced, multifaceted nature of sustainability in this chapter, underscoring its pivotal role in determining the viability of our future on Earth. We have explored frameworks like the United Nations Sustainable Development Goals (SDGs), the circular economy, and environmental, social, and governance (ESG) standards. For everyone there is an individual vision to solve for the climate change, the resource depletion, and the social inequality. Through the integration of these ideas, we have demonstrated the interrelationship between the principles of environmental integrity, economic viability, and social equity and urge for an integrated approach to sustainability development that crosses disciplinary lines.

Our scientific efforts to discover a path toward a sustainable world will serve as a pivotal turning point for our actions across all levels—government, business, and civil society. It requires new policies, new partnerships, and community involvement so that sustainability is embedded within decision-making. Since this is quite a responsibility and requires every single person on the planet to change their habits, education and awareness are critical in encouraging people to make eco-friendly choices in their lives. With that in mind, sustainability will continue to build on existing efforts and there will be ongoing research and development to better understand the challenges of sustainability and assist in creating actionable solutions. It is critical to have a multidimensional outlook which reminds us how interconnected our ecological and economic and societal conditions are with respect to the world we want to create for ourselves. It is the commitment toward sustainability that will ensure our planet's very survival for generations to come.

References

African Union. (2022). *Great green wall initiative progress report*. African Union.

Al-Chalabi, M. (2015). Vertical farming: Skyscraper sustainability? *Sustainability, 7*(4), 4987–5000.

Altieri, M. A., Nicholls, C. I., & Henao, A. (2021). *Agroecology: The ecology of sustainable food systems*. CRC Press.

Bennett, E. M., Peterson, G. D., & Gordon, L. J. (2020). Ecosystem services as a unifying concept for the sustainable development goals. *Ecosystem Services, 41*, 101099.

Bocken, N. M. P., Short, S. W., Rana, P., & Evans, S. (2022). A literature and practice review to develop a 'circular business model' framework. *Journal of Cleaner Production, 97*, 16–20.
Brundtland Commission. (1987). *Our common future*. Oxford University Press.
California Air Resources Board. (2021). *California cap-and-trade program*. California Air Resources Board. https://ww2.arb.ca.gov/our-work/programs/cap-and-trade-program
Chakir, A., Chergui, M., Medromi, H., & Sayouti, A. (2015, November). An approach to select effectively the best framework IT according to the axes of the governance IT, to handle and to set up an objective IT. In *2015 Third World Conference on Complex Systems (WCCS)* (pp. 1–8). IEEE.
Chakir, A., Chergui, M., & Andry, J. F. (2020). A Smart Updater IT Governance Platform Based on Artificial Intelligence. *Advances in Science, Technology and Engineering Systems*, 5 (5), 47–53.
Chakir, A., Chergui, M., & Andry, J. F. (2021). A decisional smart approach for the adoption of the IT green. *Environment, Development and Sustainability, 23*, 8857–8871.
City of Amsterdam. (2020). *Amsterdam's Doughnut model*. City of Amsterdam. https://www.amsterdam.nl/en/policy/
Cook, J., et al. (2021). Misinformation and climate change: A systematic review of the literature. *Environmental Research Letters, 16*(11), 121–135.
Coyle, D. (2023). The limitations of doughnut economics in a complex world. *Journal of Economics Perspectives*.
Daly, H. E., & Farley, J. (2019). *Ecological economics: Principles and applications*. Island Press.
Dinter, M., Schmid, M., & Buhl, H. U. (2022). The influence of public awareness campaigns on sustainability behavior. *Sustainability, 14*(2), 785.
Dubey, R., Bryde, D. J., & Fynes, B. (2020). Sustainable supply chain management: A systematic literature review. *International Journal of Production Economics, 214*, 182–204.
Elkington, J. (1997). *Cannibals with forks: The triple bottom line of 21st century business*. New Society Publishers.
Ellen MacArthur Foundation. (2019). *Completing the Picture: How the Circular Economy Tackles Climate Change*. Ellen MacArthur Foundation. https://www.ellenmacarthurfoundation.org/
Ellen MacArthur Foundation. (2021). *Completing the picture: How the circular economy tackles climate change*. Ellen MacArthur Foundation. https://www.ellenmacarthurfoundation.org/assets/downloads/ce100/Completing-the-Picture.pdf
European Commission. (2019). *The European Green Deal*. European Commission. https://ec.europa.eu/info/sites/default/files/european-green-deal-communication_en.pdf
European Commission. (2020). *A new circular economy action plan for a cleaner and more competitive Europe*. European Commission. https://ec.europa.eu/environment/circular-economy/
Friede, G., Busch, T., & Bassen, A. (2015). ESG and financial performance: Aggregated evidence from more than 2000 empirical studies. *Journal of Sustainable Finance & Investment, 5*(4), 210–233.
García, A., et al. (2022). The environmental footprint of lithium-ion batteries: Challenges and opportunities. *Nature Reviews Materials, 7*(5), 453–471.
Garriga, E., & Melé, D. (2023). Corporate social responsibility theories: Mapping the territory. *Journal of Business Ethics, 53*, 51–71.
Geissdoerfer, M., Savaget, P., Bocken, N. M., & Hultink, E. J. (2018). The circular economy—A new sustainability paradigm? *Journal of Cleaner Production, 143*, 757–768.
Global Reporting Initiative (GRI). (2023). *Sustainability reporting trends*. Global Reporting Initiative (GRI). https://www.globalreporting.org/
Hahn, T., et al. (2023). The challenges of measuring the triple bottom line: A review of current practices. *Sustainability*.
Hoffman, A. J. (2021). Climate change and the role of Technology in the Sustainable Transition. *Sustainability, 13*(2), 874.
Honeyman, K. (2020). *B corporations: A global movement for good*. B Lab.
ICLEI. (2023). *Local governments for sustainability*. ICLEI. https://iclei.org/

International Energy Agency (IEA). (2023). *Global EV Outlook 2023*. IEA website. https://www.iea.org/reports/global-ev-outlook-2023

International Renewable Energy Agency (IRENA). (2022). *Renewable energy and jobs: Annual review 2022*. IRENA.

IPCC. (2021). *Climate change 2021: The physical science basis*. Cambridge University Press.

IPCC. (2022). *Climate change 2022: Impacts, adaptation, and vulnerability*. Intergovernmental Panel on Climate Change.

Kearns, R. A., & Collins, D. C. (2021). Participatory budgeting and local sustainability: A comparative analysis. *Local Environment, 26*(7), 901–917.

Kirchgeorg, M., Schneider, L., & Lang, M. (2021). Circular economy: The role of innovative technologies in the transition. *Journal of Cleaner Production, 297*, 126649.

Kirchgeorg, M., et al. (2023). *Barriers to the circular economy: Insights from a systematic literature review. Resources, Conservation and Recycling*. Elsevier.

KPMG. (2023). *The future of sustainability reporting: Aligning with the SDGs*. KPMG. https://home.kpmg/xx/en/home/insights/2023/01/future-of-sustainability-reporting.html

Le Blanc, D. (2015). Towards integration at last? The sustainable development goals as a network of targets. In *DESA working paper no. 141*. United Nations.

Lea, E., & Worsley, A. (2022). Community-supported agriculture: A systematic review of the literature. *Journal of Community Engagement and Scholarship, 15*(1), 1–9.

Linder, M., & Williander, M. (2023). Circular business models: A review and future research directions. In *Resources, conservation and recycling*. Elsevier.

Mazzucato, M. (2023). *The role of government in the sustainable development goals*. Public Administration Review.

Miller, J. R. (2020). Intergenerational learning: A key to cultural sustainability. *Cultural Sustainability Journal, 2*(1), 15–27.

Nykvist, B., & Nilsson, M. (2015). Rapidly falling costs of battery packs for electric vehicles. *Nature Climate Change, 5*(4), 327–332.

OECD. (2020). *Greening the economy with agriculture*. OECD Publishing.

Piketty, T. (2014). *Capital in the Twenty-First Century*. Harvard University Press.

Porter, M. E., & Kramer, M. R. (2019). *Creating shared value: Competitive advantage through social impact*. Harvard Business Review.

Pretty, J. (2020). The role of community in sustainable development. *Journal of Community Development, 55*(2), 129–143.

Rawls, J. (1999). *A theory of justice*. Harvard University Press.

Reinhardt, F. L. (2021). Innovation gaps in the sustainable transition: A policy perspective. *Journal of Business Research, 128*, 561–570.

Rojas-Romero, L. S., & Baird, L. (2021). The role of environmental education in fostering sustainable behaviors among youth: A systematic review. *Environmental Education Research, 27*(2), 270–285.

Sachs, J. D., et al. (2021). The sustainable development goals and the global goals for sustainable development: A comprehensive approach. *Global Policy, 12*(1), 1–15.

Sachs, J. D., et al. (2023). *Accelerating Progress on the sustainable development goals: The role of multi-stakeholder partnerships*. Nature Sustainability.

Santagata, W. (2019). Cultural heritage and sustainable development. *Journal of Cultural Heritage Management and Sustainable Development, 9*(3), 265–279.

Stiglitz, J. E. (2020). *Economics of the public sector*. W.W. Norton & Company.

UNESCO. (2005). *Convention on the protection and promotion of the diversity of cultural expressions*. United Nations Educational, Scientific and Cultural Organization.

UNESCO. (2018). *Cultural diversity and sustainable development*. United Nations Educational.

UNESCO. (2021). *Education for Sustainable Development: A Roadmap*. UNESCO website. https://en.unesco.org/themes/education-sustainable-development

UNFCCC. (2022). *Nationally Determined Contributions under the Paris Agreement*. UNFCCC. https://unfccc.int/sites/default/files/resource/NDC%20Synthesis%20Report%20March%202022.pdf

United Nations. (1972). *Report of the United Nations conference on the human environment*. United Nations.

United Nations. (2015). *Transforming our World: The 2030 Agenda for sustainable development*. United Nations.

United Nations Environment Programme. (2021). *Financing the sustainable development goals: A global overview*. UNEP.

Verplanken, B., & Roy, D. (2016). Empowering interventions to promote sustainable lifestyles: A review of the literature. *Environment and Behavior, 48*(3), 285–310.

WHO. (2021). *World health statistics 2021: Monitoring health for the SDGs*. World Health Organization.

Wilkinson, R., & Pickett, K. (2023). *Cities in the Doughnut: Urban planning for a sustainable future*. Ecological Economics.

Wolfert, S., Ge, L., Verdouw, C., & Bogaardt, M. J. (2017). Big data in smart farming—A review. *Agricultural Systems, 153*, 69–80.

World Bank. (2021a). *Building resilience: A global framework for climate action*. World Bank Publications.

World Bank. (2021b). *State and trends of carbon pricing 2021*. World Bank Publications.

World Bank. (2022). *Financing for sustainable development: A global perspective*. World Bank Publications.

World Economic Forum. (2023). *Global risks report 2023*. World Economic Forum.

Integrating AI Into Sustainability

Afzal Ahmed Soomro, Ashvin Gobal, and Mohamad Hanif Md Saad

1 Introduction

1.1 Background

Sustainability has become a pressing worldwide concern because of the increasing environmental problems, such as climate change, deforestation, and biodiversity loss. A varied technique that integrates sustainable practices in different sectors is required to address these interconnected problems. The following part defines the main aspects of sustainability in response to these challenges. AI plays an important role in environmental observation which includes weather pattern analysis to predict natural disasters. These improve the alertness and help to prepare to face the disaster (Begum et al., 2024).

Interdependence of Biodiversity and Climate Change

- Functional strategies for biodiversity and climate change must be highlighted at the same time as both of them are closely linked (Ambrósio et al., 2024).
- In regions such as grasslands and tropical non-forests, scenarios with integrated measures which include protected areas and sustainable land use will outstandingly improve the biodiversity results (Ambrósio et al., 2024).

Function of Corporate Social Responsibility (CSR)

- To improve sustainability and competitiveness, businesses are increasingly adopting green practices and CSR initiatives (Ali et al., 2023).

A. A. Soomro (✉)
Interdisciplinary Research Center for Industrial Nuclear Energy (IRC-INE), King Fahd University of Petroleum and Minerals (KFUPM), Dhahran, Saudi Arabia

A. Gobal · M. H. M. Saad
Department of Mechanical and Manufacturing Engineering, Universiti Kebangsaan Malaysia, Bangi, Selangor, Malaysia

A. Chakir et al. (eds.), *Green Society, Environmental Strategies and Sustainable Development*, https://doi.org/10.1007/978-3-031-94086-6_3

- To convert the endeavor into quantifiable, long-term outcomes and promote environmental conscious culture within businesses, a managerial support is very important (Ali et al., 2023).

Global Frameworks and Policies

- The definition of sustainable enhancement is fulfilling current demands without affecting future contemporaries by focusing on a balanced economic growth as well as protection to the environment (Mahanayak, 2024).
- Roadmap for addressing global challenges including climate action and biodiversity conservation is provided in international frameworks such as the Sustainable Development Goals (SDGs) (Mahanayak, 2024).

Issues such as pollution, urbanization, and income inequality still exist although a lot has been done to promote sustainability. Therefore, to guarantee a sustainable future for everybody, finding creative solutions and maintaining dedication are necessary (Mahanayak, 2024; Said et al., 2024). Besides, in order to promote sustainability across different departments, artificial intelligence (AI) is widely recognized as a revolutionizing tool.

Universally, resource management is one of the important factors in sustainability and AI technologies play a major role in filtering energy usage and resource distribution which indicates wastage decrement and enhanced effectiveness (Sohana, 2024). The implementation of AI in agriculture helps farmers to elevate harvest at the same time reducing ecological effect. This becomes a vital development in sustainable agricultural formalities (Shahvaroughi Farahani & Ghasemi, 2024).

The capability of predictive analysis is a notable implementation of AI in ecological sustainability. Accurate forecast of environmental risks is enabled using the capability of AI to define massive datasets and therefore promoting proactive decision-making (Sohana, 2024). To aid people to plan ahead for intense weather issue and decrease potential damages, machine learning models have shown a notable capability in predicting climate patterns (Shahvaroughi Farahani & Ghasemi, 2024).

Other than environmental sustainability and resource management, the implications of AI also benefit urban sustainability drastically. The aid of AI to the ecological urban enhancement has been documented by researchers which includes optimizing traffic flow and management of waste and energy consumption in the cities (Leal Filho et al., 2024a). The role of AI in creating smart cities which concurrently improves the quality of life while decreasing ecological footprints has been highlighted in several case studies (Leal Filho et al., 2024a).

Data privacy and ethical concerns are the challenges that have to be discussed to make sure a reasonable application of AI although it shows consequential opportunities for sustainability. To maximize the potential of AI, balancing innovation with responsible practices is a must to achieve sustainable development goals.

1.2 Justification for AI in Sustainability

The sustainability initiative across different sectors is improved by the efficiency of AI to process broad datasets and analyze patterns significantly. AI contributes real-time insights that instruct decision-making and resource management optimization by capitalizing technologies such as natural language processing as well as machine learning. This incorporation of AI into sustainability attempt is noticeable in various domains such as agriculture, finance, and education.

AI has transformed traditional farming methods through advanced technological solutions in agriculture sector. The resource application is enhanced by the AI systems, especially in water management while consequently decreasing chemical inputs via precision agriculture methods (Whig et al., 2024). Farmer can now improve soil health and crop productions through real-time observation and advanced data analysis while concurrently reducing their environmental footprint (Whig et al., 2024).

AI also has become a massively important tool for promoting sustainability in the financial sector. To assess environmental, social, and governance (ESG) aspects, AI tools exhibit impressive ability in analyzing extensive datasets. This will provide valuable insights into sustainable investment opportunities to the investors (Bai & Shanavas, 2024). Other than that, the risk management capabilities have been improved through the application of predictive analysis, especially regarding the climate change consequences and social inequalities (Bai & Shanavas, 2024)

AI has developed as an essential facilitator in the education sector, especially in post-secondary education organizations. The application of United Nations (UN) Sustainable Development Goals (SDGs) is supported actively by AI technologies by enhancing the campus operations and improving the research capabilities (Leal Filho et al., 2024b). The successful implementation of AI in university sustainability initiatives has been documented in numerous case studies, establishing its practical value across educational settings (Leal Filho et al., 2024b).

As been discussed in the introduction background, the problems like data privacy, algorithmic bias as well as moral interest have to be discussed although AI provides transformative opportunities for sustainability. This is important to guarantee an equal implementation (Shahvaroughi Farahani & Ghasemi, 2024; Bai & Shanavas, 2024).

Furthermore, the responses to dynamic environmental challenges can be disrupted due to traditional environmental strategies which often struggle with fixed policies and slow adaptation. The demand for more adaptive and innovative approaches that are able to respond to quick changes in environmental changes is highlighted in recent research. Diverse strategies which are able to optimize adaptability in environmental policies can be analyzed.

To develop their personnel for ecological problems, the automated systems are revolutionizing companies in this era of organizational compliance. LearnBox is an example automated tool which has resulted in notable capability (Kleshko et al., 2024). To discuss increasing environmental problems efficiently, organizations can make sure that their personnel is well-equipped by arranging centralized access to data and powerful observation of the learning development.

By current developments in adaptive algorithms, encouraging results have been created for environmental improvement. An important discovery in the ability of the system to respond to the environmental shifts has been represented by the Evolutionary Robust Policy Optimization (ERPO) method (Paul & Deshmukh, 2024). Notable performance has been shown by this to traditional reinforcement learning processes. To accommodate new environment conditions, these processes demonstrate superior ability in quickly adjusting policies.

The indigenous communities have built the adaptive strategies by illustrating the importance of traditional knowledge in environmental management. By headlining the crucial role of traditional ecological knowledge in environmental stewardship, these indigenous communities have refined complex techniques to discuss climate problems (Sarkar et al., 2023). Their strategies include the incorporation of sustainable etiquettes whereby it is both cost-efficient and comprehensive. Also, it offers valuable teachings that are able to inform and improve broader adaptation efforts.

To generate more flexible environmental policies, it is crucial to implement constructive technologies and use indigenous knowledge as traditional methods may be slow to adapt. By using multidimensional approach in the current face of swift environmental changes, the risk associated with the fixed policies can be reduced.

By the application of AI, environmental observation and alleviation efforts via prediction analysis, automation, and intelligent managerial are consequentially optimized. Via the implementation of AI technologies like Internet of Things (IoT) and remote sensing, real-world data acquisition and studies become achievable, and therefore, enhanced accuracy in environmental assessments and proactive management approaches can be achieved.

AI has presented brilliant capabilities that exceed the potential of traditional method in the field of climate modeling. By using machine learning, AI-driven models have demonstrated important advantages over the conventional general circulation models (GCMs) to study sophisticated datasets with unprecedented effectiveness. It decreases processing time by 50% which includes a 20–30% enhancement in prediction accuracy which is a remarkable result achieved through these improvements (Khadar Maideen et al., 2024). To support timely climate forecasts and effective plans to reduce global warming, the adaptive nature of these systems is the main asset (Khadar Maideen et al., 2024).

Environmental observation has been transformative through AI automation in particular applications such as bioremediation whereby AI systems enhance pollutant degradation by automating both data gathering and analysis procedures as well as adjusting environmental parameters in real time (Akeem Akinkunmi Akintola, 2024). For instance, this enhancement in air quality monitoring has been represented by the "AI Know You" (AIKU) model whereby it demonstrates and exemplifies markedly enhanced accuracy and responsiveness compared to conventional observation techniques by focusing on the capability of AI in dynamic environmental assessment (Manongga et al., 2024).

By implementation of AI in organizational managerial, substantial advantages for sustainability initiatives have been produced. AI systems can enhance both sustainability and operational effectiveness by helping organizations to effectively study carbon emissions and waste throughout their supply chains and aiding in instant adjustments (Ameh, 2024). Significant enhancements in resource allocation and environmental impact mitigation through the enhancement of AI technologies have been documented in various case studies. This is shown in their practical value in achieving sustainability goals (Ameh, 2024).

In other words, challenges like data quality, model transparency, and the risk of overfitting remain crucial concern while AI offers revolutionary potential. This must be discussed to completely realize its benefits in environmental management (Khadar Maideen et al., 2024; Alotaibi & Nassif, 2024).

1.3 Research Questions

This chapter is systematically constructed around the following main research questions, which guide the analysis of the capacity of AI in sustainability. The research questions that arise from the previous discussion include:

- How can AI-driven technologies enhance sustainability practices across various industries?
- How do AI improve green engineering principles to mitigate environmental impacts?
- What ethical concerns that must be discussed when applying AI in sustainability?

1.4 Research Objectives

Based on the research questions above, this chapter is intended to:

- Analyze the role of AI in discussing sustainability challenges across multiple industries.
- Investigate how AI achieves green engineering principles to reduce environmental effects.
- Address probable challenges which include ethical considerations and the digital gaps which could slow down AI integration in sustainability.

1.5 Chapter Structure

Subsequent to this section, the next sub-chapter focuses on the literature review on the implementations of AI in sustainability. The third sub-chapter addresses different AI-driven sustainability methods in agriculture, energy, transportation, and industrial manufacturing. The fourth sub-chapter discusses about the ethical concerns and the final sub-chapter provides the conclusion of this chapter with some information into the future of AI-driven sustainability agendas.

2 Literature Review

2.1 Introduction

Sustainability initiatives have prevailed as an encouraging method for discussing critical environmental challenges with the integration of artificial intelligence (AI). AI offers innovative result for environmental monitoring, resource improvement, and energy efficiency. Real-time data collection and studies have enhanced air and water quality assessments, climate modeling, and biodiversity tracking (Alotaibi & Nassif, 2024). Resource application and waste reduction in industries such as energy and manufacturing have been improved through AI-driven systems. This results in decreased carbon emissions and cost savings (Khan Burki et al., 2024; Jebbor et al., 2024). However, we need to address the importance for responsible deployment and robust governance frameworks because there are challenges when applying AI technologies, such as the energy consumption of the AI, data privacy interest, and ethical issues (Shahvaroughi Farahani & Ghasemi, 2024; Alotaibi & Nassif, 2024). The contributions of AI to sustainability are uncovered in this chapter together with the limitations and impact for the future progress.

2.2 AI in Environmental Monitoring

The environmental supervising by improving data accuracy, processing, and predictive abilities has remarkably revolutionized through AI-powered satellite imaging and remote sensing technologies. The ultimate contribution of these enhancement is to facilitate comprehensive ecosystem management and pollution detection. This ensures that better resource management and conservation efforts are provided.

Innovative solutions for detecting and analyzing pollutants have been offered to transform environmental monitoring through the application of AI and the Internet of Things (IoT). When combined IoT networks with AI technologies, real-time observation of hazardous substances is enabled. This notable improves the capability to track and respond to environmental pollutants (Arowolo et al., 2024). This is important for the effectiveness of the natural resource management as the collaboration facilitates a more distinct understanding of ecological processes. Researchers and policymakers can obtain up-to-date insights into environmental changes by leveraging IoT sensors and AI-driven analytics which can enable proactive procedures to reduce risks and preserve ecosystems.

Besides, the abilities of environmental monitoring have been enhanced further with the aid of deep learning algorithm. Significance in the analysis of remote sensing images has been enhanced up to 90% accuracy in urban settings by methods such as conventional neural networks (CNNs) (Liu & Zhang, 2024). The biodiversity observation and conservation efforts are made extremely useful by the ability of the models to outshine in the aspect of object detection and scene understanding. For example, deep learning algorithms applications have been vital in tracking wildlife population, determining habitat changes, and evaluating ecosystem health (Pettorelli et al., 2024). These developments emphasize the ability of AI to deal with difficult environmental issues by offering accurate and useful data (Fig. 1).

The comprehensive application of AI in environmental surveillance is still restricted by multiple issues, although with these notable developments. However, better assessment criteria and the model tolerance are the obstacles that still persist which hinders the progress of the developments (Tasneem & Islam, 2024). The reliability and effectiveness of AI systems in various environmental conditions are vital to ensure the broad application of AI in environmental surveillance. Interdisciplinary cooperation and the development of homogenized frameworks are the challenges

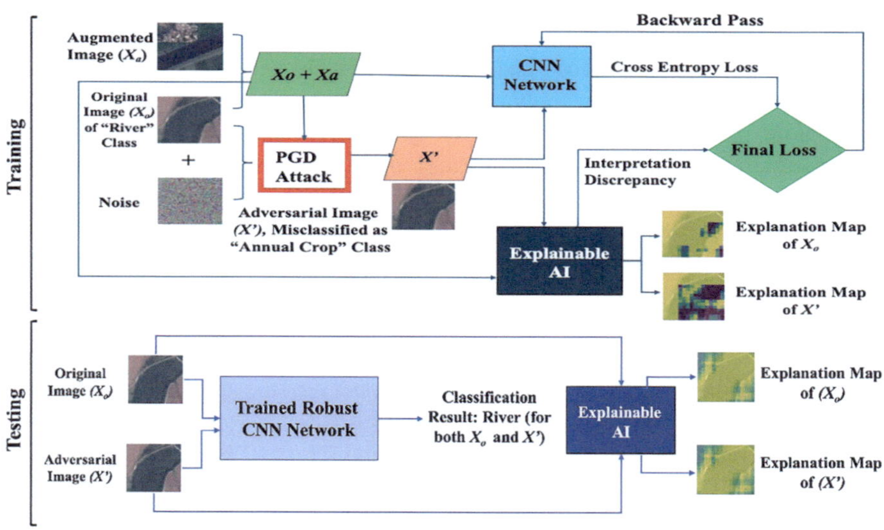

Fig. 1 Remote sensing using CNN (Tasneem & Islam, 2024)

that need to be highlighted to examine the performance of the AI models. These are the important limitations that when it is solved, it is able to be the key to unlock the maximum capability of AI and IoT in the environmental surveillance, clearing the path for more sustainable and resilient ecosystems.

Additionally, NASA's AI-driven Earth monitoring systems are paving the way to track environmental changes like deforestation, climatic trends, and ocean health. To contribute a comprehensive and timely data which is vital for environmental management and conservation efforts, these monitoring systems leverage enhanced technologies such as AI, remote sensing, and satellite imagery. Table 1 shows the summarized application and challenges of AI and Earth Monitoring Technologies.

2.3 AI in Green Engineering

The main concepts of green engineering focus on the designing methods which can mitigate waste and energy consumption, developing sustainability across a broad range of industries. These concepts systematically discuss the environmental problems of the world by supporting ecologically friendly, commercially feasible, and socially conscious behaviors. The understanding and application of green engineering encourage a new generation of experts committed to sustainable development through the implementation of green technologies into engineering education.

Green engineering is a revolutionary idea to design and manufacture, which emphasizes sustainability by mitigating environmental effect. The mitigation of waste at the source via effective design and process improvement is one of the main principles of green engineering. Industries can notably decrease material wastage and environmental hazard by applying strategies like lean manufacturing and circular economy models (Amol et al., 2024). Other vital principle is the energy efficiency, which facilitates the implementation of undepletable source of energy and energy-effective technologies into industrial methods. This will not only reduce the overall energy usage but also decreases the dependence on fossil fuels which can provide a cleaner and more sustainable future (Sukla et al., 2024). The application of sustainable materials in product design plays a major role in the environmental impact reduction. Sustainable materials which include biodegradable polymers and recycled materials promote resource conservation and reducing ecological footprints (Kushwaha et al., 2024). Table 2 gives a concise of the applications as well as the benefits.

To prepare the next generation of engineers and scientists to address global sustainability challenges, it is important to implement green engineering principles into educational programs. Students will be equipped with necessary skills to integrate sustainable practices in their future careers when importance is given to curriculum development. By including subjects like life cycle assessment, renewable energy systems, and sustainable design into assignments, education institutions can promote deeper understanding of green engineering concepts (Yue & Zhang, 2024).

Table 1 Application and challenges of AI and Earth Monitoring Technologies in environmental surveillance

Theme	Summary	Citations
Monitoring deforestation	Real-time data for identifying deforestation activities like illegal logging and forest fire can be provided by AI and remote sensing technologies	Haq et al. (2024), McCallum et al. (2023)
Climate patterns and ocean health	NASA'S plankton, aerosol, cloud, ocean ecosystem (PACE) mission deploys hyperspectral remote sensing to analyze oceanographic and atmospheric responses to climate change, enhancing understanding of marine carbon cycles and ecosystem effects. Aquatic ecosystems observation and water quality also aided by earth surveillance data	Gorman et al. (2019), Lisboa et al. (2024)
Applications and societal benefits	Global challenges like climate change, disaster management, and food security are supported by the earth surveillance data. It is applied for land cover mapping, carbon biomass assessment, and observation of air and water quality. AI improves these abilities for sustainable development	Kansakar and Hossain (2016), Ghamisi et al. (2024), Kumar et al. (2024)
Challenges and future directions	Challenges includes data accessibility, ethical concerns, and responsible AI practices. Assuring open data access, reducing biases, and improving precision are important for future developments in environmental surveillance	Lisboa et al. (2024), Ghamisi et al. (2024), Diniz et al. (2015)

Theme	Summary	Citations
Monitoring deforestation	Real-time data for identifying deforestation activities like illegal logging and forest fire can be provided by AI and remote sensing technologies	Haq et al. (2024), McCallum et al. (2023)
Climate patterns and ocean health	NASA'S plankton, aerosol, cloud, ocean ecosystem (PACE) mission deploys hyperspectral remote sensing to analyze oceanographic and atmospheric responses to climate change, enhancing understanding of marine carbon cycles and ecosystem effects. Aquatic ecosystems observation and water quality also aided by earth surveillance data	Gorman et al. (2019), Lisboa et al. (2024)
Applications and societal benefits	Global challenges like climate change, disaster management, and food security are supported by the earth surveillance data. It is applied for land cover mapping, carbon biomass assessment, and observation of air and water quality. AI improves these abilities for sustainable development	Kansakar and Hossain (2016), Ghamisi et al. (2024), Kumar et al. (2024)
Challenges and future directions	Challenges includes data accessibility, ethical concerns, and responsible AI practices. Assuring open data access, reducing biases, and improving precision are important for future developments in environmental surveillance	Lisboa et al. (2024), Ghamisi et al. (2024), Diniz et al. (2015)

Moreover, students can be provided with hands-on projects which focus on green technologies. This can improve their practical experience and creativity enhancement. As an example, students can apply theoretical knowledge to real-world problems by designing projects like solar-powered devices or developing waste-reduction prototypes can reinforce their understanding of sustainability (Yue & Zhang, 2024).

Table 2 Applications and benefits of green engineering principles

Principle	Description	Benefits	Citations
Reduction of waste	Decreasing waste at the source through effective design and process enhancement	Reduced cost of material, environmental pollution decreased	Amol et al. (2024)
Energy efficiency	Application of renewable energy and energy-efficient technology to reduce consumption	Decreased carbon emissions, decreased energy costs	Sukla et al. (2024)
Sustainable materials	Integrating eco-friendly materials into product design	Resource conservation, mitigate environmental degradation	Kushwaha et al. (2024)

Education programs can efficiently prepare students to handle pressing environmental challenges by combining theoretical knowledge with practical application. The consolidation of green engineering principles into education will benefit not only individual learners but also helps to provide comprehensive goal of creating a sustainable future.

2.4 *The Function of AI in Smart Cities*

The progression of intelligent urban areas Smart cities are characterized by enhanced resource utilization and improved urban sustainability, with artificial intelligence (AI) significantly contributing to this progress. Cities can optimize infrastructure management, enhance citizen experiences, and promote sustainable practices. This following section highlights the important aspect of the role of the AI in smart city development. Figure 2 shows the conventional energy cycle of urban energy.

To improve the effectiveness and sustainability, AI has developed as a revolutionizing tool to improve resource use across different urban systems. One of the main implementations is predictive maintenance whereby AI facilitates real-time analytics for infrastructure management. AI decreases resource wastage and increases the lifespan of critical assets by predicting potential failures and scheduling up-to-date maintenance (Rane et al., 2024a). Energy management is another important subject to be addressed where machine learning algorithms enhance energy usage in buildings, industries, and transportation system. These enhancements do not mitigate practical expenses only but together with that, it reduces ecological effect by reducing carbon emissions (Hadiyana & Ji-Hoon, 2024). To analyze traffic patterns to enhance flow, decrease congestion, and improve urban mortality, AI performs a vital function in traffic management where it aids to improved sustainability and habitable cities (Hadiyana & Ji-Hoon, 2024). In urban regions, AI technologies are increasingly used. AI technologies are increasingly used to enhance overall quality, particularly in public safety, where response times and safety measures are improved through advanced monitoring and predictive analysis.

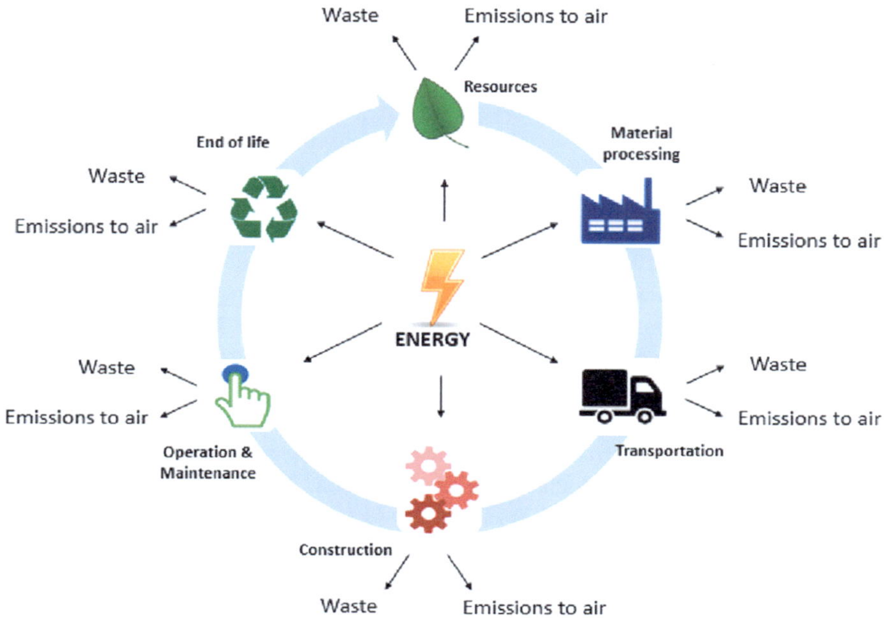

Fig. 2 Conventional energy cycle of urban energy (Haddad et al., 2021)

To assure quicker response to problems and reduce risk, this technology ultimately provides an impressive solution (Ullah et al., 2024). To simplify waste collection and recycling processes, AI also transforms waste management. By increasing recycling rates, intelligent system improves routes, decreases operational ineffectiveness, and benefits sustainability (Rane et al., 2024a). Moreover, by the applications of AI in healthcare services, urban well-being is also transformed by promoting access to medical care and improving operational effectiveness. From telemedicine platforms to predictive diagnostics, AI ensures that healthcare is more approachable, affordable, and effective for urban populations (Ullah et al., 2024).

To achieve better sustainability, adaptability and livability, cities can implement AI for resource optimization and improve the quality of life. Other than addressing current urban difficulties, these improvements also clear the path for smarter, greener, and more comprehensive cities in the future. Despite the advantages, challenges which include data confidentiality, algorithm intolerance, and the demand for moral guidelines remain vital. Discussing these issues is important to perform a responsible integration of AI in smart cities (Al-Raeei, 2024). It will be vital for sustainable urban development when innovation is balanced with ethical concerns.

3 AI-Driven Sustainability Solutions

3.1 Introduction

Artificial intelligence (AI) has become a revolutionary element in improving sustainability which provides a critical achievement to the Sustainable Development Goals (SDGs). AI enhances resource management, improves effectiveness, and encourages innovative solutions across sectors such as agriculture, energy, education, and environmental conservation by leveraging technologies such as machine learning and natural language processing. To decrease waste and increase production, AI facilitates precision farming in the agriculture sector (Shahvaroughi Farahani & Ghasemi, 2024; Tandon & Shaheen, 2024). In energy management, AI aids smart grids to enhance effectiveness and mitigate carbon footprints (Kumari et al., 2024). By personalizing learning and improving sustainability literacy, AI also provides sustainable education (Risang Baskara, 2024). Responsible deployment and interdisciplinary collaboration are heavily important as they need to address AI challenges which are stated in previous sub-chapters such as data privacy, bias, job displacement, and energy consumption of AI (Shahvaroughi Farahani & Ghasemi, 2024; Khan Burki et al., 2024). These considerations need to be discussed to ensure equitable access and balancing the contribution of AI with its societal and environmental influence for a legitimate transition toward sustainable development.

3.2 AI in Agriculture

Agriculture plays a major role in food production. However, it notably contributes to environmental challenges which include deforestation and water wastage. Although productivity is enhanced by the modern agricultural practices, it often leads to negative ecological impacts, demanding a deviation toward sustainable methods. This section presents the major environmental challenges caused by agriculture and the potential solution.

Deforestation and Habitat Loss

- The main factor of deforestation which leads to habitat loss and biodiversity mitigation is agricultural expansion (Kodaparthi et al., 2024).
- These problems become worse when crop yielding techniques such as monoculture and land conversion will weaken the ecosystem resilience (Adedibu, 2023).

Water Wastage and Pollution

- Extreme usage and pollution of water happen highly due to the usage of global freshwater in agriculture (Ayeni & Olagoke-Komolafe, 2024).
- Fertilizers and pesticide outflow contaminates river, whereas inefficient method of irrigation leads to water wastage (Kodaparthi et al., 2024; Adedibu, 2023).

Sustainable Solutions

- Sustainable methods such as agroforestry, precision farming, and organic agriculture, when applied, can reduce the environmental effects stated above (Adedibu, 2023; Ayeni & Olagoke-Komolafe, 2024).
- Besides, micro-watering and rainwater collection also improve water efficiency as well as decrease pollution (Ayeni & Olagoke-Komolafe, 2024).

Although the sustainable agriculture paves the way to solve these addressed problems, the transition of agricultural method needs vital investment in technology, policy support, and most importantly farmer education to ensure long-term sustainability and environmental health.

The optimization of irrigation, fertilization, and pest control has been notably improved by AI-driven precision farming methods. This can lead to enhanced agricultural productivity and sustainability. These methods can help farmers to make data-driven decisions that will increase resource effectiveness and crop production by leveraging sophisticated technologies like remote sensing, predictive analysis, machine learning.

By studying weather pattern and soil moisture levels to determine optimal irrigation schedules, AI-driven smart irrigation systems are transforming water management in agriculture. By using this method, water usage has reduced up to 30% which remarkably conserves this crucial natural resource (Gangwani, 2024). Furthermore, by using water only where and when it is needed, crop health and production is improved while reducing waste, all thanks to the precision irrigation methods (Raj et al., 2024). To underscore the evolutive capability of AI in modern agriculture, these innovations together not only improve water usage but also provide more sustainable and productive farming practices.

AI-enhanced fertilization practices are revolutionizing agriculture due to the use of fertilizers to improve effectiveness and sustainability. By leveraging the power of AI algorithms to analyze soil quality and nutrient requirements, precision fertilizer application can perform targeted use of fertilizers. The application of fertilizers has been reduced by 20% via this method and most importantly there is a positive impact on the production of crops as well as a reduction in waste and environmental hazard (Gangwani, 2024). Farmers are empowered with data-driven guidance together with doable information which is acquired from real-time data, ensuring the most efficient fertilization techniques customized to specific field conditions (Raj et al., 2024). These improvements not only increase resource efficiency but also provide more sustainable farming practices by focusing on the major role of AI in modern agriculture.

Pest control innovations which are driven by AI transform agriculture by improving the accuracy and effectiveness of pest management. To identify pest infestations at an early stage, AI technologies are used to automate pest detection systems which include drones and advanced imaging. This could provide an appropriate mediation which can remarkably decrease the demand for pesticide application while maintaining the quality and health of the crop (Raj et al., 2024). Moreover, by classifying pests and suggesting accurate control standard customized to specific threats, AI supports integrated pest management (IPM). This modus operandi mitigates crop losses and benefits sustainable farming protocols by reducing the dependence on chemical pesticides (Raj et al., 2024). Together, the improvements emphasize the capability of AI to revamp pest control into more competent, environmentally friendly, and resource-efficient method.

Blue River Technology, which started in 2011 and acquired by John Deere in 2017, has made itself as a pioneer in agricultural technology by adapting AI to benefit precision farming and sustainability. The company's innovative solutions which include AI-driven drones and the "See & Spray" system use computer vision and machine learning to study crop health and detect weeds with significant precision (Shamshiri et al., 2024). These technologies reduce chemical overuse by 90% at the same time enhancing yield efficiency and mitigating environmental impact by enabling precise application of herbicides (Monteiro & Santos, 2022). Blue River Technology has set a new benchmark for sustainability by addressing challenges like resource waste and labor shortages (Yeshe et al., 2022). Farmers all over the world have been empowered to adopt wiser protocols that align with global sustainability goals.

Although there are massive benefits, the need for substantial initial investments may slow down the extensive adoption of AI-driven precision farming. However, it tends to become a long-term asset which includes maximized profitability and environmental sustainability, underscoring the transformative capabilities of AI in agriculture.

3.3 AI in Energy Optimization

Energy usage is one of the important factors that leads to carbon emissions. AI systems ultimately enhance energy efficiency using predictive analysis and smart grid technologies. This improves the energy management and resource allocation. These systems can analyze historical and real-time data by integrating AI into intelligent networks. This can help to improve energy effectiveness as well as decrease operational costs.

Predictive Analytics in Energy Management

- AI uses machine learning algorithms such as linear regression and recurrent neural network (RNN) to predict energy demand which causes an increase in energy efficiency (Noviati et al., 2024).
- Predictive analytics provides better resource allocation which reduces reliance on fossil fuels and improved grid stability (Brynjolfsson & Hitt, 2000).

Smart Grid Technologies

- To enable real-time enhancement of energy distribution, smart grids are applied, which facilitate adaptability and resilience (Luo, 2024).

- Challenges such as load balancing and renewable energy have been addressed using the incorporation of multi-agent systems with machine learning, which improves managerial development (Luo, 2024).

Economic and Environmental Advantage

- AI-driven systems facilitate sustainable energy practices whereby it decreases carbon emissions and reduced energy waste (Krishna, 2024).
- The application of these technologies can lead to remarkable cost savings for consumers and organizations alike (Brynjolfsson & Hitt, 2000).

For example, Google's DeepMind AI has shown a remarkable enhancement in energy efficiency by decreasing data center cooling by 40% through real-time enhancement of energy usage (Al Munem et al., 2023). DeepMind studies huge amount of operational data to forecast and adjust cooling system requirements dynamically by using machine learning algorithms (Lazic et al., 2018). Besides reducing energy consumption, this innovative approach also contributes to substantial savings in cost and decreased carbon footprint. This demonstrates a transformative potential of AI in improving sustainability among industrial operations.

3.4 AI in Transportation

Emissions and congestions are one of the leading factors which affects sustainability in the transportation factor. To mitigate fuel consumption and emissions, AI-driven traffic management systems and autonomous vehicles remarkably enhance routes. At the same time, operational efficiency is also improved by leveraging advanced algorithms and real-time data whereby it contributes to environmental sustainability.

The application of AI revolutionizes traffic management by using predictive analysis and smart systems to improve urban mobility and sustainability. To enable the enhancement of traffic flow and reducing delays, machine learning algorithms are applied to forecast traffic congestion. For instance, to improve urban mobility, a unique framework which incorporates data from multiple sources has been presented to significantly enhance urban mobility (Poonam, 2024). Moreover, to analyze traffic patterns, Autonomous Smart Traffic Management (ASTM) systems deploy AI whereby it achieves significant outcomes such as a 50% rise in traffic flow rates and a 70% decrease in vehicle delays which contributes to more effective and sustainable urban environments (Goenawan, 2024).

In the era of autonomous vehicles, AI plays a major role in developing navigation, decision-making, and environmental sustainability. By learning from mass amount data, AI algorithms enhance routes which leads to a safer and more efficient travel while constantly enhancing performance (Nayeem et al., 2014). The incorporation of AI in autonomous vehicles supports sustainability goals by decreasing fuel consumption and reducing emissions via improved routing and reduced idle times (Mandal & Mohammed, 2024). As in real world application, Singapore has cut congestion times by 25% by leveraging AI-powered traffic monitoring as well as enhancing quality and mitigating fuel waste (Saxena, 2024). With these advances, the potential of artificial intelligence (AI) to revolutionize transportation systems for improved efficiency and environmental sustainability is evident.

3.5 AI in Industrial Sustainability

The implementation of AI in manufacturing industries is remarkably improving waste reduction and energy efficiency. This revolution aligns with the principle of Industry 5.0, where AI promotes collaboration between human and machines, enhancing workflows and reducing operational risks (Anang et al., 2024). To enhance the resource management and quality control, AI technologies like machine learning as well as predictive data are important as it leads to reduced waste and improved production efficiency (Mahesa Yahya et al., 2024). Other than that, the contribution to sustainability goals which is aided by AI-driven solutions in supply chain management leads to decrement in carbon footprint and enhanced energy distribution (Prakash Sah et al., 2024; Rane et al., 2024b).

Major Advantages of AI in Manufacturing

- **Waste Reduction:** To reduce material through better demand forecasting and stock control, AI enhances production processes (Rane et al., 2024b).
- **Energy Efficiency:** Expert system recognizes energy-saving opportunities, remarkably decreasing consumption of electricity in manufacturing operations (Ioshchikhes et al., 2024).
- **Reinforced Quality Control:** The improvement of defect detection and product quality is due to the implementation of AI whereby it reduces rework and waste (Mahesa Yahya et al., 2024).

For instance, Siemens applies AI-driven digital twins to facilitate factory operations. This leads to the reduction in energy consumption and mitigates environmental impacts (Cooke, 2021; Agostinelli Advisor Fabrizio Cumo S & Dottorato Di Ricerca, n.d.).

4 Ethical Concerns and Challenges

4.1 Introduction

Although AI provides massive potential for improving sustainability, its application comes with important ethical and technical challenges. Issues to be addressed include data bias, equitable AI access, and the environmental impact of AI model to ensure responsible AI application. The crucial ethical considerations and the demand for a balanced approach will be discussed in this sub-chapter.

4.2 *Algorithmic Bias and Data Integrity*

The reliance on massive datasets which can produce biases, inaccurate predictions, and unfair outcomes is the main challenge of AI, although it is undoubtedly powerful. Intensification of current inequalities can happen due to these biases, which can happen in multiple crucial sectors, including healthcare and finance. A multifunctional approach which focuses on understanding the sources of bias and applying efficient reduction strategies is implemented to discuss these issues.

Sources of Bias in AI Models

- **Collection of Data:** Due to inadequate representation of certain demographic groups, biased data will be generated which leads to inaccurate model predictions (Cross et al., 2024).
- **Development of Model:** Due to definite biases in labeled data and the overdependence on performance metrics, highlighting biases during model training is implied (Cross et al., 2024) (Anon, 2024).
- **Formation and Use:** The model introduces additional biases due to the user interactions with AI systems (Cross et al., 2024). This can cause the real-time performance of the AI model.

Mitigation Strategies

- **Distinct Datasets:** To mitigate bias, it is important to collect huge and representative number of datasets (Cross et al., 2024) (Nathim et al., 2024).
- **Measurement of Bias:** It is also important to operate metrics to measure bias at both datasets and model levels to help recognize and solve these issues early in the buildup process (Anon, 2024).
- **Plan for Fairness:** Applying methods such as Attribute Neutral Framework to enhance fairness without sacrificing performance can neutralize biased attributes in data (Hu et al., 2024).

Problems sustain although notable endeavors have been taken to comprehend and mitigate bias in AI. It requires continuous efforts to provide a balance between these conflicts by researching on the importance of trace-off between justice and model performance (Nathim et al., 2024). It is important to ensure a diverse and representative data acquisition to develop AI-driven sustainability efforts.

4.3 *The Digital Divide*

AI has the ability to drive crucial developments in emerging countries; however, its advantages are usually limited by technological imbalances. To make sure equitable access to AI, policies must focus on improving infrastructure, capacity building, as well as ethical governance. This method can ease sustainable development and connects the current technological gap.

Key Challenges in AI implementation

- **Infrastructure Limitations:** Many emerging countries have deficits in necessary digital infrastructure whereby it often inhibits the application of AI (Mutasa et al., 2024; Folorunso et al., 2024).
- **Skill Gaps:** There is lack of skilled professionals who are capable to leverage AI technologies and prioritize targeted education and training programs (Khan et al., 2024; Aderibigbe et al., 2023).
- **Regulatory Frameworks:** Insufficient governance structure often led to ethical considerations and data privacy problems. This is usually addressed in emerging nations (Folorunso et al., 2024; Ejjami, 2024).

Policy Recommendations

- **Comprehensive Frameworks:** It is important to develop comprehensive AI policies that discuss about local needs and promote ethical governance (Folorunso et al., 2024; Ejjami, 2024).
- **Public–Private Partnerships:** Resource allocation and innovation in AI can be strengthened by collaborations between governments and private sectors (Aderibigbe et al., 2023).

The potential of AI to revolutionize economies and enhance quality of life in the emerging nations is huge although there are significant challenges. However, the digital divide may continue to widen without strategic interventions, intensifying existing inequalities.

4.4 Energy Consumption of AI Models

There are worries about how AI systems may affect the environments because it needs a lot of processing power. Therefore, it is necessary to research into low-energy AI models and sustainable computing techniques. This research is important to reduce the carbon footprint which is linked with AI technologies.

Energy Consumption of AI Systems

- AI models, especially generative AI (GenAI) models such as ChatGPT, use substantial energy. For example, a single interaction with ChatGPT consumes about 2.9 watt-hours, notably more than traditional search tasks (Saklani & Singh, 2024).
- Deep reinforcement learning (DRL) models also generate high energy consumption, with different carbon footprints depending on the training environment and hardware used (De Araújo & Outtagarts, 2024).

Sustainable AI Practices

- Energy efficiency in AI systems can be improved by applying green cloud computing strategies. This will focus on improving resources use and integrating clean energy solutions (Saklani & Singh, 2024).

- Studies show a potential 28.7% mitigation in carbon dioxide emissions through sustainable practices by optimizing automated machine learning (AutoML) for energy efficiency (Castellanos-Nieves & García-Forte, 2024).

Policy and Future Directions
- To reduce the contradicting effects of AI energy utilization, policymakers are recommended to incentivize energy-effective AI models and bankroll in renewable energy infrastructure (Khan Burki et al., 2024).
- The further reduction in environmental footprint technology via the development of eco-friendly materials foe electronics with the support of AI (Esho et al., 2024).

It is important to take into account the wider consequences of AI's role in sustainability while focusing on the crucial low-energy AI models which include its potential to enhance energy consumption across various sectors.

4.5 Ethical AI Governance

To ensure that AI technologies align with sustainability goals, transparent regulations are important, concurrently, reducing civil and economic inequalities. The incorporation of tough regulatory frameworks can improve the explainability and accountability of AI systems, creating trust and equitable access.

Importance of Transparency in AI
- **Explainability Requirements:** It is important to understand decision-making processes in AI systems. However, current regulations lack these provisions for AI explainability (Chavula et al., 2024).
- **Right to Explanation:** To prevent biases and promote fairness in decision-making, users must have access to information about how AI algorithms operate (Shin et al., 2022).

The Function of AI in Sustainable Advancement
- **Addressing Inequalities:** To improve financial inclusion and poverty alleviation efforts, AI can aid in targeted interventions for marginalized communities (Muchokore & Kaur, 2024).
- **Environmental Implementation:** Toward contributing to environmental sustainability, AI technologies can enhance renewable energy use and improve waste management (Baig et al., 2024).

Regulatory Challenges
- **Gaps in Current Frameworks:** Environmental impacts of AI such as energy consumption and greenhouse gas emission are usually failed to be addressed by the current regulations (Alder et al., 2024).
- **Need for Comprehensive Policies:** To ensure equitable access to AI advantages, effective governance requires interdisciplinary collaboration and the establishment of ethical guideline (Muchokore & Kaur, 2024).

There is a risk that absurdly strict regulations could restrain innovation and limit the potential advantages of AI technologies although transparent regulations are crucial for aligning AI with sustainability goals. It is still a challenge to balance regulation with the demand for innovation.

5 Conclusion and Future Perspective

5.1 Summary of Key Findings

By offering innovative solutions to environmental problems and resource management, AI and green technologies are highly crucial in promoting sustainability across various sectors. Businesses can analyze massive data to make qualitative decision that aligns with sustainability by incorporating AI, hence enhancing resource use and reducing environmental impact.

AI for Sustainability

- **Resource Management:** AI technologies study sophisticated dataset which is related to resource utilization, enhancing water and energy use, handling raw material supplies efficiently, and decreasing waste. As a method to reduce resource depletion and excess waste, predictive analytics can predict resource demand accurately.
- **Sustainable Development Goals (SDGs):** From mitigating poverty and hunger. to promote clean energy and battling against climate change, AI helps track and analyze progress toward sustainability objectives. By facilitating the development of informed decisions to reduce the environmental impact through improved energy efficiency and enhanced manufacturing.
- **Green Initiatives:** AI helps in green initiatives by enhancing resource use, reducing waste, and improving efficiency. Other than that, AI also promotes the creation of sustainable urban settings via smart city solutions that control traffic flow, reduce congestion, increase public transportation, and improve air quality. Implementation of AI-driven waste management system facilitates sustainable agriculture practices and improves recycling efforts and decreases landfill waste by analyzing information on soil quality, climate sequence, and yield productions.
- **Product Development:** By stimulating design that uses fewer resources and generates lower waste, GenAI enables the design and development of effective, cost-efficient, and eco-friendly product. In terms of architecture, AI helps in designing the energy-efficient buildings with less ecological footprints, enhancing insulation, lighting, and energy consumption.
- **Supply Chains:** Besides reducing environmental impact and facilitating corporate social responsibility, AI identifies sustainable suppliers, enhances logistics for lesser emissions, and ensures ethical practices.
- **Monitoring and Awareness:** To ensure alignment with sustainability commitments and engaging consumers, AI tools observe progress toward environmental

objectives by providing information on sustainable practices and encouraging eco-friendly behavior.

By using innovations like solar panels, wind turbines, electric vehicles, and smart grids, the green technology and sustainability sector is revolutionizing industries. By aiding businesses to mitigate their ecological effects, enhance productivity, and meet administrative standards, AI-powered recycling, IoT, and big data are generating sustainability across various sectors. To ensure sustainability and responsibility is taken into account, the implementation of AI must consider the overall environmental effect. Renewable energy sources which function to power the data center is also paving its way forward through the application of green energy in AI infrastructure.

5.2 *Call to Action for Stakeholders*

The most effective way to utilize the immense capability of AI responsibly and comprehensively is through the corporation between governments, researchers, and industries. The advancements and implementation of AI bring up crucial ethical concerns which include bias, privacy, and accountability to ensure equitable and responsible use. AI models can sustain societal biases if trained on flawed data; however, it is also crucial to organize clear accountability for errors or harm which will be caused by AI due to their needs on massive amount of individual information which establishes considerations about data confidentiality as well as data protection. Ensuring open-accessibility in AI advancements is vital for all communities and not only privileged groups to prevent digital gaps whereby the challenges include affordability, equitability, and the demand of underprivileged populations which include individuals with disabilities or inadequate access to technology.

Governments can build administrative frameworks and bankroll in research to encourage ethical AI advancement and connect accessibility gaps. Researchers must play major role in constructing ethical guidelines in order to reduce bias and advancing explainable AI to improve transparency. By establishing responsible AI etiquette like working together with researchers and implementing necessary efforts, industrial AI technologies can be affordable and comprehensive. Hence, stakeholders can utilize the capability of AI together while ensuring it benefits society responsibly without any compromission.

5.3 *Future Directions*

Through the energy-demanding training and operation of large models, the urgency for low-energy AI models to improve sustainability is highlighted. When an enhanced algorithms and specialized hardware are developed, AI becomes more sustainable at the same time, it also supports edge computing applications on power-constrained devices and these help researchers to mitigate computational needs and reduce carbon emissions. Other than that, it is also importance to reduce the digital gaps and empower underprivileged communities by escalating the accessibility of AI. Open-source tools and AI education initiatives are some examples which can equalize access so that people in underprivileged countries can use AI to solve local issues like food, poverty, and illness. Moreover, to promote innovation and provide complete solutions, integrative cooperations are important. The incorporation of AI in healthcare is able to enhance diagnostics techniques, whereas in agriculture the AI application can power precision farming techniques. We can ensure AI is developed and deployed sustainably, inclusively by prioritizing the research in low-energy models, accessibility, and interdisciplinary efforts to benefit the humanity.

References

Adedibu, P. A. (2023). *Ecological problems of agriculture: Impacts and sustainable solutions*. ScienceOpen. https://doi.org/10.14293/PR2199.000145.v1

Aderibigbe, A. O., Ohenhen, P. E., Nwaobia, N. K., Gidiagba, J. O., & Ani, E. C. (2023). Artificial intelligence in developing countries: Bridging the gap between potential and implementation. *Computer Science & IT Research Journal, 4*, 185–199.

Agostinelli Advisor Fabrizio Cumo S, Dottorato Di Ricerca In *Energia E Ambiente PhD Program in energy and environment XXXVI Cycle Optimization and Management of Microgrids in the Built Environment based on Intelligent Digital Twins*.

Akeem Akinkunmi Akintola. (2024). AI-driven monitoring systems for bioremediation: Real-time data analysis and predictive modelling. *World Journal of Advanced Research and Reviews, 24*, 788–803.

Al Munem, A., Hossain, M. S., Prito, R. H., Tuhin, R. A., Reza, A. W., & Arefin, M. S. (2023). A deep reinforcement Learning framework for reducing energy consumption of server cooling system. In *Lecture notes in networks and systems* (pp. 32–42). Springer.

Alder N, Ebert K, Herbrich R, Hacker P. AI, climate, and transparency: Operationalizing and improving the AI act. 2024

Ali, A. Z., Ameer, K., Faisal, I., & Rauf, S. (2023). Forging sustainable development: exploring the impact of green initiatives and corporate social responsibility. *Sustainable Trends and Business Research, 1*, 76–81.

Alotaibi, E., & Nassif, N. (2024). Artificial intelligence in environmental monitoring: in-depth analysis. *Discover Artificial Intelligence, 4*, 4.

Al-Raeei, M. (2024). The smart future for sustainable development: Artificial intelligence solutions for sustainable urbanization. *Sustainable Development, 33*, 508.

Ambrósio, G., Doelman, J. C., Schipper, A. M., Stehfest, E., & van Vuuren, D. (2024). Global sustainability scenarios lead to regionally different outcomes for terrestrial biodiversity. *Environmental Research Letters, 19*, 104055.

Ameh, B. (2024). Digital tools and AI: Using technology to monitor carbon emissions and waste at each stage of the supply chain, enabling real-time adjustments for sustainability improvements. *International Journal of Science and Research Archive, 13*, 2741–2757.

Amol, M., Babar, C., Rohit, M., Shaha, D. (2024). A review paper on principles of green engineering.

Anang, A. N., Obidi, P. O., Mesogboriwon, A. O., Obidi, J. O., Kuubata, M., & Ogunbiyi, D. (2024). The role of artificial intelligence in industry 5.0: Enhancing human-machine collaboration. *World Journal of Advanced Research and Reviews, 24*, 380–400.

Anon. (2024). *Demographic bias in machine learning: measuring transference from dataset bias to model predictions.*

Arowolo, M. E., Aaron, W. C., Kugbiyi, A. O., et al. (2024). Integrating AI enhanced remote sensing technologies with IOT networks for precision environmental monitoring and predicative ecosystem management. *World Journal of Advanced Research and Reviews, 23*, 2156–2166.

Ayeni, O., & Olagoke-Komolafe, O. E. (2024). Environmental impact of modern agricultural practices: Strategies for reducing carbon footprint and promoting conservation. *International Journal of Management & Entrepreneurship Research, 6*, 3082–3095.

Bai, R. T., & Shanavas, R. J. A. (2024). Sustainable finance and use of artificial intelligence in investment decision making. *International Journal of Advanced Research (Indore), 12*, 1212–1218.

Mirza Abdul Aleem Baig, Huang J, Xu J. 2024 *AI for United Nations sustainable development goals.*

Begum, A., Naim, A., & Sabahath, A. (2024). *The impact of AI on sustainability* (pp. 99–113). IGI Global.

Brynjolfsson, E., & Hitt, L. M. (2000). Beyond computation: Information technology. *Organizational Transformation and Business Performance., 14*, 23.

Castellanos-Nieves, D., & García-Forte, L. (2024). Strategies of automated Machine Learning for energy sustainability in green artificial intelligence. *Applied Sciences (Switzerland), 14*, 14.

Chavula, P., Addy, A., Kayusi, F., Turyasingura, B., & Mensah, G. B. (2024). Demystifying the black box: A technical analysis of transparency requirements in clinical AI systems. *Africa Journal for Public Health Medicine & Nursing, 2024*, 14–31.

Cooke, P. (2021). Image and reality: 'Digital twins' in smart factory automotive process innovation—Critical issues. *Regional Studies, 55*, 1630–1641.

Cross, J. L., Choma, M. A., & Onofrey, J. A. (2024). Bias in medical AI: Implications for clinical decision-making. *PLOS Digital Health, 3*, e0000651.

De Araújo, L., & Outtagarts, A. (2024). Sustainability of AI-Energy consumption estimation of an mMTC admission control. In *7th conference on cloud and internet of things (CIoT)* (pp. 1–7). IEEE.

Diniz, C. G., Souza, A. A. D. A., Santos, D. C., et al. (2015). DETER-B: The new Amazon near real-time deforestation detection system. *IEEE Journal of Selected Topics in Applied Earth Observations and Remote Sensing, 8*, 3619–3628.

Ejjami, R. (2024). The AI and big data equitable transformation theory (ABETT): a pathway to inclusive development and technological empowerment. *International Journal for Multidisciplinary Research, 6*, 1.

Esho, A. O.-O., Aderamo, A. T., & Olisakwe, H. C. (2024). Sustainability-driven electrical engineering optimizing energy efficiency through ai and developing eco-friendly electronics. *Global Journal of Research in Engineering and Technology, 2*, 039–048.

Folorunso, A., Olanipekun, K., Adewumi, T., & Samuel, B. (2024). A policy framework on AI usage in developing countries and its impact. *Global Journal of Engineering and Technology Advances, 21*, 154–166.

Gangwani, N. (2024). AI-driven precision agriculture: optimizing crop yield and resource efficiency. *Computer, 6*, 1.

Ghamisi, P., Yu, W., Marinoni, A. et al. (2024) Responsible AI for earth observation.

Goenawan, C. R. (2024). *ASTM: Autonomous smart traffic management system using artificial intelligence CNN and LSTM*

Gorman, E. T., Kubalak, D. A., Patel, D., et al. (2019). *The NASA Plankton, Aerosol, Cloud, ocean Ecosystem (PACE) mission: An emerging era of global, hyperspectral Earth system remote sensing †*. SPIE.

Haddad, A., Hammad, A., Castro, D., Vasco, D., & Soares, C. A. P. (2021). Framework for assessing urban energy sustainability. *Sustainability (Switzerland), 13*, 13.
Hadiyana, T., & Ji-Hoon, S. (2024). ITEJ information technology Engineering Journals AI-driven urban planning: Enhancing efficiency and sustainability in smart cities. *ITEJ, 9*, 23–35.
Haq, B., Jamshed, M. A., Member, S., et al. (2024). Tech-driven Forest conservation: Combating deforestation with internet of things, artificial intelligence, and remote sensing. *IEEE Internet of Things Journal, 11*, 24551–24568.
Hu, L., Li, D., Liu, H., et al. (2024). Enhancing fairness in AI-enabled medical systems with the attribute neutral framework. *Nature Communications, 15*, 8767.
Ioshchikhes, B., Frank, M., & Weigold, M. (2024). A systematic review of expert Systems for Improving Energy Efficiency in the manufacturing industry. *Energies, 17*, 4780.
Jebbor, I., Benmamoun, Z., & Hachmi, H. (2024). Revolutionizing cleaner production: The role of artificial intelligence in enhancing sustainability across industries. *Journal of Infrastructure, Policy and Development, 8*, 8.
Kansakar, P., & Hossain, F. (2016). A review of applications of satellite earth observation data for global societal benefit and stewardship of planet earth. *Space Policy, 36*, 46–54.
Khadar Maideen, A. A., Mohammed Nawaz Basha, S., & Afsal Basha, V. A. (2024). Effective utilisation of AI to improve global warming Mitigation strategies through predictive climate modelling. *International Journal of Data Informatics and Intelligent Computing, 3*, 43–52.
Khan Burki, A., Normen Ahamed Mafaz, M., Ahmad, Z., Zulfaka, A., & Yazid Bin Isa, M. (2024). Artificial Intelligence and environmental sustainability: insights from PLS-SEM on resource efficiency and carbon emission reduction. *OPSearch American Journal of Open Research, 3*, 277.
Khan, M. S., Umer, H., & Faruqe, F. (2024). Artificial intelligence for low income countries. *Humanities and Social Sciences Communications, 11*, 1422.
Kleshko, I. I., Tynchenko, V. S., Grigorev, D. A., Fedorova, A. V., & Yushkova, L. V. (2024). RETRACTED: Automation of registration and training processes for more efficient adaptation of employees to environmental challenges. *E3S Web of Conferences, 549*, 08004.
Kodaparthi, A., Kondakindi, V. R., Kehkashaan, L., et al. (2024). *Environmental conservation for sustainable agriculture* (pp. 15–45). Springer.
Krishna, P. M. (2024). *Green energy revolution: AI-powered systems for sustainable power management chapter overview* (Vol. 4). National Education Services.
Kumar, C. S., Mohan, C. R., Kumar, B. S., Singh, N., Lavanya, P., & Banerjee, D. (2024). Novel deep Learning approaches to environmental management with sustainability. In *2024 international conference on communication, computer sciences and engineering (IC3SE)* (pp. 1484–1489). IEEE.
Kumari, S., Kumari, A., & Waoo, A. A. (2024). AI for sustainable development: innovations & applications. *ShodhKosh: Journal of Visual and Performing Arts, 5*, 5.
Kushwaha, A. K., Raha DrS, Kushwaha, M., & Kurup DrPS. (2024). *Environmental engineering*. AMKCORP Research Technologies Private Limited.
Lazic, N., Lu, T., Boutilier, C., et al. (2018). Data center cooling using model-predictive control. *Advances in Neural Information Processing Systems, 31*, 1.
Leal Filho, W., Mbah, M. F., Dinis, M. A. P., et al. (2024a). The role of artificial intelligence in the implementation of the UN sustainable development goal 11: Fostering sustainable cities and communities. *Cities, 150*, 105021.
Leal Filho, W., Ribeiro, P. C. C., Mazutti, J., et al. (2024b). Using artificial intelligence to implement the UN sustainable development goals at higher education institutions. *International Journal of Sustainable Development & World Ecology, 31*, 726–745.
Lisboa, F., Brotas, V., & Santos, F. D. (2024). Earth observation—An essential tool towards effective aquatic ecosystems' management under a climate in change. *Remote Sensing, 16*, 2597.
Liu, T., & Zhang, B. A. (2024). UAV-Based Remote Sensing Image Automatic Monitoring System Empowered by Artificial Intelligence. In *International conference on intelligent algorithms for computational intelligence systems (IACIS)* (pp. 1–5). IEEE.

Luo D. (2024) *Enhancing smart grid efficiency through multi-agent systems: A Machine Learning approach for optimal decision making*. https://www.preprints.org/manuscript/202411.0687/v1.
Mahanayak, B. (2024). Meaning and concept of sustainable development for the protection of environment and the role of India. *World Journal of Advanced Research and Reviews, 22*, 1460–1465.
Mahesa Yahya, L., Hidayat, D., & Yuli, V. A. (2024). *Application of artificial intelligence to improve production process efficiency in manufacturing industry article info abstract* (Vol. 2, p. 223).
Mandal, J., & Mohammed, I. (2024). Implementation of AI transportation routing in reverse logistics to reduce CO2 footprint. *International Journal of Supply Chain Management, 9*, 1–12.
Manongga, D., Rahardja, U., Sembiring, I., Aini, Q., & Wahab, A. (2024). Improving the air quality monitoring framework using artificial intelligence for environmentally conscious development. *HighTech and Innovation Journal, 5*, 794–813.
McCallum, I., Walker, J., Fritz, S., et al. (2023). Crowd-driven deep Learning tracks Amazon deforestation. *Remote Sensing, 15*, 15.
Monteiro, A., & Santos, S. (2022). Sustainable approach to Weed Management: The role of precision Weed Management. *Agronomy, 12*, 118.
Muchokore, F., & Kaur, D. G. (2024). Integration of artificial intelligence into sustainable development goals in India. *International Journal of Scientific Research in Engineering and Management, 08*, 1–9.
Mutasa, E. T., Dhiwwale, C., & Gopal, S. S. A. (2024). Artificial intelligence in developing economies: Unpacking Business innovations, prospects, and challenges. *International Journal of Academic Research in Business and Social Sciences, 14*, 63.
Nathim, K. W., Hameed, N. A., Salih, S. A., Taher, N. A., Salman, H. M., & Chornomordenko, D. (2024). Ethical AI with balancing bias mitigation and fairness in machine learning models. In *36th conference of open innovations association (FRUCT)* (pp. 797–807). IEEE.
Nayeem, T., Kumar, S., Rathore, R., & Bhardwaj, A. (2014). AI system for autonomous vehicles. *International Journal for Multidisciplinary Research, 6*, 1.
Noviati, N. D., Maulina, S. D., & Smith, S. (2024). Smart grids: Integrating AI for efficient renewable energy utilization. *International Transactions on Artificial Intelligence (ITALIC), 3*, 1–10.
Paul, S., Deshmukh, J. V. (2024). *Survival of the fittest: Evolutionary adaptation of policies for environmental shifts*
Pettorelli, N., Williams, J., & Schulte to Bühne H, Crowson M. (2024). Deep learning and satellite remote sensing for biodiversity monitoring and conservation. *Remote Sensing in Ecology and Conservation, 11*, 123.
Poonam, K. M. (2024). AI-driven predictive analysis for urban traffic management: A novel approach. *International Journal of Innovative Science and Research Technology (IJISRT), 2024*, 2270–2277.
Prakash Sah, B., Begum, S., Rahman Bhuiyan, M., & Shahjalal, M. (2024). The role of AI in promoting sustainability within the manufacturing supply chain achieving lean and green objectives. *Academic Journal on Business Administration, 4*, 79.
Raj, P., Gayathri, N., & Kathrine, G. J. W. (2024). *Artificial intelligence for precision agriculture*. Auerbach Publications.
Rane, N. L., Rane, J., Paramesha, M., & Kaya, Ö. (2024a). Artificial intelligence, machine learning, and deep learning for enabling smart and sustainable cities and infrastructure. In *Artificial intelligence and industry in society 5.0*. Deep Science Publishing.
Rane, N. L., Kaya, Ö., & Rane, J. (2024b). Artificial intelligence, machine learning, and deep learning applications in smart and sustainable industry transformation. In *Artificial intelligence, Machine Learning, and deep Learning for sustainable industry 5.0*. Deep Science Publishing.
Risang Baskara, F. X. (2024). Generative AI as an enabler of sustainable education: Theoretical perspectives and future directions. *British Journal of Teacher Education and Pedagogy, 3*, 122.
Said, M. A., Hassan, F. U., Islam, M. S., et al. (2024). *Environmental sustainability* (pp. 1–15). IGI Global.

Saklani, S., & Singh, D. (2024). Minimizing carbon emissions by improving water and energy use efficiencies in AI servers: A green cloud computing strategy for sustainable artificial intelligence systems. *International Journal of Innovative Science and Research Technology (IJISRT), 9*, 1822–1824.

Sarkar, M., Gorai, P., & Modak, B. K. (2023). *Indigenous strategies and adaptive approaches to scrabble recent climate crisis in two districts (Bankura and Purulia) of West Bengal, India* (pp. 75–101). Springer.

Saxena, R. R. (2024). *Artificial intelligence in traffic systems.*

Shahvaroughi Farahani, M., & Ghasemi, G. (2024). How artificial intelligence plays a role in achieving sustainable development goals? *Sustainable Economies, 2024*, 66.

Shamshiri, R. R., Shafian, S. A., & Hameed, I. (2024). *Precision agriculture—emerging technologies*. IntechOpen.

Shin, D., Kee, K. F., & Shin, E. Y. (2022). Algorithm awareness: Why user awareness is critical for personal privacy in the adoption of algorithmic platforms? *International Journal of Information Management, 65*, 102494.

Sohana, A. M. (2024). *Harnessing Technology for Environmental Sustainability: Utilizing AI to tackle global ecological challenges* (Vol. 2, p. 49). JAIGS.

Sukla, B., MrR, J., DrS, B., Jasmine, M. M., & Sundaram, S. (2024). Innovations in sustainable engineering: A review of green technologies. *Educational Administration: Theory and Practice, 30*, 3667–3675.

Tandon, N., & Shaheen, M. (2024). *AI-driven innovations for environmental sustainability across sectors* (pp. 429–454). IGI Global.

Tasneem, S., & Islam, K. A. (2024). Improve adversarial robustness of AI Models in remote sensing via data-augmentation and explainable-AI methods. *Remote Sensing, 16*, 3210.

Ullah, A., Quddusi, S., & Haider, I. (2024). Incorporation of artificial intelligence in enhancing quality of life in smart cities. *American Journal of Artificial Intelligence, 8*, 48–54.

Whig, P., Sharma, P., Aneja, N., Elngar, A. A., & Silva, N. (2024). *Artificial intelligence and Machine Learning for sustainable development*. CRC Press.

Yeshe, A., Gourkhede, P. H., Vaidya Vasantrao, P. H., Marathwada, N., & Vidyapeeth, K. (2022). Blue river technology: futuristic approach of precision farming. *Just Agriculture, 2*, 1.

Yue, C., & Zhang, Z. (2024). Integration of green techniques into undergraduate-level chemical design courses for practising the concept of sustainable development and developing green education. *European Journal of Education, 59*, 59.

The Role of Policy in Environmental Strategies

Khalid Balar

1 Introduction

To address the global sustainability challenge, it is critical to understand how political structures and environmental policies interact. This chapter investigates the role of political systems in promoting sustainable development on a global, national, and local scale (Sachs, 2015). We use both scientific research and local experience to demonstrate the importance of informed, ethical policymaking (Larrue, 2017). This chapter presents real-life scenarios that provide practical advice and forward-looking measures for achieving long-term environmental goals. Political frameworks dictating environmental action serve as the foundation for sustainability efforts. International treaties, such as the Paris Agreement and the United Nations Sustainable Development Goals (SDGs) (OECD, 2019), establish the framework for coordinated global action. Although these frameworks prioritize mutual accountability and international collaboration, they frequently face issues such as carbon taxation, renewable energy subsidies, and industrial emission reduction regulations (Moomaw, 2012).

Municipal and regional policies at the local level have a significant impact on the implementation of sustainability measures. The separation of community and government through waste management programs, urban green spaces, and other initiatives highlights the importance of grassroots action (Ostrom, 1990). Local initiatives frequently demonstrate a thorough understanding of specific environmental and social conditions, highlighting the importance of including diverse perspectives in broader policy discussions. This chapter emphasizes the importance of combining scientific research with local knowledge when developing effective policies

K. Balar (✉)
Ain Chock Faculty of Legal, Economic and Social Sciences, Hassan II University Casablanca, Casablanca, Morocco
e-mail: lncs@springer.com

A. Chakir et al. (eds.), *Green Society, Environmental Strategies and Sustainable Development*, https://doi.org/10.1007/978-3-031-94086-6_4

(Lascoums, n.d.). From climate modeling to resource management, scientists have access to important data and insights that can help with decision-making. Simultaneously, folklore and indigenous knowledge provide a complex understanding of ecosystems and sustainable practices developed over centuries (Reid, 2016). By combining knowledge systems, policymakers can create culturally sensitive and scientifically sound strategies. Despite its potential, this integration does not come without challenges. A significant gap between academic proficiency and local perspective can stymie collaboration. Political and institutional factors frequently obstruct the inclusion of underrepresented voices in policymaking (Biermann, 2008). These issues can only be addressed by actively participating in dialogue and creating open and accessible learning environments. Practical examples highlight the advantages and disadvantages of environmental policies. The renewable energy transition in Denmark demonstrates how wind energy can be used as a political priority to drive progress toward sustainability (Nations, 2015). Through reforestation and carbon neutrality programs, Costa Rica can enhance biodiversity and climate by supporting forest conservation efforts through legislative measures. Meanwhile, China's ambitious Green Belt and Road Project demonstrates how difficult it is to strike a balance between industrial development and environmental goals (Wilfried et al., 2014). This chapter emphasizes the importance of taking progressive, ethical, and comprehensive approaches to addressing long-term sustainability concerns. Environmental policies can be made more efficient and transparent by utilizing technology such as smart grids, AI-powered resource management, and blockchain-based carbon credits (Géraldine & Fano, n.d.). Green practices can be encouraged through financial instruments such as green bonds and carbon markets, which provide additional incentives. To ensure that today's decisions do not jeopardize the future, ethical policy must prioritize ecological integrity and a sense of intergenerational justice (Rockström et al., 2009). A comprehensive and long-term solution requires the involvement of stakeholders from various sectors, including government, private industry, civil society, and local communities.

This chapter ends by emphasizing the role of politics in shaping environmental policy at all levels. Policymakers can tackle difficult issues like sustainability by integrating scientific and local information, learning from global case studies, and implementing new and ethical measures. This section encourages daring and collaborative initiatives to build a strong green society.

2 The Nexus Between Sustainable Development and Environmental Legislation

Sustainable development has emerged as one of the most pressing global challenges of our time, necessitating a delicate balance between economic growth, social justice, and environmental preservation (UNEP, 2021). At this point, environmental policymaking is critical in shaping government, corporate, and community strategies, behaviors, and commitments. Legislation that addresses pressing environmental issues while also promoting sustainable practices is critical to societal well-being. The interdependence of social, economic, and environmental issues requires the implementation of sophisticated and well-integrated policies (Bank, 2010). Climate change, resource depletion, and biodiversity loss are not isolated problems; they are part of a larger environmental puzzle that includes poverty, economic inequality, and international trade.

Sustainable development has emerged as one of the most pressing global challenges of our time, necessitating a delicate balance between economic growth, social justice, and environmental preservation (Meadows et al., 1972). At this point, environmental policymaking is critical in shaping government, corporate, and community strategies, behaviors, and commitments. Legislation that addresses pressing environmental issues while also promoting sustainable practices is critical to societal well-being. The interdependence of social, economic, and environmental issues requires the implementation of sophisticated and well-integrated policies. Climate change, resource depletion, and biodiversity loss are not isolated problems; they are part of a larger environmental puzzle that includes poverty, economic inequality, and international trade (IPCC, 2022).

3 Key Concepts in Environmental Policymaking

Environmental policy is an important component of sustainability initiatives, as it provides the necessary legislation and regulations to address environmental issues while also taking into account social and economic goals (Anne et al., n.d.). This section explains the basic concept of environmental policy as well as the various levels of implementation, such as national, international, and local. The study also looks into the impact of these systems on corporate behavior and promotes community participation in sustainability initiatives.

3.1 Fundamental Principles

3.1.1 Definition of Environmental Policymaking

The primary goal of environmental policymaking is to develop and implement legislative and regulatory actions that will reduce environmental degradation and promote sustainable development (Simms, 2010). Governments, organizations, and communities use these policies to address pressing issues such as climate change, pollution, resource depletion, and habitat destruction, all while balancing human health and environmental well-being.

3.1.2 The Triple Bottom Line

The triple bottom line, which prioritizes economic growth, social justice, and environmental protection, is an essential component of sound environmental policymaking (Ostrom, 1990). Sustainability necessitates that policymakers consider the trade-offs and synergies between these three dimensions, ensuring that none is pursued at the expense of the others. This comprehensive approach is critical for ensuring environmental strategies' inclusiveness and long-term resilience.

3.2 Policies Adopted at the National, International, and Local Levels

3.2.1 National Policies

National governments are responsible for developing and enforcing environmental regulations that affect industries' and individuals' choices (Reid, 2016). Examples include taxes that penalize activities or products with a negative environmental impact, such as carbon taxes on fossil fuels. These measures generate revenue for sustainability projects while encouraging cleaner practices. Carbon pricing refers to the implementation of policies that penalize carbon emissions while encouraging businesses to adopt low-carbon technologies (Silvia et al., 2014). Finally, there are energy projects that use financial aid to accelerate the transition to cleaner renewable energy sources such as solar or wind power.

3.2.2 International Policies

The need for coordinated efforts to address global environmental issues is evident (Biermann, 2008). The key international frameworks include: The Paris Climate Agreement is an international treaty aimed at limiting global temperature rise by reducing greenhouse gas emissions.

The Kyoto Protocol, a precursor to the Paris Agreement, seeks to set binding emission reduction targets for developed countries. Universal framework: UN Sustainable Development Goals (SDGs), which include specific targets for climate action, clean energy, and consumption (OECD, 2019). The accords provide direction, establish standards, and encourage international cooperation to address global environmental challenges.

3.2.3 Local Policies

Local governments and communities play an important role in developing policies that are appropriate for their specific regions (Ostrom, 1990). Examples of community-based initiatives that promote sustainability include local recycling programs and conservation projects. Urban environmental management policies, such as waste management systems, green space initiatives, and sustainable public transportation projects, play a central role in municipal governance. Localized initiatives, which address specific environmental challenges within a community or region, frequently supplement national and international policies.

3.2.4 Impact on Corporate Behavior and Community Involvement

Environmental policies have a direct impact on corporate practices, as they establish regulatory guidelines, incentive programs, and penalizing fines (Jean-Pierre & Dominique, 2019). For example, carbon pricing or environmental taxes compel businesses to adopt greener practices such as energy efficiency measures or sustainable sourcing. Furthermore, subsidies and tax cuts encourage the development of clean technologies and renewable energy. Community-level policies are intended to encourage individuals and groups to participate actively in sustainability initiatives (Reid, 2016). The promotion of local waste management initiatives or urban greening projects fosters a sense of ownership and responsibility in residents. Furthermore, participatory governance approaches that involve communities in policy formulation and implementation improve the effectiveness and legitimacy of these interventions.

This section focuses on the fundamental principles and policy measures that define the complex role of environmental policy in promoting sustainability. Furthermore, it emphasizes the importance of coordinating policies at all levels to create a cohesive and productive framework for dealing with environmental issues.

4 Opportunities and Challenges of Environmental Policies

Environmental policies are critical for addressing ecological crises and promoting sustainable development (James et al., n.d.). However, their effectiveness is dependent on how well opportunities are capitalized and challenges are mitigated. This section critically evaluates the dual aspects of environmental legislation, focusing on both its potential and the challenges that policymakers face.

4.1 Opportunities Brought by Environmental Policies

4.1.1 Advancing Corporate Environmental Responsibility

Environmental policies are a powerful tool for promoting corporate environmental responsibility (Makarius et al., 2020). Regulations such as carbon pricing, emission standards, and sustainability reporting requirements force businesses to adopt greener practices. Furthermore, incentives such as tax breaks and subsidies for renewable energy and clean technologies encourage businesses to invest in environmentally friendly innovations. These measures not only reduce environmental damage, but they also boost a company's reputation and competitiveness in an increasingly environmentally conscious market.

4.1.2 Empowering Local Communities

Policies aimed at encouraging local participation enable communities to contribute to sustainability initiatives (Lascoums, n.d.). Community-driven waste management programs, conservation projects, and local renewable energy cooperatives are some examples. By involving communities in decision-making and implementation, these policies increase public ownership, foster trust, and generate more effective and contextually relevant solutions.

4.1.3 Integration of Technology

The integration of cutting-edge technology opens up transformative opportunities for environmental policies (Géraldine & Fano, n.d.). Smart cities, for example, use IoT-enabled infrastructure to optimize resource utilization, reduce emissions, and improve urban living conditions. Green technology innovations such as solar energy, electric vehicles, and AI-driven resource management systems improve sustainability efforts' efficiency and scalability. These technological advancements make environmental policies more adaptable, data-driven, and effective.

4.2 Challenges Faced by Policymakers

4.2.1 Conflicts of Interest Between Economic Growth and Environmental Protection

Policymakers frequently face the challenge of balancing competing demands for economic growth and environmental protection (Aguinis et al., 2011). Mining and fossil fuels, for example, are frequently incompatible with sustainability objectives. Crafting policies that address both short-term economic needs and long-term environmental preservation necessitate delicate negotiation and novel solutions.

4.2.2 Lack of Coordination Between Policy Levels

Disparities and misalignment at the global, national, and local policy levels impede the effectiveness of environmental strategies (Johan & Frank, n.d.). While international agreements establish broad objectives, national policies may lack the commitment or resources to carry them out. Similarly, local governments may struggle to align their initiatives with larger frameworks due to funding constraints or policy gaps. This lack of cohesion reduces the overall effectiveness of environmental policies.

4.2.3 Political Resistance and Stakeholder Conflicts

Environmental policies frequently face opposition from political actors and stakeholders with conflicting interests (Anne et al., n.d.). Lobbying by anti-regulation industries, political polarization on climate issues, and public skepticism about the economic costs of sustainability measures all pose significant challenges. Conflicts among government bodies, corporations, and civil society groups further complicate policy development and implementation.

4.2.4 Balancing Scientific Research and Local Knowledge

Another critical challenge is integrating scientific research with local knowledge during the policymaking process (Larrue, 2017). While scientific data is essential for making evidence-based decisions, local communities provide practical, context-specific knowledge that is critical for successful implementation. Striking a balance between these two knowledge systems is frequently difficult, particularly in diverse or marginalized areas where trust and communication gaps may exist.

5 Case Studies: Global Examples of Environmental Policy Implementation

Environmental policies are not implemented in a vacuum; they are influenced by a wide range of political, economic, and social factors that determine their success or failure (Johan & Frank, n.d.). This section looks at real-world examples to highlight both the opportunities and challenges of environmental policymaking. Drawing on global experiences, it emphasizes the critical importance of adaptable, inclusive, and context-sensitive approaches to achieving sustainable development goals.

5.1 Successful Policy Initiatives

Denmark has emerged as a global leader in renewable energy, particularly wind power (Nations, 2015). Its success is based on decades of strategic policies that prioritize environmental sustainability while maintaining economic growth. The

Danish government has taken innovative steps, such as providing subsidies for renewable energy projects, setting ambitious emission reduction targets, and fostering strong public–private partnerships. Denmark's approach is notable for its emphasis on community participation, as citizen-owned wind farms have received widespread public support and increased trust in environmental policies. These efforts have allowed Denmark to generate nearly half of its electricity from wind power, demonstrating how integrated policies and technological innovation can drive transformative change. Costa Rica provides another example of successful environmental policy.

The country has reversed deforestation trends, thanks to its innovative Payments for Environmental Services (PES) program, which rewards landowners for conserving forests and protecting biodiversity (Reid, 2016). This initiative, combined with a commitment to carbon neutrality, has established Costa Rica as a global model for sustainable development. Over 99% of the country's electricity comes from renewable sources like hydropower, wind, and solar energy. Furthermore, Costa Rica's legal protections for biodiversity and extensive network of protected areas, which cover approximately 25% of its land, demonstrate the effectiveness of aligning national policies with global sustainability goals.

5.2 *Challenges in Policy Implementation*

The implementation of environmental policies, while critical to achieving sustainability goals, is frequently fraught with difficulties, particularly in developing countries (James et al., n.d.). Many countries in sub-Saharan Africa bear the dual burden of combating environmental degradation while also addressing pressing socioeconomic priorities such as economic growth and poverty alleviation. This delicate balance is complicated by limited financial resources, inadequate institutional frameworks, and competing interests. Kenya's ambitious efforts to restore the Mau Forest, a critical water catchment area, demonstrate these challenges. The forest, which supports biodiversity, agriculture, and water security, has been severely harmed by illegal logging, agricultural encroachment, and uncontrolled settlement. Although policies for reforestation and conservation have been implemented, their success has been hampered by insufficient funding, poor enforcement, and competing land-use demands.

These obstacles highlight the inherent difficulty of prioritizing ecological preservation in regions where immediate socioeconomic needs often take precedence. China, by contrast, offers an example of both the potential and the limitations of environmental policymaking in the context of rapid industrialization and urbanization (Wilfried et al., 2014). The country has made notable progress in advancing green technologies and integrating sustainability into its development agenda. Initiatives such as the national carbon trading system and significant investments in renewable energy have positioned China as a global leader in certain aspects of environmental innovation. Solar and wind energy capacity has grown exponentially, demonstrating the government's ability to mobilize resources and drive large-scale change. However, these efforts are frequently undermined by competing priorities and structural challenges. For example, despite progress in renewable energy, China remains heavily reliant on coal, which continues to drive high levels of air and water pollution. Rapid urban expansion and industrial growth exacerbate these issues, creating a complex dynamic in which economic imperatives often overshadow environmental goals. One of the most significant challenges to China's environmental governance is the inconsistent implementation of policies across its provinces. While national directives are frequently ambitious, local governments may lack the capacity or political will to carry them out effectively. This disparity reflects a larger challenge of balancing top-down policy objectives with the realities of diverse local contexts. In some cases, local governments prioritize short-term economic benefits, such as industrial output and job creation, over long-term sustainability goals. This misalignment highlights the critical need for more integrated and enforceable strategies that align national ambitions with local capacities and priorities. Both Kenya and China demonstrate the multifaceted challenges of environmental policymaking in diverse socioeconomic contexts. In Kenya, the primary challenge is addressing immediate socioeconomic needs while developing the institutional capacity required to implement and sustain environmental initiatives. Meanwhile, China's experience highlights the tension between rapid economic growth and environmental preservation, as well as the challenges of translating national policies into actionable and effective local measures. Despite their differences, these cases highlight a common need for policies that are both well-designed and adaptable to specific local conditions. The examples of Kenya and China highlight the importance of international cooperation in addressing these issues.

Developing countries, such as Kenya, frequently require external assistance in the form of financial assistance, technology transfer, and capacity building to effectively implement environmental policies (Moomaw, 2012). Similarly, China's ability to transition to a greener economy could benefit from increased international cooperation, particularly in areas such as knowledge sharing and cross-border climate initiatives. Both cases demonstrate that, while environmental policymaking can be an effective tool for sustainability, its success is dependent on the ability to navigate economic, social, and institutional complexities through innovation, inclusivity, and global collaboration.

5.3 Analysis: Lessons Learned

The case studies of Denmark, Costa Rica, Kenya, and China offer invaluable insights into the complex dynamics and diverse outcomes of environmental policymaking in various socioeconomic and ecological contexts (Biermann, 2008). These examples show how proactive strategies, community involvement, and well-structured incentives can lead to sustainability success while also highlighting the challenges presented by competing priorities, limited resources, and systemic constraints.

Proactive planning and a strong commitment to renewable energy have helped Denmark establish itself as a global sustainability leader. Complexities are associated with innovation, inclusivity, and global cooperation.

The government's active collaboration with businesses and communities has ensured widespread support for its energy transition, making Denmark a model for combining economic development and environmental preservation (Nations, 2015). Costa Rica has another compelling success story, particularly in terms of reforestation and carbon neutrality. Costa Rica has made remarkable progress in restoring its ecosystems while also strengthening its economy, thanks to policies that prioritize forest conservation, sustainable tourism, and renewable energy. Financial incentives, such as payments for ecosystem services (PES), have encouraged landowners to participate in reforestation initiatives, transforming sustainability into a national priority. The country's emphasis on inclusivity and environmental education has boosted community engagement, demonstrating the value of grassroots involvement in achieving policy objectives (Reid, 2016).

Kenya and China, however, show how difficult it is to balance environmental goals with economic and social priorities. In Kenya, efforts to combat deforestation and promote renewable energy are frequently hampered by financial and technological constraints (Moomaw, 2012). While initiatives like the Green Belt Movement have had some success in combating desertification and empowering women, systemic issues like poverty and limited infrastructure complicate the implementation of larger sustainability policies. These challenges highlight the need for international support in the form of funding, technology transfer, and capacity building for developing nations to achieve their environmental goals. One of the biggest economies in the world, China must balance maintaining its fast industrial growth with reducing its substantial environmental impact (Wilfried et al., 2014). To address these issues, the government has put in place ambitious policies, such as strict pollution controls and investments in renewable energy. But the extent of urbanization and industrialization has resulted in enduring problems like water and air pollution. Furthermore, striking a balance between environmental preservation and economic growth frequently leads to competing priorities, highlighting the significance of flexible and context-sensitive policymaking. As China's experience shows, attaining sustainability necessitates strong enforcement mechanisms and a clear alignment of national priorities with environmental objectives, even with significant resources.

A common thread across these case studies is the critical importance of context-sensitive approaches (Johan & Frank, n.d.). Each country's unique socioeconomic and ecological conditions necessitate tailored strategies that reflect local realities while aligning with global sustainability goals. For instance, policies that succeed in high-income nations like Denmark may not be directly transferable to developing countries such as Kenya without significant adaptation and support. Policymakers must also recognize the dynamic nature of environmental challenges, requiring frameworks that are both flexible and resilient to future uncertainties.

The importance of international cooperation in removing technical and financial obstacles, especially for developing countries, is another important lesson (Biermann, 2008). Global programs like the Green Climate Fund and the Paris

Climate Agreement offer a forum for information exchange, resource mobilization, and group action. These mechanisms aid in bridging the gap between ambition and implementation by encouraging international collaboration, allowing nations to more successfully pursue sustainability. It is clear from looking at these examples from around the world that environmental policymaking is both an art and a science. In order to develop strategies that address current environmental challenges and foster future resilience, policymakers must strike a careful balance between innovation, inclusivity, and adaptability. These lessons emphasize that success depends on the ability to balance local needs with global imperatives, and they provide a road map for creating effective policies that promote sustainable development. These case studies remind us of the transformative power of thoughtful, contextual, and cooperative policymaking as countries continue to struggle with the challenges of environmental stewardship.

6 The Need for Collaborative and Inclusive Policymaking

An inclusive and cooperative approach to policymaking is increasingly necessary as environmental issues continue to worsen (Lascoums, n.d.). Policies pertaining to the environment must be developed with the varied needs, viewpoints, and experiences of different stakeholders in mind. The most effective way to shape these policies is through a participatory process that involves input from a variety of actors, each of whom offers unique insights that together guarantee the development of comprehensive and long-lasting strategies. First and foremost, the development and implementation of environmental policies are greatly influenced by government agencies. But governments cannot solve complicated environmental problems like pollution, biodiversity loss, and climate change on their own. Businesses and the private sector, whose operations may have both beneficial and detrimental effects on the environment, must cooperate with them (Makarius et al., 2020). Businesses can spur innovation by implementing sustainable production techniques, and governments can encourage green practices through regulatory frameworks by involving the private sector. A sustainable economy is increasingly seen as requiring green technologies and corporate social responsibility (CSR) programs. The synergistic approach fostered by this collaboration allows for a greater environmental impact by leveraging the strengths of both private sector initiatives and public policy.

Important viewpoints and firsthand experience are offered by civil society organizations and non-governmental organizations (NGOs), which enhance the policymaking process (Lascoums, n.d.). NGOs frequently lead the charge in promoting environmental justice because they speak for underrepresented groups and ecosystems that might otherwise go unnoticed in policy debates. They play a crucial role in the policymaking process by conducting research, monitoring environmental policies, and influencing public opinion. Since local communities bear the brunt of environmental problems, they also offer priceless insights into the realities of implementing policies. Many times, traditional knowledge that has been handed down through the generations is also held by local communities, offering long-lasting sustainable practices.

When scientific research and traditional/local knowledge are combined, more comprehensive policies that are not only successful but also suitable for the culture and society can result. Institutions of higher learning and science provide crucial evidence-based information that guides policy decisions (Larrue, 2017). Using scientific data is essential for comprehending the implications of various policy options as environmental issues become more complex. Through thorough monitoring and evaluation, research institutions also contribute to the assessment of the effectiveness of current policies.

However, science alone is not enough. Policies that fail to incorporate the lived experiences and indigenous knowledge of local communities often overlook the nuances of environmental challenges (Jean-Pierre & Dominique, 2019). By bridging the gap between scientific research and local wisdom, policymakers can ensure that their strategies are more adaptable and context-specific, addressing both global and localized environmental concerns. Collaboration between these diverse stakeholders ensures that environmental policies are inclusive, equitable, and context-specific. An inclusive policymaking process promotes social equity by ensuring that all voices particularly those of vulnerable populations are heard and taken into account (Sachs, 2015). In many cases, marginalized groups, such as indigenous communities or low-income households, are disproportionately affected by environmental degradation. Engaging these groups in the policy process is essential for creating policies that address their specific needs and protect their rights. Moreover, policies designed through a participatory approach are more likely to enjoy broader public support, as they reflect the values and priorities of the people they aim to serve.

In conclusion, the need for collaborative and inclusive policymaking cannot be overstated (Anne et al., n.d.). Only by engaging a diverse array of stakeholder's governments, businesses, NGOs, local communities, and scientific institutions, we can develop effective and equitable environmental policies. This approach ensures that policies are not only grounded in scientific evidence but also socially relevant and culturally sensitive. By fostering collaboration, we create a foundation for sustainable development that is inclusive, adaptive, and capable of addressing the pressing environmental challenges of our time.

7 Innovative and Ethical Policy Solutions for Long-Term Sustainability

A sustainable future depends on finding creative and moral policy solutions to the urgent environmental issues of our day (Rockström et al., 2009). Although they may provide momentary respite, short-term fixes frequently overlook the underlying systemic problems that sustain environmental deterioration. Policy approaches must incorporate forward-thinking strategies that strike a balance between the urgency of today's crises and the need for lasting change in order to be sustainable over the long term. In order to ensure a resilient and sustainable future, this section promotes drastic policy solutions that match immediate actions with long-term environmental objectives.

7.1 Radical Policy Approaches for Sustainability

Integrating radical policy approaches that have the power to change current systems is one of the most important ways to promote long-term sustainability (Simms, 2010). These regulations ought to prevent further damage in addition to attempting to lessen present environmental issues. For instance, carbon taxes, emissions trading programs, and resource efficiency requirements are some of the policies that governments can enact to drastically move industries toward sustainable practices. Businesses are encouraged by such policies to embrace green technologies and abandon environmentally damaging practices, like excessive use of fossil fuels or deforestation. A more sustainable and just world may result from these radical policies that match economic incentives with environmental objectives (Silvia et al., 2014).

Furthermore, these policies need to include mechanisms for adaptation that can change over time (Johan & Frank, n.d.). Designing policies that are adaptable and responsive to new scientific discoveries, technological breakthroughs, and changes in societal demands is crucial given the dynamic nature of environmental challenges. This flexibility guarantees that policies stay applicable and efficient when conditions shift, offering long-term direction while resolving immediate environmental emergencies.

7.2 Innovative Strategies for Sustainable Development

The foundation of any effective sustainability strategy is innovation (Wilfried et al., 2014). Technological developments have enormous potential to improve environmental monitoring and management, especially in areas like artificial intelligence (AI) and data analytics. AI, for instance, can be used to track ecosystems in real time and forecast environmental hazards like pollution, deforestation, and the effects of climate change (Géraldine & Fano, n.d.). Governments and organizations can make better, more proactive decisions to protect natural resources and enhance policy outcomes by using AI to collect and analyze massive amounts of environmental data.

In addition to technological innovation, public engagement plays a crucial role in achieving long-term sustainability (Jean-Pierre & Dominique, 2019). Educating the public about the importance of sustainable practices and their individual impact on the environment can foster widespread behavioral change. Policymakers can use incentives, awareness programs, and public campaigns to encourage environmentally friendly actions, such as reducing waste, conserving water, or switching to renewable energy sources. These initiatives not only raise awareness but also create a collective sense of responsibility, empowering citizens to contribute to sustainability efforts on a larger scale. Another important tactic for guaranteeing sustainability over the long run is to fortify international cooperation (Biermann, 2008). Global in scope, environmental problems like ocean pollution, biodiversity loss, and climate change cannot be adequately addressed by individual countries acting alone. The sharing of resources, technologies, and best practices can be facilitated by international cooperation through programs like multilateral agreements, global climate funds, and knowledge-sharing platforms. By working together, nations can combine their resources, share the cost of environmental problems, and produce long-lasting results that are advantageous to all.

7.3 The Role of Ethical Principles in Environmental Policymaking

Ethical principles serve as a compass to shape policies that protect ecological integrity and future generations' rights (Rockström et al., 2009). These principles act as a guide for decision-making by ensuring that environmental strategies are not only successful but also equitable, just, and sustainable.

7.3.1 Ecological Integrity: Preserving the Balance of Natural Systems

The ecological integrity principle, which places a strong emphasis on the preservation and protection of ecosystems, is fundamental to moral policymaking (Rockström et al., 2009). Policies must preserve the balance of vital ecological processes, including soil fertility, water cycles, and biodiversity, and stop the deterioration of natural systems in order to preserve ecological integrity. A symbiotic relationship between society and nature is fostered when policymakers prioritize ecological integrity, ensuring that human activities do not surpass the planet's capacity to regenerate resources and absorb waste. Recognizing the inherent uncertainties of environmental systems, policies that uphold ecological integrity frequently adopt a precautionary approach.

7.3.2 Intergenerational Equity: Responsibility to Future Generations

Intergenerational equity is another fundamental tenet of moral policymaking, arguing that the current generation has a moral duty to protect the environment for coming generations (Meadows et al., 1972). This principle highlights the long-term effects of environmental degradation, challenging the predominance of short-term thinking in policy decisions. It emphasizes that the ability of future generations to meet their needs and enjoy a healthy and thriving planet should not be compromised by current actions. In order to achieve intergenerational equity, policies must put sustainability ahead of short-term financial gains. For example, switching to renewable energy sources from fossil fuels shows a dedication to long-term environmental health and guarantees a stable climate for future generations. In a similar vein, policies that support waste reduction, water conservation, and reforestation are investments in the welfare of future generations.

7.3.3 Ethical Policymaking: A Just and Inclusive Approach

Strategies are also guaranteed to be inclusive and equitable when ethical principles are incorporated into environmental policymaking (Sachs, 2015). Ethical policies take into account the needs of marginalized groups, such as indigenous peoples or low-income communities, who are disproportionately impacted by environmental issues. Policies can promote social cohesion and bridge disparities by addressing environmental justice, guaranteeing that sustainability initiatives benefit all. Furthermore, incorporating different viewpoints into the policymaking process is morally supported by ethical principles. Policymakers, scientists, and local communities work together to develop collaborative approaches that support the development of culturally sensitive and context-specific policies. By utilizing a diverse range of expertise and experiences, this inclusivity not only increases the legitimacy of policies but also improves their efficacy.

7.3.4 From Principles to Action: The Role of Policymakers

Decision-makers must take a comprehensive approach that strikes a balance between social, economic, and environmental priorities in order to convert moral principles into workable policies (Anne et al., n.d.). Transparency, accountability, and flexibility must also be embraced in the formulation and application of policies (James et al., n.d.). Policymakers are encouraged by ethical frameworks to carefully weigh trade-offs, making sure that immediate gains don't come at the expense of sustainability over the long run (Simms, 2010).

Investing in green infrastructure or putting in place carbon pricing mechanisms are two examples of how moral values can influence observable results (Silvia et al., 2014). These laws promote actions that uphold ecological integrity and intergenerational equity by coordinating financial incentives with environmental objectives (OECD, 2019).

7.4 *Conclusion: A Road Map for Sustainable Policy Development*

7.4.1 Conclusion: Toward a Sustainable and Resilient Future

Innovative and moral policy solutions are essential for tackling the complex issues of environmental sustainability and building a resilient and prosperous global society (UNEP, 2021). Given the severity of the environmental crisis, drastic, game-changing measures that go beyond small tweaks are required (Aguinis et al., 2011). Realizing that the choices made today will have a significant impact on future generations, governments, organizations, and communities must adopt radical policy approaches that close the gap between short-term demands and long-term sustainability goals (Meadows et al., 1972). Aligning short-term activities with long-term goals is essential to this change.

In addition to reducing current environmental risks, policies must lay the groundwork for long-term ecological equilibrium and social justice (Bank, 2010). Examples of strategies that can simultaneously address present environmental pressures and pave the way for future resilience include switching to renewable energy, putting circular economic models into practice, and embracing sustainable urban planning techniques (Nations, 2015). This dual emphasis guarantees that sustainability initiatives are both realistic and progressive, offering solutions that change to meet new problems.

It is impossible to overestimate the importance of technological innovation in this undertaking (Géraldine & Fano, n.d.). Unprecedented opportunities to monitor environmental changes, optimize resource use, and develop scalable solutions for sustainability are presented by advancements in artificial intelligence, big data, and green technologies (Wilfried et al., 2014). For instance, renewable energy technologies like solar and wind power have already shown their potential to lessen dependency on fossil fuels, and AI-driven models can forecast climate risks, allowing policymakers to create adaptive strategies (Moomaw, 2012). Accelerating the shift to a low-carbon and resource-efficient future requires promoting investment in R&D as well as the implementation of cutting-edge technologies.

The impact of these initiatives is further increased through international cooperation and public involvement (Biermann, 2008). Individuals, communities, and nations must all be committed to sustainable development; it cannot be accomplished in a vacuum (Ostrom, 1990). People can actively participate in environmental solutions by being informed about the value of sustainability, receiving incentives for eco-friendly behavior, and supporting grassroots projects (Jean-Pierre & Dominique, 2019). At the same time, enhancing global cooperation through accords such as the Paris Climate Accord and global climate funds guarantees that resources, information, and skills are distributed fairly among countries, especially helping those most at risk from environmental problems (OECD, 2019).

By establishing ethical foundations for these policy solutions, sustainability initiatives are guaranteed to be both equitable and successful (Rockström et al., 2009).

In order to preserve biodiversity and ecosystem health, the ecological integrity principle requires that human activities adhere to the planet's natural boundaries (IPCC, 2022). Policies founded on this idea put human–nature harmony first, encouraging a mutualistic relationship that is advantageous to both. In a similar vein, intergenerational equity calls on decision-makers to take the long-term effects of their actions into account and highlights our moral obligation to protect the environment for coming generations (Sachs, 2015). By serving as a moral compass, ethical frameworks direct the creation and application of laws that support equity, justice, and inclusivity.

The interdependence of environmental, social, and economic issues is becoming more and more obvious as the twenty-first century goes on (Makarius et al., 2020). In order to achieve integrated solutions, policymakers must embrace a comprehensive viewpoint that recognizes the intricacy of these problems (Johan & Frank, n.d.). They are able to navigate this complexity with clarity and purpose by utilizing cutting-edge tools, interacting with a variety of stakeholders, and abiding by ethical requirements (Larrue, 2017).

In summary, achieving a sustainable and resilient future requires a lot of work and visionary leadership, but it is not an easy or problem-free journey (Lascoums, n.d.). The foundation of this journey is made up of creative and moral policies that provide a thorough road map for tackling the urgent environmental problems of our day (Reid, 2016). We can build a future where humanity flourishes by cooperating, embracing technology, and promoting justice and equity.

References

Aguinis, H., Gottfredson, K. R., & Wright, A. T. (2011). Best-practice recommendations for estimating interaction effects using meta-analysis. *Journal of Organizational Behavior., 32*, 1033.

Anne, M. S., Sarah, J., & Steven, J. (n.d.). Supporting actionable science for environmental policy. *Frontiers in Conservation Science*.

Bank, W. (2010). *World development report: Development and climate change*. Bank, W.

Biermann, F. (2008). Global environmental governance: Taking stock, moving forward. *Annual Review of Environment and Resources, 33*, 277–294.

Géraldine , F., & Fano , A. (n.d.). "Les stratégies environnementales des organisations internationales dans les pays en développement: continuité ou ruptures ?".

IPCC. (2022). *"Climate Change 2022: Impacts, Adaptation and Vulnerability" Contribution of Working Group II to the Sixth Assessment Report of the Intergovernmental Panel on Climate Change*. IPCC.

James, P., Dave, H., & Julia, E. (n.d.). *The future of 'environmental' policy in the Anthropocene: Time for a paradigm shift*. Environmental Politics.

Jean-Pierre, G., & Dominique, B. (2019). *"Politiques environnementales et changements techniques : éléments pour une approche renouvelée"*.

Johan , S., & Frank, W. (n.d.). Strategic niche management and sustainable innovation journeys: Theory, findings, research agenda, and policy.

Larrue, C. (2017). Recherche et politiques publiques environnementales : quelles interactions ? *Natures Sciences Sociétés, 25*, S12–S17.

Lascoums, P. (n.d.). "Les politiques environnementales: évolutions et perspectives".

Makarius, E., Mukherjee, D., Fox, D. J., & Fox, K. A. (2020). Rising with the machines: A sociotechnical framework for bringing artificial intelligence into the organization. *Journal of Business Research, 120*, 262–273.

Meadows, D., Randers, L., & Meadows, D. L. (1972). *The limits to growth.* De Gruyter.

Moomaw, W. R.-C. (2012). Renewable energy and energy efficiency in developing countries: Contributions to reducing global greenhouse gas emissions. *Journal of Environment & Development*, 193–222.

Nations, U. (2015). *Transforming our world: The 2030 agenda for sustainable development.* Nations, U.

OECD. (2019). *Environmental policy toolkit for green growth.* OECD.

Ostrom, E. (1990). *Governing the commons: The evolution of institutions for collective action.* Cambridge University Press.

Reid, C. E. (Jun 2016). "The case for 'Environment in all Policies': Lessons from the 'Health in All Policies' approach in Public Health".

Rockström, J., Steffen, W., Noone, K., et al. (2009). A safe operating space for humanity. *Nature, 461*, 472–475.

Sachs, J. D. (2015). *The age of sustainable development.* Columbia University Press.

Silvia, A., Enrico, B., Tomasz, K., & Vera, Z. (2014). *Do environmental policies matter for productivity growth? Insights from new cross-country measures of environmental policies.* OECD.

Simms, A. J. (2010). *Growth isn't possible: Why we need a new economic direction.* New Economics Foundation.

UNEP. (2021). *Making peace with nature: A scientific blueprint to tackle the climate, biodiversity, and pollution emergencies.* UNEP.

Wilfried, L., Tilman, A., Anna, P., & Georgeta, V. (October 2014). *Green industrial policy: Managing transformation under uncertainty.*

Innovations in Green Technology

Muhammad Hussain, Tieling Zhang, and Aziza Chakir

1 Introduction

A wide range of inventions intended to reduce environmental effect and promote sustainability are included in green technology, also known as sustainable technology. These technologies are intended to reduce pollution, conserve resources, and slow down global warming. Green technology advancements are revolutionizing a number of industries, including manufacturing, transportation, energy, and agriculture (Lidder et al., 2025). Innovations in this field are not merely technological advancements; they represent a paradigm shift toward more sustainable practices across various sectors, including energy, agriculture, and urban infrastructure. As noted by Pang et al. (2021), innovation is a key driver for industries aiming to achieve sustainable growth, facilitating both breakthrough and incremental technological progress. This is particularly evident in sectors such as agriculture, where advancements like biotechnology and precision farming are revolutionizing traditional practices, enhancing productivity while reducing ecological footprints (Sott et al., 2021a, 2021b). The integration of digital technologies, such as Global Navigation Satellite Systems (GNSS), further exemplifies how modern innovations can optimize resource use and improve efficiency in agricultural practices (Sott et al., 2021a).

As the global community struggles with the effects of industrialization, this chapter emphasizes the urgent need to embrace green technologies, which are essential for reducing ecological harm and promoting economic growth. Kaluarachchi (2021) emphasizes that integrating smart and green infrastructure is

M. Hussain (✉) · T. Zhang
School of Mechanical, Materials, Mechatronic and Biomedical Engineering, University of Wollongong, Wollongong, NSW, Australia

A. Chakir
Faculty of Law, Economics and Social Sciences, Hassan II University, Casablanca, Morocco

A. Chakir et al. (eds.), *Green Society, Environmental Strategies and Sustainable Development*, https://doi.org/10.1007/978-3-031-94086-6_5

crucial for cities to meet their long-term sustainability objectives. Wu et al. (2018), who emphasize the significance of technological innovation, market demand, and governmental policies in propelling the growth of new energy vehicle industries, corroborate this cooperative viewpoint. In addition to promoting the adoption of green technologies, these synergies also create an atmosphere that encourages creativity.

The authors highlight the critical role of advancements in energy efficiency, storage solutions, and renewable energy sources such as solar, wind, and hydroelectric power in driving technological progress. These innovations are delivering tangible benefits, including reduced carbon emissions and the expansion of employment opportunities within the growing green economy (Wang et al., 2022a; Saunila et al., 2018).

In the energy sector, advancements in renewable sources are making clean energy more affordable and widely available. Improvements in energy storage, including next-generation batteries and smart grid systems, are enhancing the reliability and effectiveness of renewable energy solutions. Similarly, the transportation industry is undergoing major changes with the adoption of electric vehicles (EVs), hydrogen fuel cells, and biofuels, all of which contribute to lower greenhouse gas emissions and decreased dependence on fossil fuels (Sabet & Farooq, 2025). Additionally, the implementation of smart transportation systems and infrastructure improvements is supporting more efficient and sustainable urban mobility.

1.1 Background

In light of environmental change and sustainable development, the introduction of green technology marks a significant turning point in the adoption of global environmental concerns. Innovations in this sector are mainly motivated by the necessity to lower carbon emissions and improve energy efficiency. In this regard, renewable energy technology innovation, or RETI, is essential. It includes a range of technologies that facilitate the transition from fossil fuels to renewable energy sources. These global issues are not only threatening ecosystems but also affecting human health, economic stability, and social well-being, Table 1 shows key environmental challenges and its overall impact.

The empirical evidence presented by Usman et al. (2023) underscore the crucial function of environmental innovations in generating new renewable energy sources while concurrently meeting traditional energy demand. Enhancing the effectiveness of using renewable energy requires a combined focus on innovation and consumption control, especially in light of the European Union's aggressive targets to cut greenhouse gas emissions by 2050 (Sipa & Gorzeń-Mitka, 2021). The research by Koval et al. (2023), who assess the economic and environmental aspects of energy generation through renewable methods and predict a major impact on sustainable development, further supports the EU's commitment to developing renewable energy technologies.

Table 1 Key environmental challenges and their impacts

Environmental issue	Description	Examples and impacts
Climate change	Caused by a rise in greenhouse gas (GHG) emissions brought on by industrial operations, deforestation, and the burning of fossil fuels	Sea level rise, extreme weather, biodiversity loss, food shortages, community dislocation, rising global temperatures, and an increase in natural disasters
Resource depletion	Over-extraction and consumption of vital resources like water, fossil fuels, minerals, and arable land due to population growth and increased consumption	Soil erosion, deforestation, exhaustion of fossil fuels (coal, oil, and natural gas), increased carbon emissions, and the urgent need for cleaner, renewable alternatives
Environmental degradation	Deterioration of the natural environment caused by pollution, habitat destruction, and overuse of natural resources	Land degradation, ocean pollution, biodiversity loss, reduced carbon sequestration from deforestation, endangered wildlife, and disrupted water cycles

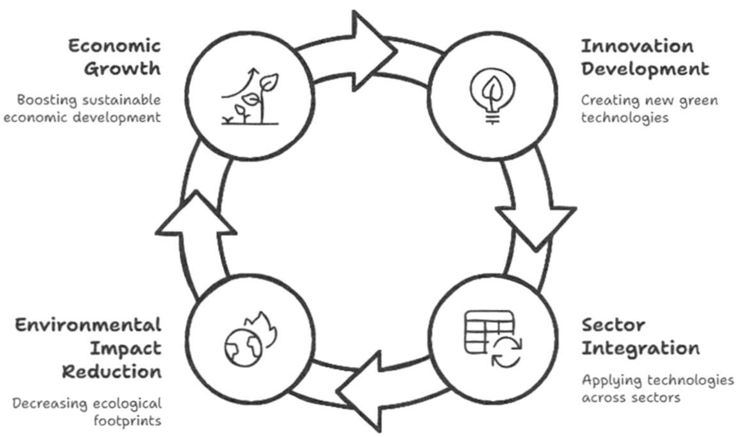

Fig. 1 Cycle of green technology integration

1.2 *The Role of Green Technology*

It is becoming more widely recognized that integrating green technology into many industries is essential to attaining sustainable growth and addressing environmental problems. Green technology encompasses many innovations to reduce environmental impacts while promoting economic growth. Figure 1 shows cycle of green technology integration.

The relationship between environmental performance and economic results in the corporate realm is intricate. Wang et al. (2022b) explore how patents for green technology can offer companies sustainable competitive advantages, improving their public image and, in turn, their economic performance. This finding highlights the dual advantages of innovating in green technology, as illustrated in Fig. 2, where companies can achieve both environmental and economic objectives.

It is crucial to consider the impact of market competition on the progress of green technology. Song and Wang (2018) argue expansion of green technology may mitigate some environmental impact but detracts from gaining a competitive edge in markets with low labor cost and weak environmental regulations. This became paramount the need for a balanced solution that considers factors of both economic competitiveness and environmental sustainability. Furthermore, the role of human capital in stimulating green technology innovation is substantial. Zhang and Li (2023) highlight that educational and health-related human capital significantly influences green technology innovation, suggesting that investments in human capital are crucial for fostering an environment conducive to sustainable technological advancements. This perspective aligns with the broader understanding that a skilled workforce is essential for successfully implementing green technologies.

2 Current Landscape of Green Technology

The current landscape of green technology is showed by a dynamic interaction between innovation, regulatory frameworks, and market forces, all aimed at promoting sustainable development and mitigating environmental challenges. As global awareness of climate change and ecological degradation intensifies, adopting green technologies has become imperative for economic growth and environmental sustainability. Green technology, often synonymous with green innovation, refers to technological advancements that aim to reduce environmental harm while promoting resource efficiency. Chen et al. (2022) define green innovation as technological

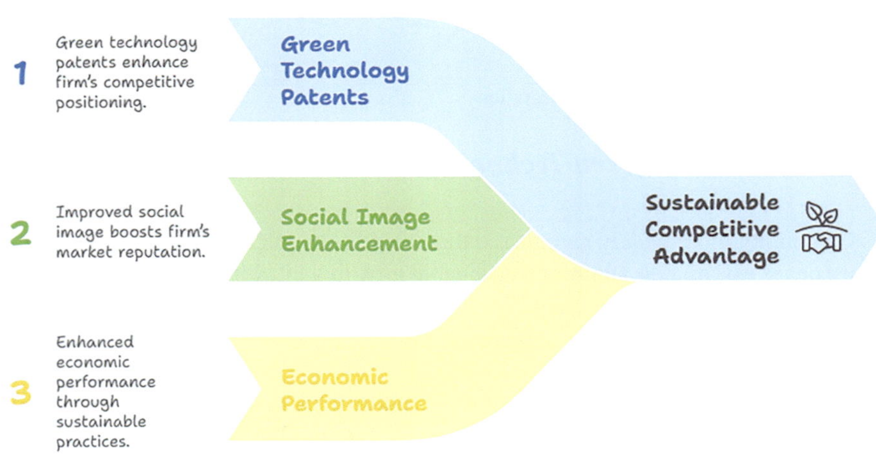

Fig. 2 Dual benefits of green innovation

advances that minimize waste, reduce pollution, and conserve energy, thereby addressing the contrary effects of climate change. This definition underscores the importance of integrating sustainability into technological development, which is necessary for succeeding long-term ecological aims.

Zhang et al. (2022) present evidence from China and show that strong environmental regulation can induce green innovation by pushing firms to adopt cleaner technologies. They could be seen as a compliance mechanism, and a driver of innovation, which pushes enterprises toward more sustainable practices.

But such green technology is also essential in their business and financial support mechanisms are also indispensable to progress. Green finance is a branch of holistic financial practices that paves the way for investment into sustainable technologies. Jiang et al. (2022) emphasize that green finance can significantly enhance enterprises' capacity to innovate by providing the necessary funding for research and development.

This is echoed by Feng et al. (2023), explaining the importance of a strong green financial system, as illustrated in Fig. 3, helped by certain instruments (green bonds, climate finance, etc.), to move toward sustainable technologies.

The digital economy is proving to be a transformative influence in the field of green technology innovation. Ye et al. (2023) emphasize that the digital economy enhances the efficiency and effectiveness of green technology initiatives by facilitating data-driven decision-making and encouraging collaboration among stakeholders. This convergence of technologies not only fosters innovation but also aids in the development of a low-carbon, circular production system, aligning economic growth with environmental sustainability. Additionally, the current landscape of green technology is increasingly influenced by the recognition of the importance of human capital and institutional support. El-Kassar and Singh (2019) point out that organizational commitment and human resource practices are essential for effectively harnessing green innovation. This indicates that cultivating a culture of sustainability within organizations is just as crucial as the technological advancements themselves.

Fig. 3 Role of green finance

2.1 Renewable Energy Sources

The global shift to renewable energy has been fueled by the urgent need to reduce carbon emissions and slow down climate change. As indicated in Fig. 4, solar, wind, and hydroelectric power are among the most well-known renewable energy sources and have become important contributors to the decarbonization of the energy sector.

2.2 Energy Efficiency and Storage

As renewable energy adoption grows, the need for efficient energy storage and distribution systems becomes increasingly vital. Energy efficiency measures and storage advancements ensure a stable and reliable power supply while reducing energy waste.

2.2.1 Technology for Batteries

Energy storage capabilities have been greatly increased by advancements in battery storage technology, especially in the areas of solid-state and lithium-ion batteries. By storing extra energy produced by renewable sources for later use, large-scale battery systems aid in balancing the supply and demand for energy. New technologies that improve energy efficiency and dependability for grid storage include sodium-ion batteries and flow batteries.

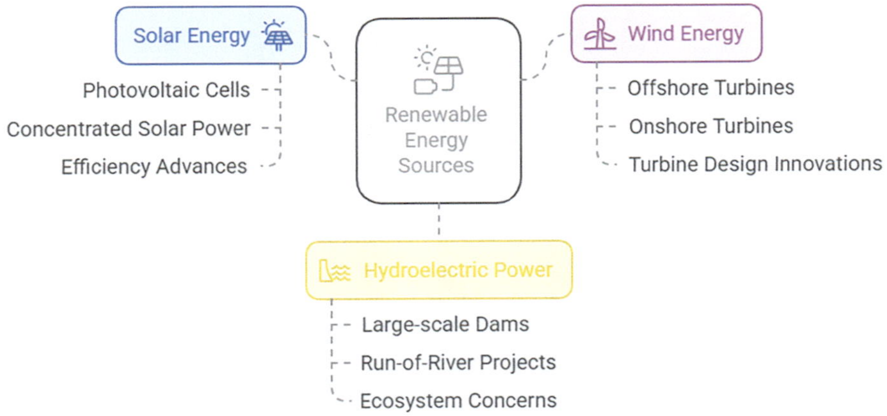

Fig. 4 Comparison of renewable energy sources and their impact on carbon reduction

2.2.2 Intelligent Grids

Digital technology is incorporated into smart grids to improve the sustainability, dependability, and efficiency of power distribution. These grids optimize electricity distribution, track energy usage, and identify problems instantly by utilizing sensors, automation, and sophisticated analytics. Energy systems may efficiently handle variations in energy production and consumption by combining renewable energy sources with smart grids, which lessens the need for backup power plants that rely on fossil fuels.

2.3 Economic Impact

The green technology industry is rapidly expanding, providing employment opportunities and driving investments in sustainable infrastructure. Some of the economic impact is shown in Table 2.

3 Role of Digital Technologies in Driving Green Innovation

Digital technologies are becoming increasingly important in fostering green innovation, particularly in addressing global environmental issues. This section looks at several digital technologies, including blockchain, the Internet of Things, and artificial intelligence (AI), and how they promote green innovation. Each subsection goes into great detail about the uses and effects of these technologies in promoting sustainability and enhancing resource management.

Table 2 Economic impact of renewable energy sector

Aspect	Description
Job creation	The renewable energy sector has become a major job creator, employing millions worldwide in roles such as solar panel installation, wind turbine maintenance, battery manufacturing, and smart grid technology development. Solar photovoltaic and wind energy have created the most jobs, with employment opportunities expected to grow as demand for clean energy solutions increases
Industry growth	The green technology sector, including renewable energy, energy storage, and efficiency solutions, has attracted significant investments from governments and private enterprises. Green finance mechanisms like green bonds and carbon credits have supported industry growth. Countries investing in renewable energy infrastructure have seen economic benefits, including reduced dependence on fossil fuel imports and increased energy security

3.1 Artificial Intelligence (AI) for Green Technology

Artificial Intelligence (AI) is essential to the development of green technology, especially in the areas of smart grids and energy management predictive analytics. Large volumes of data may be analyzed by AI systems to predict energy demand, optimize energy distribution, and boost renewable energy sources' efficiency. AI, for example, can forecast periods of high energy consumption, enabling improved load balancing and decreased energy waste (Shabrina & Andriani, 2022). AI plays a key role in waste efficiency, precision farming, and climate modeling in addition to energy management. Figure 5 AI-Driven Sustainability Cycle and its application to green technology.

By evaluating crop health, weather patterns, and soil conditions, AI systems in agriculture enable precision farming, which maximizes resource utilization, lowers chemical inputs, and boosts yields (Carbonell et al., 2016). Additionally, by anticipating garbage generation trends and streamlining collection routes, AI can optimize waste management procedures and reduce carbon emissions related to waste transportation (Yaacoub & Alouini, 2020).

Fig. 5 AI-driven sustainability cycle

3.2 Internet of Things (IoT) and Smart Systems

A key component of contemporary green innovation, the Internet of Things (IoT) allows for improved resource management, water monitoring, and energy efficiency through networked devices. Real-time energy consumption monitoring via IoT-enabled devices enables consumers to spot inefficiencies and modify their usage (Petrolo et al., 2014). Smart meters, for example, give users comprehensive information about their energy usage trends, enabling them to make more environmentally friendly decisions (Janssen et al., 2019). IoT technologies also enable water monitoring systems that minimize water consumption by detecting leaks and improving irrigation techniques (Janssen et al., 2019).

The Internet of Things is being incorporated into smart cities, smart households, and green infrastructure, making urban landscapes more sustainable ecosystems. Smart cities use IoT technology to reduce their negative environmental consequences, enhance public services, and improve transportation (Mahmud & Buyya, 2019). For example, IoT applications in smart transportation can optimize traffic flow, reduce congestion, and reduce emissions by providing real-time data on traffic conditions.

3.3 Role of Blockchain in Green Supply Chains

Blockchain technology is becoming a potent instrument for improving traceability and transparency in green supply chains, see Fig. 6.

Blockchain is crucial for enabling carbon credits and decentralized energy markets in addition to supply chain transparency. A more effective and transparent carbon trading system can be promoted by using blockchain technology to safely record and validate carbon credit transactions (Sabri & El Kamoun, 2023). In order to encourage the use of renewable energy sources, this technology also makes it possible to establish decentralized energy marketplaces where people and companies may directly exchange excess renewable energy. Incorporating blockchain technology into these systems not only improves productivity but also gives customers the ability to actively engage in the shift to a low-carbon economy.

3.4 Emerging Technologies and Trends

Emerging technologies such as quantum computing and synthetic biology hold immense potential for optimizing energy systems and creating bio-based products. Quantum computing, with its ability to process complex calculations at unprecedented speeds, can significantly enhance the modeling and optimization of energy systems (Alsamhi et al., 2019). This capability allows for more efficient energy

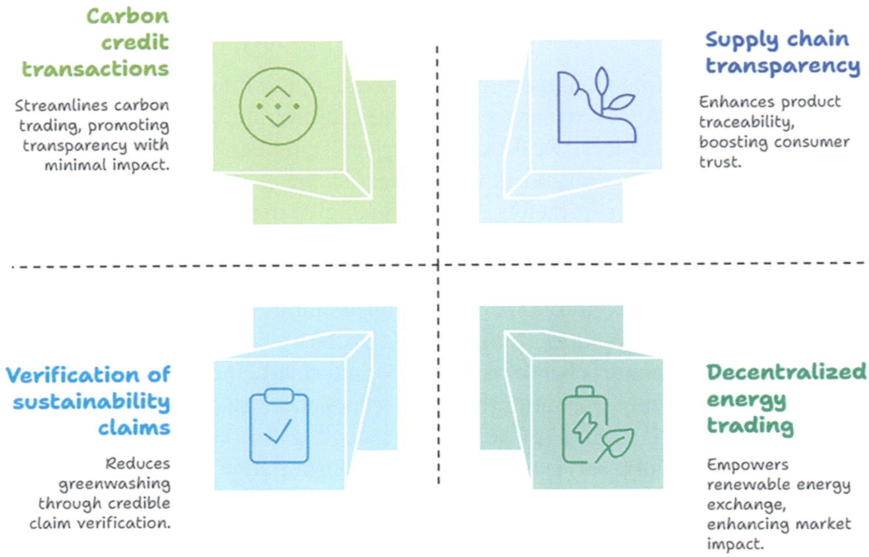

Fig. 6 Blockchain applications in green technology

distribution and the development of advanced materials that can improve energy storage and conversion (Shenkoya, 2023). As quantum technologies mature, they are expected to revolutionize various sectors, including energy, by enabling breakthroughs that were previously unattainable with classical computing methods.

Synthetic biology is another promising field that is driving green innovation by enabling the creation of bio-based products and materials. Advances in synthetic biology allow for the engineering of microorganisms to produce sustainable materials, such as bioplastics and biofuels, from renewable resources. This approach not only reduces reliance on fossil fuels but also minimizes waste and environmental impact. The integration of synthetic biology with IoT and AI technologies can further enhance the efficiency of production processes, leading to a more sustainable bioeconomy (Marine & Saluky, 2018).

4 Major Categories of Innovations in Green Technology

The major categories of innovations in green technology encompass a wide array of advancements aimed at promoting sustainability and reducing environmental impact. This section will explore these innovations in detail, focusing on renewable energy technologies, green building and construction technologies, sustainable transportation innovations, circular economy and waste management, water management and conservation, and advanced materials for sustainability.

4.1 *Renewable Energy Technologies*

Leading the charge in the shift to sustainable energy systems are renewable energy technologies. Among them, solar energy has witnessed notable advancements, especially in the areas of flexible solar panels, solar storage systems, and perovskite solar cells. Because of their high efficiency and inexpensive manufacturing costs, perovskite solar cells have become a viable substitute for conventional silicon-based cells (Adewumi et al., 2024). These cells have the potential to reach efficiencies of over 25% and can be produced utilizing less complicated procedures (Verma, 2024). The intermittency of solar power can also be addressed by improvements in solar storage technology, such as lithium-ion and next-generation batteries, which allow energy to be saved for usage during cloudy weather. Another creative way to use solar is using flexible solar panels, which may be incorporated into a variety of surfaces, such as clothes and building materials.

Innovations in wind energy are also important, especially in the creation of vertical-axis turbines, floating wind farms, and offshore wind turbines. Stronger and more reliable winds are an advantage of offshore wind farms, which increases energy output (Etukudoh et al., 2024). The potential for wind energy generation is increased by the deployment of floating wind turbines in deeper waters where conventional fixed turbines cannot be erected. For urban settings with limited space, vertical-axis turbines—which can capture wind from any direction—are being investigated (Safavi & Khoshbakht, 2024). With advancements in wave energy converters, mini-hydropower systems, and tidal energy, hydropower and marine energy technologies are developing. Wave energy converters absorb energy from surface waves, whereas tidal energy systems use the moon's and sun's gravitational pull to create power.

For small-scale energy generation, mini-hydropower systems offer a sustainable option, especially in isolated locations where larger infrastructure is not feasible (Ogunrinola et al., 2024).

4.2 *Sustainable Transportation Innovations*

Innovations in sustainable transportation are essential for lowering greenhouse gas emissions and advancing greener modes of transportation. Significant progress has been made in battery technology, fast-charging systems, and wireless charging capabilities for electric cars (EVs). Consumers now find EVs more enticing due to improved driving ranges and shorter charging periods brought forth by advancements in battery technology (Sarsah et al., 2024). In order to facilitate long-distance travel and lessen range anxiety among prospective EV users, fast-charging stations are being frequently installed. EV infrastructure is being revolutionized by wireless charging technology, which enables cars to be charged while parked or moving.

A promising substitute for trucks, buses, and heavy-duty vehicles is hydrogen fuel cells. Green hydrogen, which is created by electrolysis using renewable energy, provides a transportation fuel alternative with no emissions. Greenhouse gas emissions and urban air pollution can be considerably decreased by integrating hydrogen fuel cells into public transportation networks. With AI being essential to the development of self-driving electric cars and ride-sharing services, autonomous and shared mobility solutions are also becoming more popular. Autonomous vehicles have the ability to reduce emissions, ease traffic congestion, and improve traffic flow through efficient routing and shorter idle times. Ride-sharing services further enhance the sustainability of urban transportation by promoting the use of electric vehicles and lowering the number of car owners.

In order to reduce emissions in these industries, sustainable shipping and aviation technology are essential. The carbon footprint of air travel can be considerably decreased by developing sustainable aviation fuel (SAF) from renewable resources. To reduce the environmental impact of maritime transportation, new developments in green shipping technology are being investigated, such as energy-efficient hull designs and wind-assisted propulsion.

4.3 Circular Economy and Waste Management

A revolutionary strategy, the circular economy aims to reduce waste and encourage the continuous use of resources. Recycling technology advancements like chemical recycling, automated trash sorting, and e-waste recovery are crucial to the advancement of the circular economy. AI and robotics are used by automated waste sorting systems to increase recycling processes' accuracy and efficiency, lower contamination rates, and improve the quality of recycled materials. Plastics can be broken down into their original monomers via chemical recycling processes, producing new, superior goods. The goal of e-waste recovery systems is to lessen the environmental impact of electronic waste by removing valuable elements from discarded gadgets.

4.4 Water Management and Conservation

Particularly in light of growing water shortages and climate change, water management and conservation are essential elements of sustainable development. Access to clean water depends on advancements in desalination and water purification technology. Nanotechnology has demonstrated potential in improving filtration procedures and eliminating pollutants at the molecular level in water purification systems (Adedoja et al., 2021). Nanomaterials, such as carbon nanotubes and graphene oxide membranes, can significantly improve the efficiency of water treatment by allowing for selective permeability and higher flow rates. Solar desalination techniques, which harness solar energy to evaporate and condense water, are gaining traction as a sustainable method for producing fresh water from saline sources (Taheri et al., 2016). Additionally, biomimicry-based purification systems, which draw inspiration from natural processes, are being developed to create more efficient and environmentally friendly water treatment solutions.

Effective management of water resources also depends on water conservation technologies. In order to maximize water use in agriculture and guarantee that crops receive the proper amount of water without wasting any, smart irrigation systems make use of sensors and data analytics (Jamil et al., 2024). These devices can preserve agricultural yield while drastically lowering water usage. In both urban and rural areas, rainwater harvesting systems—which collect and store rainwater for later use—are being used more and more to augment water supplies and lessen dependency on conventional sources. These methods improve resistance to drought and water scarcity in addition to encouraging water conservation.

It is impossible to overestimate the importance of AI, big data, and IoT in managing water resources. Real-time monitoring and predictive analytics are made possible by these technologies, which facilitate better decision-making about the distribution and use of water. AI systems, for example, can forecast water consumption and spot possible shortages by analyzing past data, enabling proactive management techniques (Jamal et al., 2024; Hussain & Zhang, 2025). Additionally, big data analytics can improve coordination between the several water management stakeholders, allowing for prompt reactions to events like flooding or pollution. Water resource management can become more effective, sustainable, and sensitive to shifting environmental conditions by using these cutting-edge technology.

4.5 Advanced Materials for Sustainability

Advanced materials play a pivotal role in promoting sustainability across various sectors, particularly in clean energy, construction, and environmental remediation. Nanomaterials are increasingly recognized for their potential in clean energy applications, including energy storage, energy harvesting, and water purification. For example, nanostructured materials can enhance the performance of batteries and supercapacitors, leading to more efficient energy storage solutions. In energy harvesting, nanomaterials can be used to develop more efficient photovoltaic cells, thereby increasing the conversion efficiency of solar energy (Somoza-Norton et al., 2023). Furthermore, in water purification, nanomaterials can facilitate the removal of contaminants, making them essential for sustainable water management.

Biomaterials and bio-based innovations are also gaining prominence in the quest for sustainability. The use of algae-based products, mycelium materials, and biofabrication techniques is revolutionizing material science by providing renewable alternatives to conventional materials. Algae, for instance, can be used to produce biodegradable plastics and biofuels, reducing reliance on fossil fuels and minimizing environmental impact. Mycelium, the root structure of fungi, is being explored as a sustainable building material that is both lightweight and strong, offering an eco-friendly alternative to traditional construction materials (Somoza-Norton et al., 2023). Biofabrication techniques enable the production of materials that are not only sustainable but also tailored for specific applications, further enhancing their utility in various industries.

5 Impact of Green Technology Innovations

Innovations in green technology have a significant and diverse impact on social justice, economic progress, and environmental sustainability. This section will examine these technologies' effects on the environment, the economy, and society, emphasizing how important they are to building a sustainable future.

5.1 Environmental Impact

Reductions in carbon emissions, air pollution, and deforestation are the main indicators of the environmental impact of green technology breakthroughs (see Fig. 7). Energy-efficient appliances, sustainable farming methods, and renewable energy systems are examples of green technology that greatly reduce greenhouse gas emissions.

The efficiency of renewable energy systems is increased by advancements in energy storage technology, allowing for a more steady and dependable energy source while reducing environmental damage. Green technologies are essential for managing air pollution in addition to lowering carbon emissions. By reducing the

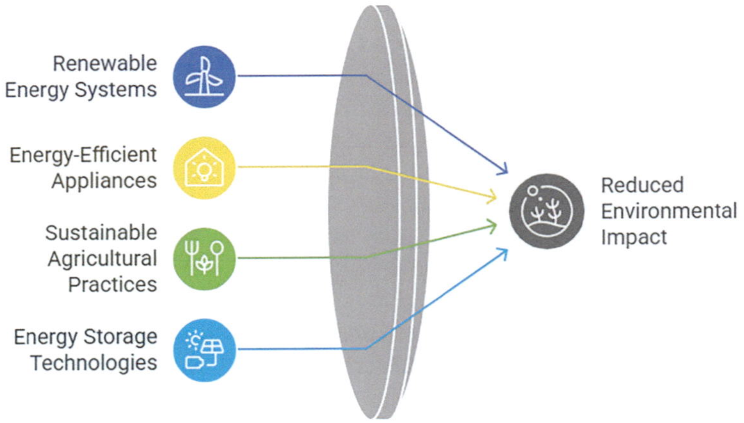

Fig. 7 Pathways to a greener future

number of conventional vehicles on the road, technologies like electric vehicles (EVs) and better public transportation networks can cut down on hazardous emissions.

Innovations in green technology also aid in adapting to climate change. For instance, by boosting biodiversity and enhancing soil health, sustainable farming methods like agroforestry and precision farming contribute to increased resilience to the effects of climate change (Saunila et al., 2018).

5.2 Social Impact

Innovations in green technology have an equally significant social impact, especially when it comes to advancing social justice and reducing poverty. By giving underserved populations access to resources and clean energy, green technology can promote social fairness.

Innovations in green technology are essential to attempts to electrify and provide energy access in rural areas. Communities that formerly relied on expensive and polluting fossil fuels might switch to greener and more economical energy alternatives by utilizing renewable energy sources. By giving communities energy independence and resilience against changes in energy prices, this shift not only lessens its effects on the environment but also gives them more power (Kwilinski et al., 2023). Green technologies can therefore promote economic growth and social inclusion in traditionally underserved areas.

6 The Future of Green Technology

Integrating cutting-edge technologies with sustainable practices will be essential as the world struggles with issues like resource depletion, climate change, and environmental degradation.

6.1 Emerging Innovations

The circular economy, green AI for environmental monitoring, and sustainable farming practices are all becoming increasingly important to new breakthroughs in green technology. Figure 8 illustrates the use of green artificial intelligence in environmental management. By highlighting the need of recycling, reusing, and reducing waste, the circular economy reduces the negative effects of production and consumption on the environment. This approach encourages businesses to design products with their entire lifecycle in mind, thereby promoting sustainability from the outset. For instance, companies are now focusing on creating products that are

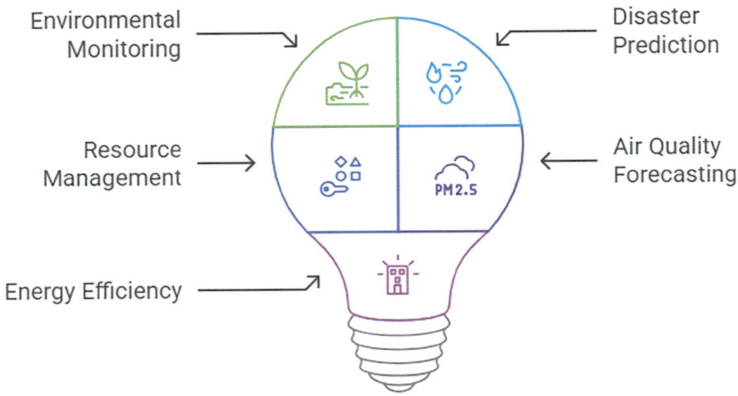

Fig. 8 The role of green AI in environmental management

easy to disassemble and recycle in order to reduce the amount of waste that ends up in landfills (Hou et al., 2023).

Green AI is becoming a key instrument for managing and monitoring the environment. Green AI can evaluate enormous volumes of environmental data to monitor ecosystem changes, forecast natural disasters, and improve resource management by utilizing machine learning algorithms and big data analytics (Wang et al., 2022a).

6.2 Role of Digital Technology

The integration of AI, IoT, and big data analytics enables real-time environmental condition monitoring, which facilitates informed decision-making for sustainability programs. IoT devices that collect energy, water, and air quality data include smart sensors and connected infrastructure, providing crucial information to regulators and businesses (Ley et al., 2016). More efficient resource management and the identification of areas in need of intervention are made possible by this data-driven approach. This data-driven approach enables more effective resource management and the identification of areas that require attention.

By examining both past and present data, artificial intelligence (AI) systems are able to predict changes in the environment, such as shifts in weather patterns or the likelihood of natural disasters. For the purpose of creating pre-emptive plans to lessen the effects of climate change, this predictive ability is crucial (Yu et al., 2022). For example, by anticipating drought situations and facilitating effective watering techniques, AI can maximize the management of water resources. By identifying trends and patterns that guide policy and regulatory frameworks, big data analytics can improve our understanding of complex environmental systems (Niu, 2024; Hussain et al., 2023).

Digital technology also plays a part in raising public knowledge of environmental issues and encouraging public participation. People can access sustainable products, track their carbon footprints, and take part in community efforts, thanks to mobile applications and web platforms.

6.3 Vision for a Sustainable Future

Governments, corporations, and communities must work together to adopt policies and practices that put sustainability first in order to make the shift to a low-carbon economy. In addition to addressing environmental issues, this cooperative strategy will boost the economy and enhance people's quality of life.

Green technology will be essential to accomplishing the Sustainable Development Goals (SDGs) of the UN in this sustainable future. Green technology can help end poverty and lessen inequality by encouraging access to clean energy, sustainable cities, and ethical consumerism (Liu & Yang, 2023). By guaranteeing that resources are recycled and reused, the incorporation of circular economy principles would further improve resource efficiency and reduce waste and environmental effect (Fang & Shao, 2022).

6.4 New Market Opportunities

The future of green jobs and the expansion of green startups are essential elements of the developing green economy. Entrepreneurs are taking advantage of the growing demand for sustainable goods and services to provide cutting-edge solutions to environmental problems (Guo et al., 2018). New markets are emerging as a result of this trend, especially in waste management, sustainable agriculture, and renewable energy.

The development of these startups is greatly aided by incubators and accelerators for green technology businesses. These initiatives assist entrepreneurs in overcoming the obstacles of introducing green inventions to the market by offering cash, resources, and mentorship. As a result, it is anticipated that the green job market would grow, providing jobs in a range of industries, from installing renewable energy to designing sustainable products.

6.5 Ethical Considerations and Challenges

While the future of green technology holds great promise, it also presents ethical considerations and challenges. Issues related to data privacy, AI use, and automation raise important questions about the implications of these technologies on society.

For instance, the collection and analysis of environmental data must be conducted transparently and ethically to protect individuals' privacy and ensure that data is used responsibly (Cui et al., 2022). Additionally, the automation of jobs in sectors such as manufacturing and agriculture may lead to workforce displacement, necessitating policies that support retraining and upskilling.

Furthermore, discussions around fair access to green technologies for developing countries are crucial. As green technology innovations advance, it is essential to ensure that these solutions are accessible and affordable for all, particularly in low-income regions (Zhang et al., 2024). Bridging the technology gap will require international collaboration and investment to support the deployment of sustainable technologies in underserved communities.

References

Adedoja, O. S., et al. (2021). Applications of nanomaterials for water quality sustainability: Present status and future trends. *International Journal of Sustainable Development and Planning, 16*, 357.

Adewumi, A., et al. (2024). Solar energy innovations in the USA: A comparative global review. *International Journal of Science and Research Archive, 11*(1), 1528–1539.

Alsamhi, S. H., et al. (2019). Greening internet of things for greener and smarter cities: A survey and future prospects. *Telecommunication Systems, 72*, 609–632.

Carbonell, P., et al. (2016). Bioinformatics for the synthetic biology of natural products: Integrating across the design–build–test cycle. *Natural Product Reports, 33*(8), 925–932.

Cui, B., et al. (2022). Evolutionary game analysis of green technology innovation under the carbon emission trading mechanism. *Frontiers in Environmental Science, 10*, 997724.

El-Kassar, A.-N., & Singh, S. K. (2019). Green innovation and organizational performance: The influence of big data and the moderating role of management commitment and HR practices. *Technological Forecasting and Social Change, 144*, 483–498.

Etukudoh, E. A., et al. (2024). Solar power integration in urban areas: A review of design innovations and efficiency enhancements. *World Journal of Advanced Research and Reviews, 21*(1), 1383–1394.

Fang, Y., & Shao, Z. (2022). Whether green finance can effectively moderate the green technology innovation effect of heterogeneous environmental regulation. *International Journal of Environmental Research and Public Health, 19*(6), 3646.

Feng, W., et al. (2023). Green finance: Current status, development, and future course of actions in China. *Environmental Research Communications, 5*(3), 035005.

Guo, Y., et al. (2018). Environmental regulation, government R&D funding and green technology innovation: Evidence from China provincial data. *Sustainability, 10*(4), 940.

Hou, J., Bai, W., & Sha, D. (2023). Does the digital economy successfully facilitate carbon emission reduction in China? Green technology innovation perspective. *Science, Technology and Society, 28*(4), 535–560.

Hussain, M., & Zhang, T. (2025). Machine learning-based outlier detection for pipeline in-line inspection data. *Reliability Engineering & System Safety, 254*, 110553.

Hussain, M., Zhang, T., & Seema, M. (2023). Adoption of big data analytics for energy pipeline condition assessment. *International Journal of Pressure Vessels and Piping, 206*, 105061.

Jamal, E., et al. (2024). *Big data analytics for the Moroccan water actors: Towards an adoption model*. In *E3S web of conferences* (Vol. 489). EDP Sciences.

Jamil, H., et al. (2024). *Data process and water resources management in Morocco: Issues and challenges*. In *E3S web of conferences* (Vol. 489). EDP Sciences.
Janssen, M., et al. (2019). Challenges for adopting and implementing IoT in smart cities: An integrated MICMAC-ISM approach. *Internet Research, 29*(6), 1589–1616.
Jiang, S., et al. (2022). Does green finance promote enterprises' green technology innovation in China? *Frontiers in Environmental Science, 10*, 981013.
Kaluarachchi, Y. (2021). Potential advantages in combining smart and green infrastructure over silo approaches for future cities. *Frontiers of Engineering Management, 8*, 98–108.
Koval, V., et al. (2023). Economic and environmental dimensions of energy production with the use of renewable technologies. *Energy Policy Journal, 28*(1), 5.
Kwilinski, A., Lyulyov, O., & Pimonenko, T. (2023). Greenfield investment as a catalyst of green economic growth. *Energies, 16*(5), 2372.
Ley, M., Stucki, T., & Woerter, M. (2016). The impact of energy prices on green innovation. *The Energy Journal, 37*(1), 41–76.
Lidder, P., Cattaneo, A., & Chaya, M. (2025). Innovation and technology for achieving resilient and inclusive rural transformation. *Global Food Security, 44*, 100827.
Liu, L., & Yang, L. (2023). The influence of coupling degree of technological innovation and industrial structure upgrading on the green economic efficiency. *Polish Journal of Environmental Studies, 32*(2), 1721.
Mahmud, R., & Buyya, R. (2019). *Modelling and simulation of fog and edge computing environments using iFogSim toolkit. Fog and edge computing: Principles and paradigms* (p. 1). Wiley.
Marine, Y., & Saluky, S. (2018). Penerapan IoT untuk Kota Cerdas. *ITEJ (Information Technology Engineering Journals), 3*(1), 36–47.
Niu, J. (2024). The mechanism of the impact of the digital economy on green technology innovation. In *Proceedings of the 5th international conference on E-commerce and internet technology, ECIT 2024, March 15–17, 2024, Changsha, China*.
Ogunrinola, I., et al. (2024). Solar-powered electric vehicles: Enhancing endurance and sustainability through photovoltaic integration. *IOP Conference Series: Earth and environmental science, 1342*, 012044.
Pang, Y., Fan, Y.-c., & Huang, X. (2021). *Innovation investment and sustainable development*. In *E3S web of conferences* (Vol. 236). EDP Sciences.
Petrolo, R., et al. (2014). Integrating wireless sensor networks within a city cloud. In *2014 eleventh annual IEEE international conference on sensing, communication, and networking workshops (SECON workshops)*. IEEE.
Sabet, S., & Farooq, B. (2025). Exploring the combined effects of major fuel technologies, eco-routing, and eco-driving for sustainable traffic decarbonization in downtown Toronto. *Transportation Research Part A: Policy and Practice, 192*, 104385.
Sabri, Y., & El Kamoun, N. (2023). Medium access in cloud-based for the internet of things based on mobile vehicular infrastructure. *TELKOMNIKA (Telecommunication Computing Electronics and Control), 21*(2), 280–289.
Safavi, M., & Khoshbakht, M. (2024). On the performance of solar thermophotovoltaics (STPVs) and wavelength-selective Thermophotovoltaics (TPVs): Case study of a high-rise residential building in a hot and semi-arid climate. *Buildings, 14*(1), 269.
Sarsah, E. A., Sunnu, A. K., & Bawa, A.-R. (2024). Assessing the techno-economic feasibility of hybrid solar-hydrogen systems: A review of recent studies. *iEnergy, 3*(1), 28–38.
Saunila, M., Ukko, J., & Rantala, T. (2018). Sustainability as a driver of green innovation investment and exploitation. *Journal of Cleaner Production, 179*, 631–641.
Shabrina, W., & Andriani, W. (2022). Systematic literature review implementation of the internet of things (IoT) in Smart City development. *Buana Information Technology and Computer Sciences (BIT and CS), 3*(2), 41–46.
Shenkoya, T., Sustainable urban development: An evaluation of the impact of IoT on sustainable development and energy management in SMART cities. 2023.

Sipa, M., & Gorzeń-Mitka, I. (2021). Assessment of the progress towards the management of renewable energy consumption in the innovativeness context—A country approach. *Energies, 14*(16), 5064.

Somoza-Norton, A., Pabigian, L., & Drap, M. (2023). Biomimetic leadership: Core beliefs for sustainable organizations. *Journal of Ecohumanism, 2*(1), 77–91.

Song, M., & Wang, S. (2018). Market competition, green technology progress and comparative advantages in China. *Management Decision, 56*(1), 188–203.

Sott, M. K., et al. (2021a). A bibliometric network analysis of recent publications on digital agriculture to depict strategic themes and evolution structure. *Sensors, 21*(23), 7889.

Sott, M. K., et al. (2021b). Agriculture 4.0 and smart sensors. In *The scientific evolution of digital agriculture: challenges and opportunities.*

Taheri, R., et al. (2016). Biodesalination—On harnessing the potential of nature's desalination processes. *Bioinspiration & Biomimetics, 11*(4), 041001.

Usman, M., et al. (2023). Investigation on the causality relationship between environmental innovation and energy consumption: Empirical evidence from EU countries. *Energy & Environment, 34*(8), 3130–3159.

Verma, V. (2024). Advancement in solar technology: evolution, generation, future prospective, and challenges-a review. *Physical Sciences, 3*, v1. https://doi.org/10.20944/preprints202407.0

Wang, M., et al. (2022a). The mixed impact of environmental regulations and external financing constraints on green technological innovation of enterprise. *International Journal of Environmental Research and Public Health, 19*(19), 11972.

Wang, Z., et al. (2022b). Threshold effects of environmental regulation types on green investment by heavily polluting enterprises. *Environmental Sciences Europe, 34*(1), 26.

Wu, J., et al. (2018). Exploring driving forces of sustainable development of China's new energy vehicle industry: An analysis from the perspective of an innovation ecosystem. *Sustainability, 10*(12), 4827.

Yaacoub, E., & Alouini, M.-S. (2020). Efficient fronthaul and backhaul connectivity for IoT traffic in rural areas. *IEEE Internet of Things Magazine, 4*(1), 60–66.

Yan, C., et al. (2022). A two-staged SEM-artificial neural network approach to analyze the impact of FinTech adoption on the sustainability performance of banking firms: The mediating effect of green finance and innovation. *Systems, 10*(5), 148.

Ye, J., Xu, W., & Zheng, J. (2023). The digital economy, spatial effects and green technology innovation: Analysis based on the spatial Durbin model. In *Proceedings of the 4th management science informatization and economic innovation development conference, MSIEID 2022, December 9-11, 2022, Chongqing, China.*

Yu, Y., Yi, Z., & Jia, J. (2022). The efficiency evolution and risks of green development in the Yangtze River Economic Belt, China. *Sustainability, 14*(16), 10417.

Zhang, J., & Li, S. (2023). The impact of human capital on green technology innovation—Moderating role of environmental regulations. *International Journal of Environmental Research and Public Health, 20*(6), 4803.

Zhang, M., Xie, W., & Gao, W. (2022). Have environmental regulations promoted green technological innovation in cities? Evidence from China's green patents. *PLoS One, 17*(12), e0278902.

Zhang, Y., Liu, X., & Yang, J. (2024). Digital economy, green dual innovation and carbon emissions. *Sustainability, 16*(17), 7291.

Corporate Responsibility and Sustainable Business Practices

Francka Sakti Lee, Johanes Fernandes Andry, Yunianto Purnomo, Kevin Christianto, Davina Valerian Cong, and Randy Efraim Tanuwijaya

1 Introduction

With regard to "how modern companies build their eco-friendly innovations," the impact of knowledge management (KM) on green innovations (GIs) can be seen as significant (Aprilia et al., 2024; Lee et al., 2024b; Sahoo et al., 2022). As the focus on sustainability deepens in many sectors, understanding how organizations share and exploit knowledge to resolve issues considering environmental impacts is of utmost importance (Andry et al., 2021; Dinata et al., 2024; Javaid et al., 2022). KM is more than simply implementing technology and information systems; it is also about creating an organizational culture (OC) that encourages an environment for collaboration, ideas exchange, and lifelong learning (Lee et al., 2024a; Vuong et al., 2022).

Knowledge management practices (KMP) in GI enhances an organization's ability to respond to critical environmental concerns (Aydın & Yildirim, 2021; Rehman et al., 2021). Firms who manage well the information in green technology, sustainable manufacturing methods, and environmentally conscious marketing are likely to obtain market leadership in innovative offerings (Awan et al., 2021). This not only provides new business avenues but also enables the business to meet its social and environmental goals (Azeem et al., 2021). In addition, the studies regarding the relationship between KM and GI are beneficial for company owners who want to adopt more sustainable approaches to operational management (Takalo et al., 2021; Tu et al., 2021).

KM within GI plays a key role in mobilizing past experiences and organizational learning (Di Vaio et al., 2021). From past success and failures, organizations can extract patterns that are worth the effort to be learnt from, but avoiding future

F. S. Lee (✉) · J. F. Andry · Y. Purnomo · K. Christianto · D. V. Cong · R. E. Tanuwijaya
Department of Information System, University of Bunda Mulia, Jakarta, Indonesia
e-mail: flee@bundamulia.ac.id

A. Chakir et al. (eds.), *Green Society, Environmental Strategies and Sustainable Development*, https://doi.org/10.1007/978-3-031-94086-6_6

mistakes and learning how to solve new problems (Wang et al., 2021a; Yousaf, 2021). Previous knowledge management fosters the interaction of departments, partners, and other stakeholders (Demir et al., 2023; Jayaraman et al., 2023). It means bringing together the perspectives, experiences, and knowledge of different kinds that organizations can source knowledge from (employees, partners, and external stakeholder). In doing so, it creates space to nurture stretched and therefore systemic/sustainable innovations that can provide a framework for future and present challenges in one shot. When all put together, it increases the organization's capacity to generate creative solutions that are effective and sustainable in the long term (Shan et al., 2021; Yang & Wang, 2021).

This research seeks to explore several key discussions, including how the implementation of effective KM practices drives green innovation and contributes to achieving Sustainable Development Goals (SDG) (Cui et al., 2020; Frempong et al., 2021) and how knowledge-based leadership plays a crucial role in bridging the gap between green KM and organizational sustainability by creating a strong connection between the integration of sustainable practices and the efficient management of environmental knowledge. This leadership cultivates a culture centered on environmental awareness, encourages continuous learning to foster innovation, and guides strategic decisions to align business goals with long-term sustainability objectives (Ahakwa et al., 2021; Firman, 2023), how effective leadership facilitates the seamless integration of green KM practices within an organization, ensuring that sustainability initiatives align with corporate goals and industry best practices and how companies can design and implement comprehensive knowledge management strategies that leverage technology, cross-functional collaboration, and data-driven insights to enhance green innovation and achieve superior sustainability performance in a rapidly evolving business landscape (Aftab et al., 2022).

This research offers new insights by integrating the concepts of green knowledge management and GI to explore their intricate interconnections in promoting sustainability (Al-Faouri, 2023; Wang et al., 2021b). Additionally, it examines knowledge-based leadership as a mediating factor that links the adoption of green technologies with company performance outcomes (Imran et al., 2021; Roh et al., 2021). Study holds both academic and practical significance, enhancing the existing literature while providing actionable strategies for companies to strengthen their competitive advantage and sustainability efforts (Wang & Juo, 2021). Furthermore, its findings can inform public policies that encourage the widespread adoption of environmentally friendly practices, contributing to global sustainability initiatives (Song et al., 2020).

2 Literature Review

Knowledge management refers to the use of intellectual asset of an organization to capture, store know how, and share knowledge to meet strategic goals (Wibowo et al., 2021). It plays a central role in managing organizational knowledge by ensuring that critical knowledge and expertise are effectively utilized. One of the primary reasons for implementing knowledge management is the significant risk of losses when knowledge is not properly managed (Mohsin et al., 2020). Although knowledge management offers numerous benefits, its implementation can be challenging due to factors such as a weak knowledge-sharing culture, inadequate technological infrastructure, and insufficient leadership support. However, the benefits of knowledge management (improved organizational performance and competitiveness among others) outweigh these challenges (Shahzad et al., 2021).

Green innovation (GI): Green innovation encompasses a wide range of innovations on different levels from the creation and diffusion of energy-efficient products and processes, pollution prevention programs up to improvements in recycling waste, the incorporation of environmental considerations into the design of products by companies and gradual development of a holistic environmental management system, which collectively strive to facilitate sustainability leading environmental benefits for the long run and decrease ecological impact of firms and industries on nature (Alshammari, 2023). Its main objective is to minimize harmful environmental effects, such as air, soil, and water contamination, along with greenhouse gas emissions that drive climate change (Abbas & Sağsan, 2019). Additionally, GI focuses on conserving finite natural resources, such as water, energy, and raw materials, by adopting sustainable and efficient technologies.

In recent years, green innovation has garnered significant attention as it provides practical solutions to urgent environmental challenges, such as global warming and ecological degradation, which pose serious risks to humanity (Bai et al., 2021). This type of innovation not only enhances corporate value and performance but also involves the development of new technologies—both hardware and software—related to products and production processes that improve energy efficiency, reduce pollution, promote waste recycling, and support corporate environmental initiatives (Wang et al., 2016). While green innovation encounters various challenges, its long-term benefits in fostering a more sustainable and environmentally friendly future far outweigh its limitations (Ahmed et al., 2022).

3 Methodology

Method employed follows a SLR process, as outlined by Alghamdi et al. (2023) and illustrated in Fig. 1. The research began with the identification of 191 records from electronic databases, including Google Scholar (n = 143), Scopus (n = 43), and PubMed (n = 2). In the first screening phase, 96 records were excluded for various reasons: lack of an abstract or full-text, non-English language, focus on nonhuman research, absence of a joint among management of knowledge and green newness, or the format being reviews, meta-analyses, letters, book chapters, or reports. After this initial exclusion, 95 studies remained.

These studies then underwent a second screening phase, during which duplicates were removed, resulting in 24 unique records. The full-text articles of these 24

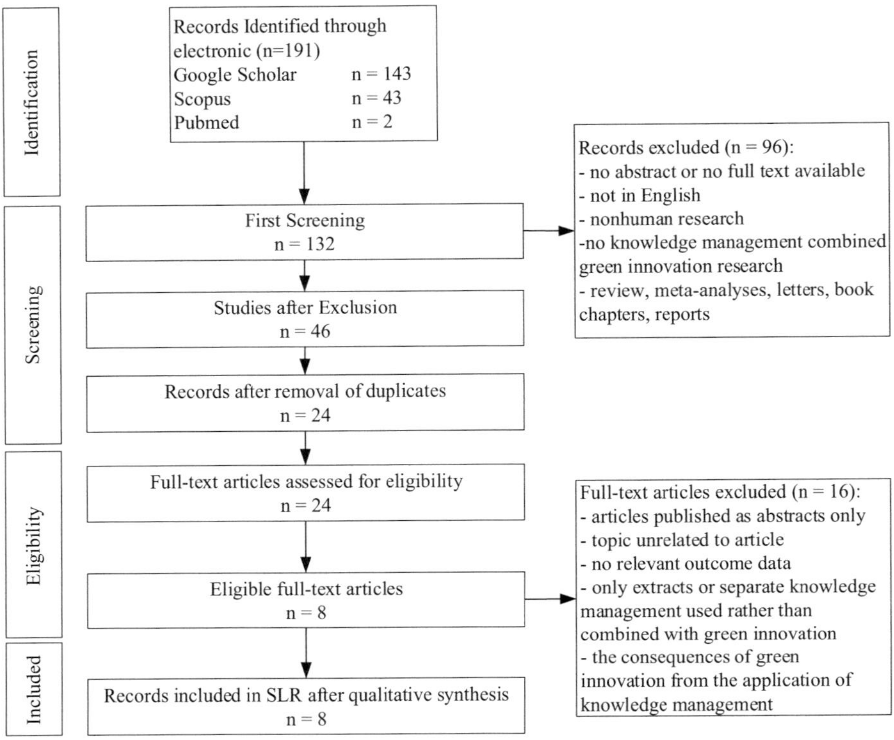

Fig. 1 Methodology of SLR (Ali et al., 2018)

studies were assessed for eligibility based on specific criteria. In this step, 17 articles were excluded for reasons such as being published only as abstracts, irrelevance to the topic, lack of relevant outcome data, the independent use of either knowledge management or green innovation without exploring their combination, or a sole focus on the consequences of green innovation stemming from knowledge management applications. Ultimately, eight full-text articles met the eligibility criteria and were included in the qualitative synthesis. The articles chosen as an input for this review are the bases of systematic review literature analysis laid down on the complex interrelation, organizational culture (OC), and knowledge management (KM) with green innovation (GI). Such studies explore the interplay and the effect of these critical drivers to show their co-efficiency in influencing sustainable practice in organizations. Taken together, the articles provide a more enriching appreciation of how OC, KM, and GI intersect to push the boundless of innovation and enable sustainable development goals. The detailed search highlights the importance of an intense and solidly entrenched organizational culture in terms of enabling and advancing adoption of successful KM strategies that lead to GI. Focus on these elements is key to enable a huge enabler in creating more sustainable, socially responsible, and environmentally sustainable business practices—cornerstones for long-term success and the deliverance of sustainability objectives of the enterprise.

3.1 Comparative Article of Record Result SLR

In order to get a richer understanding of how organizational culture (OC) and KM impact GI, related studies across different dimensions need to be compared (Di Vaio et al., 2021). Comparative analysis can identify similarities and differences in research questions, methods, and results to offer a more encompassing view of the topics. This section develops a systematical comparison of different papers, focused on presenting the title, model of research, aims, research question, study object, and results. We hope to stress the variety of approaches and comments that these various studies offer, in order to increase our comprehension of the OC–KM–GI relationship (Lam et al., 2021).

4 Analysis and Results

4.1 Comparative Analysis of Articles

4.1.1 What Does a Successful KM Culture Look Like?

Abbas and Sağsan (2019) defined successful KM culture as an organization where knowledge is sourced, shared, and applied to drive innovation and sustainable performance (Cui et al., 2020). This calls for high-level management commitment that represents resource allocation, setting goals, and also in being primarily involved with KM initiatives (Shahzad et al., 2020).

A strong knowledge management (KM) culture is also reflected in the presence of accessible learning conditions that empower the workforce to share lessons learned from mistakes and to sustain their competencies and skills (Alshammari, 2023). This can be practiced through both formal/institutional as well as informal knowledge-sharing mediated by tools which could have multiple dimensions.

Technology is a key element in developing a successful KM culture. Implementing the right technologies—such as KM platforms, collaboration tools, and document management systems—helps make knowledge management more structured and widely accessible to all employees, ultimately enhancing the efficiency and effectiveness of knowledge sharing.

Ultimately, establishing a KM culture is essential for creating value that aligns with the organization's goals. This means that the knowledge acquired and transferred, as well as its utilization, must align with the organization's strategic objectives to enhance performance, drive innovation, and support long-term sustainability (Abbas & Sağsan, 2019).

4.1.2 Organizational Culture Analysis

A very good positive toward GI is the KM process. Furthermore, as for corporate sustainable performance (Ahmed et al., 2022; Alshammari, 2023), GI is highly related to individual sustainability. The incorporation of KM and GI is equally important in promoting the sustainable organization (Alshammari, 2023). Wang et al. (2016) is a key mediating variable with KM, GI, and corporate sustainability at the knowledge-based leadership.

Abbas and Sağsan (2019) indicated that the ecosystem pathway between knowledge management and green technological innovation technologies is a main driver for sustainability in different dimensions of sustainability.

Knowledge-based leadership amplifies the influence of KM on green innovation by linking these together and making sure that the practices actually get integrated and are applied within the organization.

Organizational success in driving knowledge sharing across employees lifts operational, economic, and non-economic performance (Ahmed et al., 2022). Further, better collaboration between heterogeneous stakeholders contributes to environmental innovations (Cui et al., 2020). This results not just in increasing the competitive edge but also supports the performance from the environmental sustainability and societal well-being. The amalgamation of KM and green innovation, when done right, creates a high impact synergy that ensures sustainable and resilient growth for tomorrow.

4.1.3 Culture at the Foundation of KM

Organizational culture is essential in influencing the overall performance of an organization. It can significantly influence whether an organization succeeds or fails. A strong and adaptive culture fosters innovation, collaboration, and long-term growth (Abbas & Sağsan, 2019). Dynamic companies perceive knowledge as a valuable tool for enhancing customer satisfaction (Abbas & Sağsan, 2019) and gaining a competitive advantage (Wang et al., 2016). Over the past two decades, knowledge management (KM) has been a key focus in business, recognized as essential for developing strategies, innovative offerings, and solutions (Qasrawi et al., 2017), along with overseeing operational workflows (Yusr et al., 2014). Effective KM enhances innovation and organizational efficiency (Bolisani & Bratianu, 2018).

Consequently, many organizations consider KM a strategic resource to outperform their competitors. Organizational culture is a fundamental factor in understanding this phenomenon (Alshammari, 2023). Knowledge creation is a dynamic process that involves continuous interaction and the development of innovative ideas through the integration of both tacit and explicit knowledge, allowing for the generation of new insights and solutions within an organization (Maravilhas & Martins, 2019). Organizations must gather knowledge from multiple sources, such as suppliers, employees, and customers, to adapt to evolving customer preferences and dynamic business environments (Qasrawi et al., 2017). This enables organizations to leverage their strengths and address weaknesses (Shahzad et al., 2020).

It is essential for employees, particularly in relevant departments, to continuously acquire new knowledge (Ahmed et al., 2022). Learning organizations encourage active employee participation in various discussions, enabling managers to view challenges from different perspectives and develop effective solutions (Abbas & Khan, 2023). The application of acquired knowledge in key areas is crucial for improving organizational processes and overall performance.

4.1.4 The Effects of Culture on Individuals

The adoption of knowledge management can greatly influence organizational culture, especially in the realm of green innovation, where it serves a vital function in organizational durability and leadership grounded in knowledge (Wang et al., 2016). Embracing practices and technologies in knowledge management that facilitate eco-friendly innovation has a substantial impact on various aspects of organizational sustainability (Abbas & Sağsan, 2019). Knowledge-based leadership mediates and strengthens the role of KM in facilitating GI (Wang et al., 2016).

A culture of knowledge sharing among employees fosters a collaborative and innovative work environment (Ahmed et al., 2022), leading to improvements in operational and economic performance as well as non-economic outcomes for the company (Ahmed et al., 2022). By adopting environmentally friendly practices and enhancing resource efficiency, companies can reduce operational costs and improve their brand image (Ahmed et al., 2022). These improvements have a positive impact on different stakeholders: (customers, suppliers, and investors) (Shahzad et al., 2020).

An organizational culture that supports knowledge sharing and green innovation also enhances a company's competitiveness (Ahmed et al., 2022). Organizations that adapt to environmental demands and demonstrate strong social responsibility are more likely to achieve a sustainable competitive advantage (Ahmed et al., 2022). Therefore, fostering an organizational culture that integrates KM and GI not only promotes long-term sustainability but also strengthens global competitiveness and success (Abbas & Sağsan, 2019).

4.1.5 KM Can Change Organizational Culture

Knowledge management (KM) does not necessarily require a complete transformation of an organization's culture; instead, it can enhance the current culture to promote green innovation (GI) and sustainable growth. Wang et al. (2016) demonstrated that a robust green organizational culture can enhance the connection between green KM and GI, supporting the attainment of SDG. Similarly, Al-Faouri (2023) highlighted the importance of knowledge-driven leadership in facilitating the connection between eco-friendly KM and the sustainability of organizations, suggesting that KM can facilitate cultural transformation through effective leadership initiatives.

However, the effective implementation of KM depends on the commitment of top management and the active participation of employees at all levels. Organizations must cultivate an environment that prioritizes knowledge sharing, continuous learning, and innovation to fully harness KM's potential in advancing green innovation and sustainability.

Furthermore, Shahzad et al. (2021) indicated that for sustainability projects an organizational culture focused on collaboration and learning is more conducive than any other kind of organizational culture (2021). This implies that KM can be harnessed to strengthen and create new cultural values that are conducive for sustainability practices.

Overall, knowledge management (KM) plays a crucial role in enhancing organizational culture and promoting green creativity, which in turn supports sustainable development. It is limited to strong leadership commitment and buy-in across the board, culture of knowledge sharing, continuous learning, and innovation from employees.

4.1.6 Transformation to a Knowledge-Sharing Culture to Green Innovation

Alshammari (2023) Research constructs the significance of knowledge management (KM) in enhancing sustainability performance through research. Organizations that adopt efficient KM practices where this could range from sharing knowledge to creation and utilization have been observed to have better sustainability gains. Furthermore, green innovation (GI) serves as an intermediary in this connection, as KM promotes green innovation, which in turn enhances sustainability performance.

Ahmed et al. (2022) highlight the positive leverage of GI on sustainable development. Manufacturing and industrial companies are encouraged to incorporate green processes into their production methods to meet sustainability goals. However, it is equally important to foster knowledge-sharing practices, as leveraging the collective knowledge of employees enables organizations to achieve their sustainability objectives.

Al-Faouri (2023) illustrates that green knowledge management (GKM), which includes obtaining, retention, distribution, utilization, and development of eco-friendly information, significantly boosts organizational sustainability. In Jordan's IT sector, GKM has been effectively employed to promote green technologies and engage in environmental and social initiatives. Additionally, knowledge-based leadership plays a vital role in helping businesses to develop sustainable practices.

Research by Shahzad et al. (2021) suggests that the knowledge management process (KMP), which includes its fundamental aspects—knowledge gathering, distribution, and utilization—fuels environmentally friendly innovation, subsequently influencing sustainability performance across environmental, economic, and social dimensions. Moreover, organizational agility positively affects both green innovation and sustainability outcomes (Rahim et al., 2023).

In line with these findings, Wang et al. (2016) demonstrate that GKM enhances an organization's ability to achieve GI and sustainable development goals (SDGs). Green innovation emerges as a key predictor of corporate sustainable development, with an organizational green culture further enhancing the connection between GKM and environmental innovation in accomplishing SDGs.

Collectively, these studies suggest that fostering a knowledge-sharing culture is essential for promoting green innovation and improving sustainability performance. This cultural shift enables organizations to enhance their innovation capabilities, meet sustainability targets, improve environmental and economic outcomes, empower employees, build customer trust and loyalty, and make more informed decisions.

According to Shahzad et al. (2021), Knowledge Management Practices (KMP)—a holistic process that includes key components such as knowledge gathering, distribution, and utilization—directly drive environmentally sustainable innovation, which in turn positively influences sustainability performance across environmental, economic, and social dimensions. Organizational agility also has a positive influence on green innovation and sustainability performance (Rahim et al., 2023).

Consistent with these findings, the study by Xu and G. Wang et al. (2016) shows that Green Knowledge Management (GKM) supports organizations in achieving Green Innovation (GI) and Sustainable Development Goals (SDGs). Green innovation emerges as a pivotal determinant in corporate sustainability, accompanied by green organizational culture toward the mediating effect at GKM as a facilitator of environmental innovation toward SDGs.

These studies indicate that to encourage green innovation and enhancement of sustainability performance, a knowledge-sharing culture needs to be built up. This cultural liberalization allows organizations to increase the innovation capability of their organization, fulfill sustainability targets, improve environmental and economic results, empower employees, build customer trust and loyalty, and make wise choices.

4.1.7 Influence of Green Innovation on Organizational Culture

Mohsin et al. (2020) and Wang et al. (2016) highlight the role of the Knowledge Management Process (KMP) in driving Green Innovation (GI). Understanding in more detail the key elements of KMP-KA, KD, and KAP, as has been shown, is closely related to the promotion on green innovation (Alshammari, 2023; Shahzad et al., 2020). In sum, the combined knowledge base provides a solid foundation to comprehend fundamental organizational functions like money management, human resource grooming, and customer preferences which are integral elements for the support of any sustainability initiatives.

This is essential because knowledge acquisition (KA) is vital for employees to work together and contribute continuous improvements in all systems (Abbas & Sağsan, 2019). This method bolsters organizational effectiveness in general by offering employees requisite comprehension and skills to adopt and implement environmental sustainability initiatives. KA which establishes a learning culture and sustainability for the organization leads to innovative outcomes enabling the organization to fulfill its green innovation agendas.

Knowledge dissemination (KD) ensures the dissemination of sometimes key green practices and technologies throughout the organization. KDP provides a high quality of knowledge sharing of environmentally conscious solutions that are directly transferred to all employees; updated information enabling them timely implementation of knowledge. This goes to increasing the uptake of green innovations and focusing the organization in pursuit of its sustainability targets.

According to Ahmed et al. 2022, Knowledge Application (KAP) enables organizations to integrate the acquired and shared knowledge into their daily operations. Sound KAP enables mainstreaming of green technologies and sustainability initiatives by companies through higher efficiency material usage and reduced environmental impact. Lastly, KAP contributes to green highly initiative monitoring and improvement based upon feedback and lessons learned from the past to enhance the sustainability of that effort.

The significance of knowledge management in shaping corporate sustainability initiatives cannot be overstated. A well-structured KMP ensures that sustainability practices and green innovations are seamlessly embedded into the organizational culture (Alshammari, 2023). This not only supports environmental objectives but also fosters an adaptive and innovative workplace that fosters continuous learning and active contribution to sustainability goals (Ahmed et al., 2022).

Green Innovation (GI) significantly impacts Organizational Culture (OC) through Knowledge Management Practices (KMP). Companies that effectively implement KMP undergo cultural transformations that enhance sustainability, collaboration, and ongoing innovation. As shown in Table 1, these initiatives ultimately generate lasting advantages for both the organization and the environment (Ahmed et al., 2022; Wang et al., 2016).

Table 1 Mapping article result

Process	Variable	Article
What does a successful knowledge management culture look like?	Management commitment, open and collaborative learning environment, role of technology, value creation for the organization	Shahzad et al. (2021) and Wang et al. (2016)
Organizational culture analysis	Impact of KM processes on GI, the impact of GI on CSP, the role of knowledge-based leadership	Abbas and Sağsan (2019), Ahmed et al. (2022), Alshammari (2023) and Cui et al. (2020)
Culture at the foundation of km	Organizational culture as a valuable resource, knowledge creation, knowledge acquisition from various sources, learning organization	Abbas and Sağsan (2019), Ahmed et al. (2022) and Wang et al. (2016)
The effects of culture on individuals	Impact of KM on OC, GI, and sustainability, knowledge-sharing culture, environmentally friendly practices, resource efficiency	Abbas and Sağsan (2019) and Wang et al. (2016)
KM can change organizational culture	KM enhances existing culture, top management commitment, employee involvement, environment that encourages knowledge sharing	Shahzad et al. (2020), Shahzad et al. (2021) and Wang et al. (2016)
Transformation to a knowledge sharing culture to green innovation	Knowledge management improves sustainability performance, green innovation as a mediator, effective knowledge management practices, use of green technology	Alshammari (2023), Ahmed et al. (2022), Cui et al. (2020) and Shahzad et al. (2020)
Influence of a green innovation on organization culture	The influence of KM processes on GI, dimensions of KM (KA, KD, KAP), advanced knowledge, consistent improvement across departments, dissemination of information on green practices and technologies, implementation of knowledge in daily practices	Abbas and Sağsan (2019), Ahmed et al. (2022), Shahzad et al. (2020, 2021) and Wang et al. (2016)

5 Discussion

The findings of this report strongly align with the themes explored in four of the eight reviewed journals, highlighting the interconnection between knowledge management (KM), green innovation (GI), and sustainable development. One study conducted by Ahmed et al. (2022) extensively examines the significant influence of GI on achieving sustainable development while also highlighting the crucial intermediary role that knowledge management plays in facilitating and enhancing this relationship. In another work, Alshammari (2023) investigated the effects of KM on sustainability performance, specifying a green innovation as a critical mediator to foster environmental sensitive strategies and sustainability-embedded long-term organizational success. A different study (Shahzad et al., 2021) shows the effect of knowledge management process on corporate GI through sustainable development practices as one important contributor. Abbas and Sağsan (2019) examined the consequences of KM practices on GI and its consequences to corporate sustainable development.

These results emphasize the importance of knowledge management as part of an organizational culture that embraced working together, learning continuously and innovating. They offer a broad overview illustrating how effectively integrated knowledge management practices can drive green innovation, enhance organizational responsiveness, and ultimately support the achievement of long-term Sustainable Development Goals (SDGs) across various industries and business sectors. This adds to the significance of findings in their study that highlight the relationship between OC, KM dimension, and GI as a dynamic process sought by practitioners aiming for optimized business strategies within environment and social responsibility nexus.

These lessons apply beyond to the private sector and for improving the broader use of green in industries as public policy. To enhance sustainability, by weaving KM-driven green innovation into organizational strategies, companies can make significant contributions in terms of global environmental and social well-being.

6 Results

Research conducted by (Ahmed et al., 2022; Alshammari, 2023; Shahzad et al., 2020, 2021; Wang et al., 2016) has consistently pointed out the intuitive correlation between knowledge management (KM) and GI with corporate sustainability. Green innovation benefits from practices like sharing/exchanging, creating, and applying knowledge directly to sustainability performance. Moreover, knowledge-based leadership is shown to be an overarching link between green knowledge management and sustainable conduct overall in a manner underlining its strategic importance.

The results further reveal the significance of cultivating an organization culture where both sustainable innovation and knowledge exchange take foot. A culture of collaboration and lifelong learning is conducive to innovation, as well as increases both the efficiency of operations (efficiency), and how well business success enhances various economic and non-economic impacts. Integrating Knowledge Management (KM) practices and Green Innovation (GI) principles into the core of an organization's culture and operational strategies not only helps companies meet and exceed sustainability goals, but also enhances their competitive advantage, fosters more frequent innovation, and builds a strong foundation for long-term sustainability and stability in an increasingly conscious and rapidly evolving market.

In short, the evidence suggests that a GI and sustainable development is a key driver for KM and has most likely a strong basis of organizational culture. Importantly, when organizations do it right they can get an edge and also help advance global efforts to address sustainability. The ideas provide concrete advice for both enterprises and policy makers on how to apply their KM practices to foster green innovation addressing environmental problems while supporting organizational and societal goals.

7 Conclusion and Recommendations

7.1 Conclusion

Knowledge management (KM) is one of the key drivers for green innovation (GI) and a source of improving sustainability performance of organizations. The effectiveness of KM practices—such as knowledge sharing, creation, and utilization—has been shown to catalyze green innovation, significantly improving firms' green performance as measured by various assessment approaches. Finally, KM, knowledge-based leadership and KM-in-theory to facilitate practices supporting green innovation are indispensable factors for maximizing benefits to GI and sustainability.

This review puts forward that KM is reinforced to the positive end of green innovation and influences other several performance dimensions of an organization such as operational, economic, and non-economic outcomes as well. A company that thrives in a green innovation–friendly environment, where collaboration promotes knowledge, sharing, tends to achieve stronger competitiveness. When sustainability is aligned with their goal-oriented KM strategies, organizations have the potential to deliver on sustainability, gain sustainability edge, and contribute effectively to global sustainable development.

7.2 Recommendations

Drawing from the conclusions, the following recommendations are provided:

- Enhancing Implementation of KM: Companies need to build KM systems and infrastructure to develop effective ways in which knowledge is shared as well as created, captured, and utilized.
- Skills for Knowledge-Based Leadership: Faculties should develop leadership training and skill enhancement programs to equip leaders with the ability to systematically manage and utilize knowledge in support of green innovations and sustainable practices.
- Foster Organizational Culture That Fosters Collaboration, Knowledge Sharing, and Green Innovation Principles: Through training, internal communications, and staff recognition programs to literally solicit ideas for green innovation.
- Incorporating Knowledge Management Into Sustainability Action Plans: Toward sustainable knowledge-based embedding in strategies of organizations to enable the accessibility and use of relevant knowledge for sustainable development as such. Additional study is needed to determine the impacts of the knowledge management on eco-innovative behavior and organizational sustainability on

varying industries and in cross-cultural context. Additional future research could leverage richer KM models to promote green innovation and sustainability.

References

Abbas, J., & Khan, S. M. (2023). Green knowledge management and organizational green culture: An interaction for organizational green innovation and green performance. *Journal of Knowledge Management, 27*, 1852–1870.

Abbas, J., & Sağsan, M. (2019). Impact of knowledge management practices on green innovation and corporate sustainable development: A structural analysis. *Journal of Cleaner Production, 229*, 611–620.

Aftab, J., Abid, N., Cucari, N., & Savastano, M. (2022). Green human resource management and environmental performance: The role of green innovation and environmental strategy in a developing country. *Business Strategy and the Environment*, 1782–1798.

Ahakwa, I., Yang, J., Tackie, E. A., & Asamany, M. (2021). Green human resource management practices and environmental performance in Ghana: The role of green innovation. *SEISENSE Journal of Management, 4*, 100–119.

Ahmed, S., Naz, F., Abbas, Z., & Hummaira Batool, K. (2022). Impact of green innovation on sustainable development with mediating effect of knowledge management. *Journal of Business and Management Studies, 1*, 1–16.

Al-Faouri, A. H. (2023). Green knowledge management and technology for organizational sustainability: The mediating role of knowledge-based leadership. *Cogent Business & Management, 10*, 2262694.

Alghamdi, A. M., Pileggi, S. F., & Sohaib, O. (2023). Social media analysis to enhance sustainable knowledge management: A concise literature review. *Sustainability, 15*.

Ali, Y., Nazir, M., Ahmad, N., & Zakaria, N. H. (2018). Social media for knowledge-sharing: A systematic literature review. *Telematics and Informatics, 32*, 72–113.

Alshammari, K. H. (2023). Unlocking sustainable success in Saudi Arabia: The interplay of knowledge management and green innovation. *Archives of Business Research, 11*, 1–21.

Andry, J. F., Deny, Lee, F. S., & Liliana, L. (2021). Evaluation of the human resource information system with COBIT 5 and ITIL V3 (case study: Pharmaceutical company). *International Journal of Engineering and Information Systems, 5*, 123–129.

Aprilia, K., Lee, F. S., & Geasela, Y. M. (2024). Design of Enterprise Architecture using Zachman framework at private school in Center Jakarta. *Journal of Computer Science, 20*, 1059–1068.

Awan, U., Arnold, M. G., & Gölgeci, I. (2021). Enhancing green product and process innovation: Towards an integrative framework of knowledge acquisition and environmental investment. *Business Strategy and the Environment, 30*, 1283–1295.

Aydın, A. E., & Yildirim, P. (2021). Spatial effect of environmental regulation on green innovation efficiency: Evidence from prefectural-level cities in China. *Journal of Cleaner Production, 286*, 125032.

Azeem, M., Ahmed, M., Haider, S., & Sajjad, M. (2021). Expanding competitive advantage through organizational culture, knowledge sharing and organizational innovation. *Technology in Society, 66*, 101635.

Bai, Y., Wang, J. Y., & Jiao, J. L. (2021). A framework for determining the impacts of a multiple relationship network on green innovation. *Sustainable Production and Consumption, 27*, 471–484.

Bolisani, E., & Bratianu, C. (2018). The emergence of knowledge management. In *Emergent knowledge strategies* (pp. 23–47).

Cui, R., Wang, J., Xue, Y., & Liang, H. (2020). Interorganizational learning, green knowledge integration capability and green innovation. *European Journal of Innovation Management, 24*, 1292–1314.
Demir, A., Budur, T., Omer, H. M., & Heshmati, A. (2023). Links between knowledge management and organisational sustainability: Does the ISO 9001 certification have an effect? *Knowledge Management Research and Practice, 21*, 183–196.
Di Vaio, A., Palladino, R., Pezzi, A., & Kalisz, D. E. (2021). The role of digital innovation in knowledge management systems: A systematic literature review. *Journal of Business Research, 123*, 220–231.
Dinata, D. F., Lee, F. S., Geasela, Y. M., Everlin, S., & Purnomo, Y. (2024). Website-based educational application to help MSMEs in Indonesia develop. *Journal of Computer Science, 20*, 742–750.
Firman, A. (2023). Knowledge management implementation and human resource development on employee performance. *Jurnal Manajemen Bisnis, 10*, 221–234.
Frempong, M. F., Mu, Y., Adu-Yeboah, S. S., Hossin, M. A., & Adu-Gyamfi, M. (2021). Corporate sustainability and firm performance: The role of green innovation capabilities and sustainability-oriented supplier–buyer relationship. *Sustainability, 13*, 10414.
Imran, M., Arshad, I., & Ismail, F. (2021). Green organizational culture and organizational performance: The mediating role of green innovation and environmental performance. *Jurnal Pendidikan IPA Indonesia, 10*, 515–530.
Javaid, M., Haleem, A., Singh, R. P., Suman, R., & Gonzalez, E. S. (2022). Understanding the adoption of industry 4.0 Technologies in improving environmental sustainability. *Sustainable Operations and Computers, 3*, 203–217.
Jayaraman, K., Jayashree, S., & Dorasamy, M. (2023). The effects of green innovations in organizations: Influence of stakeholders. *Sustainability, 15*, 1133.
Lam, L., Nguyen, P., Le, N., & Tran, K. (2021). The relation among organizational culture, knowledge management, and innovation capability: Its implication for open innovation. *Journal of Open Innovation: Technology, Market, and Complexity, 7*, 66.
Lee, F. S., Andry, J. F., & Setiawansyah. (2024a). *Buku Ajar Tata Kelola Teknologi Informasi*. PT. Kamiya Jaya Auatic, Maluku Utara.
Lee, F. S., Hartono, H., Andry, J. F., Chakir, A., & Jessica. (2024b). E-learning to increase Services in Vocational High Schools Using ISO 9126. *Ingénierie des Systèmes d'Information, 29*, 2275–2282.
Maravilhas, S., & Martins, J. (2019). Strategic knowledge management a digital environment: Tacit and explicit knowledge in fab labs. *Journal of Business Research, 94*, 353–359.
Mohsin, S., Qu, Y., Javed, S. A., Zafar, A. U., & Rehman, S. U. (2020). Relation of environment sustainability to CSR and green innovation: A case of Pakistani manufacturing industry. *Journal of Cleaner Production, 253*, 119938.
Qasrawi, B. T., Almahamid, S. M., & Qasrawi, S. T. (2017). The impact of TQM practices and KM processes on Organisational performance: An empirical investigation. *International Journal of Quality & Reliability Management, 34*, 1034–1055.
Rahim, R. A., Ong, M. H. A., Wahab, L. N. A., & Anwar, N. (2023). Knowledge management strategies and green innovation practices: Empirical evidence from the Malaysian public sector. In *ASIA Proceedings of Social Sciences* (p. 11).
Rehman, S. U., Kraus, S., Shah, S. A., Khanin, D., & Mahto, R. V. (2021). Analyzing the relationship between green innovation and environmental performance in large manufacturing firms. *Technological Forecasting and Social Change, 163*, 120481.
Roh, T., Lee, K., & Yang, J. Y. (2021). How do intellectual property rights and government support drive a firm's green innovation? The mediating role of open innovation. *Journal of Cleaner Production, 317*, 128422.
Sahoo, S., Kumar, A., & Upadhyay, A. (2022). How do green knowledge management and green technology innovation impact corporate environmental performance? Understanding the role of green knowledge acquisition. *Business Strategy and the Environment, 32*, 551–569.

Shahzad, M., Qu, Y., Zafar, A. U., Rehman, S. U., & Islam, T. (2020). Exploring the influence of knowledge management process on corporate sustainable performance through green innovation. *Journal of Knowledge Management, 24*, 2079–2106.

Shahzad, M., Qu, Y., Zafar, A. U., & Appolloni, A. (2021). Does the interaction between the knowledge management process and sustainable development practices boost corporate green innovation? *Business Strategy and the Environment, 30*, 4206–4222.

Shan, S., Genç, S. Y., Kamran, H. W., & Dinca, G. (2021). Role of green technology innovation and renewable energy in carbon neutrality: A sustainable investigation from Turkey. *Journal of Environmental Management, 294*, 113004.

Song, W., Yu, H., & Xu, H. (2020). Effects of green human resource management and managerial environmental concern on green innovation. *European Journal of Innovation Management, 24*, 951–967.

Takalo, S. K., Tooranloo, H. S., & Parizi, Z. S. (2021). Green innovation: A systematic literature review. *Journal of Cleaner Production, 279*, 122474.

Tu, Y., Wu, W., Production, S., Wu, W., & Production, S. (2021). How does green innovation improve enterprises' competitive advantage? The role of organizational learning. *Sustainable Production and Consumption, 26*, 504–516.

Vuong, Q. H., Le, T. T., La, V. P., Nguyen, H. T. T., Ho, M. T., Van Khuc, Q., & Nguyen, M. H. (2022). Covid-19 vaccines production and societal immunization under the serendipity-mindsponge-3D knowledge management theory and conceptual framework. *Humanities and Social Sciences Communications, 9*, 1–12.

Wang, C. H., & Juo, W. J. (2021). An environmental policy of green intellectual capital: Green innovation strategy for performance sustainability. *Business Strategy and the Environment, 30*, 3241–3254.

Wang, S., Abbas, J., Sial, M. S., Alvarez-Otero, S., & Cioca, L.-I. (2016). Achieving green innovation and sustainable development goals through green knowledge management: Moderating role of organizational green culture. *Journal of Innovation & Knowledge, 25*, 1–9.

Wang, H., Khan, M. A. S., Anwar, F., Shahzad, F., Adu, D., & Murad, M. (2021a). Green innovation practices and its impacts on environmental and organizational performance. *Frontiers in Psychology, 11*, 1–15.

Wang, Y., Yang, Y., Fu, C., Fan, Z., & Zhou, X. (2021b). Environmental regulation, environmental responsibility, and green technology innovation: Empirical research from China. *PLoS One, 16*, 1–23.

Wibowo, D. Y. A., Muljono, P., & Sumertajaya, I. M. (2021). Analysis of the effect of knowledge management, competency, and innovation on employee performance. *Jurnal Aplikasi Manajemen, 19*, 804–811.

Yang, Y., & Wang, Y. (2021). Research on the impact of environmental regulations on the green innovation efficiency of Chinese industrial enterprises. *Polish Journal of Environmental Studies, 30*, 1433–1445.

Yousaf, Z. (2021). Go for green: Green innovation through green dynamic capabilities: Accessing the mediating role of green practices and green value co-creation. *Environmental Science and Pollution Research, 28*, 54863–54875.

Yusr, M. M., Mokhtar, S. S. M., Othman, A. R., & Sulaiman, Y. (2014). Does interaction between TQM practices and knowledge management processes enhance the innovation innovation performance? *International Journal of Quality & Reliability Management, 28*, 2–3.

Education for Sustainability

Rubina Ashmi Nabin

1 Introduction

Change starts with education. Education for sustainability can be applied for purposes of development. The world needs to be a better place to live in. The problem that needs to be addressed is linked to the means to achieve this. The UNESCO is actively researching on this aspect, and the Canadian Commission of UNESCO has set goals to be achieved by the year 2030 (Commission, UNESCO Inititatives-Education for Sustainable Development, n.d.-b).

> Education for sustainability develops the knowledge, skills, values and world-views necessary for people to act in ways that contribute to more sustainable patterns of living. It enables individuals and communities to reflect on ways of interpreting and engaging with the world. Sustainability education is futures-oriented, focusing on protecting environments and creating a more ecologically and socially just world through informed action. Actions that support more sustainable patterns of living require consideration of environmental, social, cultural and economic systems and their interdependence.
>
> *Source: Australian Curriculum, Assessment and Reporting Authority (ACARA), viewed on the Australian Curriculum website on 21/10/2015.*

The four pillars of education for sustainability according to Scott and Gaugh (Scott, 2003) are:

1. Learning to know
2. Learning to do
3. Learning to live together
4. Learning to be

In his book *Education for Sustainability*, Paul Clarke (n.d.) comes up with seven possibilities about sustainability as follows:

R. A. Nabin (✉)
Department of Business, Canadore Stanford College, Mississauga, ON, Canada

A. Chakir et al. (eds.), *Green Society, Environmental Strategies and Sustainable Development*, https://doi.org/10.1007/978-3-031-94086-6_7

Possibility one: Reinventing our relationship with the environment. Possibility 2: Is education purposeful if the purpose is sustainable living. Possibility 3: How can communities make it possible for schools to live with uncertainty. Possibility 4: Open-source living when sustainability is a way of life. Possibility 5: Can we make schools of sustainability. Possibility 6: The ways in which cities are functioning toward sustainability. Possibility 7: Our societal knowledge-related sustainability.

The research gap between the four pillars and the seven possibilities could be bridged by a mapping technique that this risk research follows.

2 Theoretical Framework

Linked hypothesis will relate to the literature review contents assuming the possibilities with possible and impossible using null and alternate (Vinod Chandra, n.d.). For instance,

Possibility One: Rethinking our relationship with our environment.

E1H0: No change.
E1H1: Change adapted.

Possibility Two: Is the current educational scenario enough for us to understand the impact of sustainable actions?

The concept of evaluating the scenario of science, technology, engineering, and mathematics streams of learning.

E2H0: Assuming that education is fit for purpose if the purpose is sustainable living.
E2H1: Assuming that education is not fit for purpose if the purpose is sustainable living.

Possibility Three: How can communities help schools to live with uncertainty?

Analyzing the effects of cross-cultural systems in the society of education.

E3H0: If communities help schools to live with uncertainty.
E3H1: If communities do not help schools to live with uncertainty.

Possibility Four: Open-source living—When sustainably is the way of life?

Adaptation of place-based education.

E4H0: Open-source living is adapted when sustainability is a way of life.
E4H1: Open-source living is not adapted when sustainability is a way of life.

Possibility Five: Is it possible to create schools of sustainability?

UNESCO's Education for Sustainability (USD) has adapted eight competencies (Deborah Nusche, 2024).

E5H0: We can promote sustainability through imparting education.
E5H1: We cannot promote sustainability through imparting education.

Possibility Six: The future of cities—does it start in one's mind?

Possibilities and difficulties to the urban fix (Fig. 1)

E6H0: The urban fix being made to achieve sustainable cities and sustainable minds.
E6H1: The urban fix not being made to achieve sustainable cities and sustainable minds.

Possibility Seven: Our great work education for sustainability.

Means and ways to make education for sustainability work better—how does individual knowledge and perception contribute to this possibility?

E7H0: Our great work is adapted as education for sustainability.
E7H1: Our great work is not adapted as education for sustainability.

For each of the stated concept of possibility the adaptation (Deborah Nusche, 2024), a multi-level approach is being adapted (Fig. 2).

The change and adaptability must start with current scenarios working from internal aspects to external factors which would in turn help us reach the goals faster.

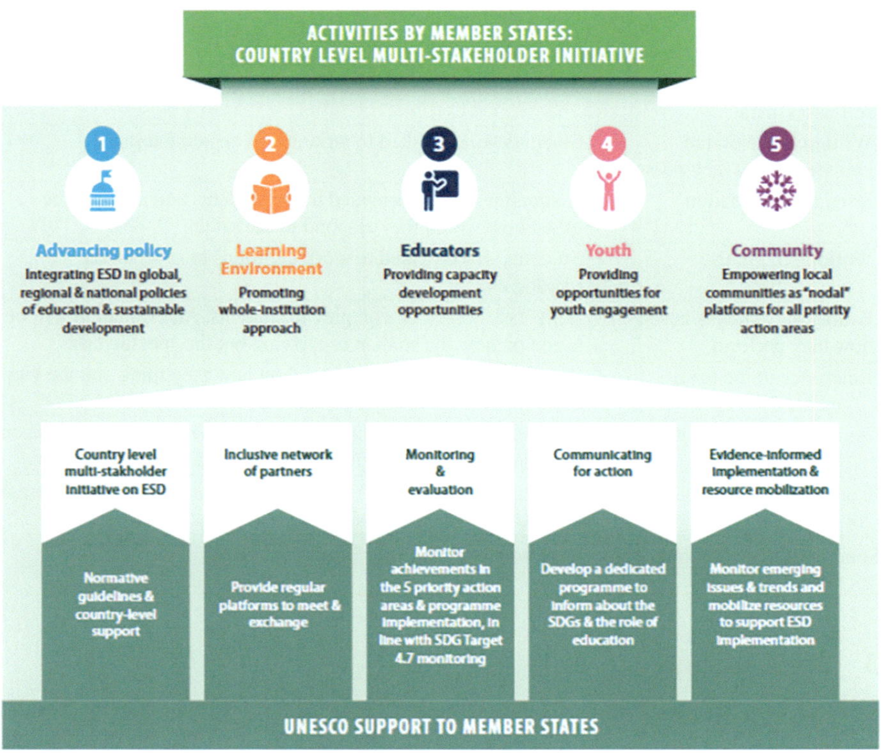

Fig. 1 ESD for 2030 framework. (Source: UNESCO, 2020)

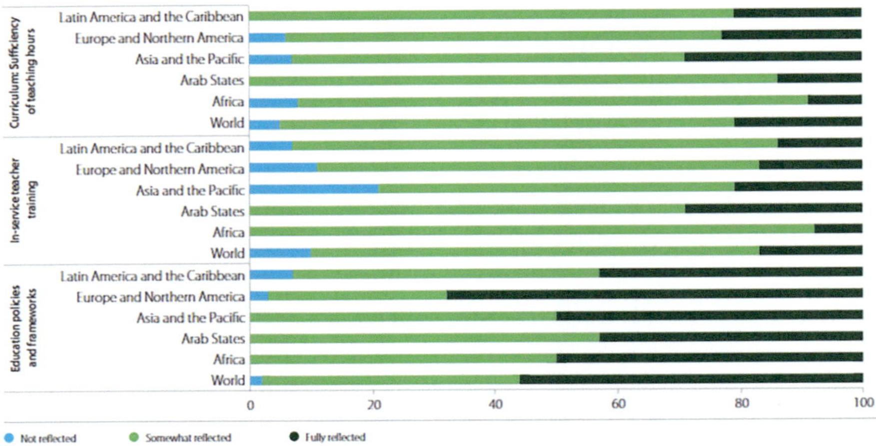

Fig. 2 Countries reflecting sustainability. (Source: 2019 Global Education Monitoring Report (UNESCO, 2019), p. 205)

Table 1 Tools and descriptions to assess sustainability

Tool	Description
Criterion-based self-assessment	Students will be asked to rate based on their own experience
Writing based on the knowledge that they posses	Students will be asked to write based on some topics
Providing a situation	A situation will be provided to the students and they will be asked how would they respond to the same
Working in groups	Students will be asked to work in groups to reflect their ideologies
Evaluating students based on how they perform	Some tasks could be complex to students. An evaluation will be based on how the student is carrying out the specific task
Relevance to the team	Students will be evaluated based on how they think and the way their approach is to solve a problem
Formative assessments	Students will be asked to take formative assessments in between the terms having some questions related to the costs
Complete assessment of the course	Students will be assisted based on different criteria both internal and external factors considered

Source: (European Commission: Directorate-General for Education, 2022)

3 Methodology

This qualitative study is going to be based on grounded theory (Merriam & Tisdell, n.d.). The primary data collection would be web analytics and the secondary data collection would be through published articles (Vinod Chandra, n.d.) (Table 1). The following table shows few tools in this context.

4 Data Analysis

Lots of cognitive models and analysis have been done in the past, mainly the inspiration comes from analysis approaches (Siddhartha Bhattacharyya, 2021).

Interpretation strategy is to learn such cognitive models and propose an effective model for education for sustainability.

Correlation and regression analysis will be done for the data; this will lead us to identify the *R* value with the assumption as if $\alpha > 0.5$, there will be a relationship between the variables. This relationship could be one of the following (Vinod Chandra, n.d.) (Table 2):

(a) Strong positive correlation.
(b) Weak positive correlation.
(c) Strong negative correlation.
(d) Weak negative correlation; the research will arrive at a regression equation.

Table 2 Age-wise understanding of students

Preliminary years (4–8 years old)	Primary (8–12 years old)	Lower secondary (12–15 years old)	Upper secondary (over 15 years old)
Feeling at home	Exploring the possibilities with some guidance from outside	Exploring the possibilities with one's own knowledge	Knowing how to apply the so far learned possibilities into worldly actions
Understanding the environment is protected us	Feeling comfortable in the place that we are in	Feeling comfortable in the place that we are in and willing to take care of that	Sense of action
Having that sense to take care of the world that we live in	Feeling safe in the world that we live in	Understanding that we are a part of the environment, so we need to take care of it	Ability to speak about the environment and its issues with others
Trying to understand this structure and organization about how the world works	Taking care of the environment	Ability to make others understand to take care of the environment	Connecting with people with same goals about environment and taking it one step ahead at a time
Asking questions about the world	Inhibits sense of questioning and deeper connections with improvised levels of understanding	Being curious about what could be their contributed to the same	Instills a sense of doing and accountability
Enhances the sense of equity, diversity, and inclusion	Enables them to adapt to the changing environment	Makes it possible for them to adjust to the change and smoothens the process of transformation	Makes it possible for them to excel in the environment

Source: (Lausselet, 2022)

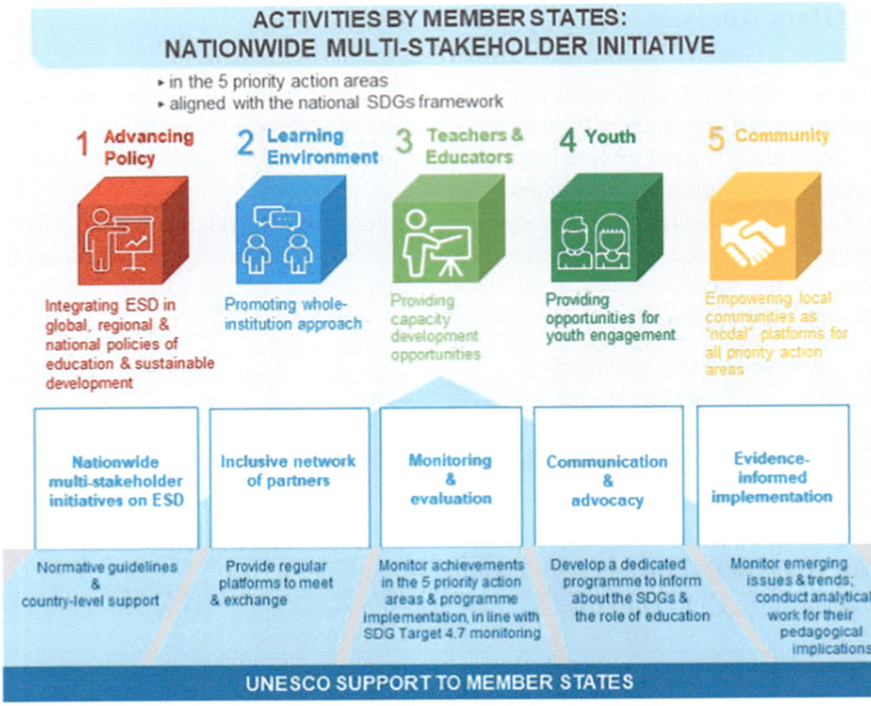

Fig. 3 Understanding the stakeholders at various levels (UNESCO, 2019)

The bond grows to be performative through being inclusive, adaptive, and transformative. However, if any of the steps is missed or not done properly, it will affect the results (Fig. 3).

The United Nations Educational, Scientific and Cultural Organization provides various frameworks on a deeper level by helping the communities come together and understand the education for sustainable developments; they do have various activities in which various member states do participate and they also have a nationwide initiative which brings together all the stakeholders on disregard. They have identified five main steps: the first step is formulating the different policies, the second step is to promote unacceptable educational environment, the third step is when the educators and the facilitators including teachers come into picture, the fourth step is involving the students who have to adapt to the changes in the education, and the fifth step is where the society comes together.

5 Cognitive Mapping Technique

As this is a qualitative study, cognitive mapping is chosen as a technique as it explores the depth of person-to-person differential thinking (Miles, 1994) in relation to the said hypothesis. Let's depict the education for sustainability using the method adopted (Wheeldon, 2009) (Figs. 4 and 5).

Fig. 4 Method percentile of ESD. (Source: (UNESCO n.d.)

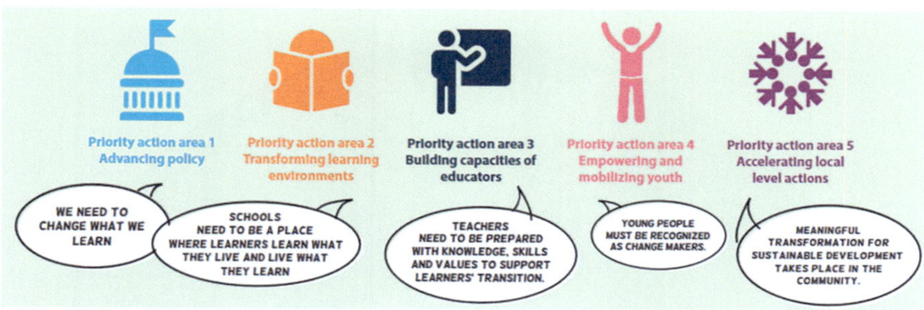

Fig. 5 ESD for 2030. (Source: Berlin Declaration on Education for Sustainable Development, 2021)

In order for us to achieve EST or education for sustainable development, we need to understand how to prioritize different action areas. Change is the only constant in life. We have to start by changing how our methodologies of learning in educational institutions have to be. Ideally, it must be a better place where it would be possible for those who learn to know, live and to practice what they learn. The educators have to be prepared with adequate skills needed to support this transition for those who are learning. The standards must be set promoting the valuable knowledge sharing of such actions. This will eventually create better awareness in this society.

> We now have the perfect vehicle for emphasizing sustainability education with the introduction of the Australian Curriculum. We must be clever by giving teachers solutions not more work. Children share their learning and reinforce the fact that we are all in this together. Our future really is in their hands.
>
> Andrew Best, Principal in NSW schools from 1991–2015

Ecosystem is made possible with the combination from the big ideas. These require certain skills. To apply the acquired knowledge, actions at different stage of dispositions in the learning classroom needs their own instructional practices. A best ecosystem is not just the one that exists, but it is the one that is cultivated with respect to people and the society. It takes a lot of steps and it also involves educational institutions formulating and practicing the sustainable environment. The classroom is where knowledge begins and it is important to import knowledge when it comes to sustainability as and when possible. People do have capabilities, especially the young minds, they have great ideas. If we can kinder them in the right manner they can be made very much responsible. By promoting their actions to apply the knowledge that they possess (Fig. 6).

The below figures are some examples of students interpretation when it comes to sustainability. This involves different forms of nature. It is so amazing how they are able to go at different levels from complex to simple. It helps us to understand that we all do need just that mindset shift or the paradigm shift to think from what is impossible to think that nothing is impossible (Fig. 7).

The above story map is an example of a student's perception on how sustainability is seen from that student's point of view in a trip taken.

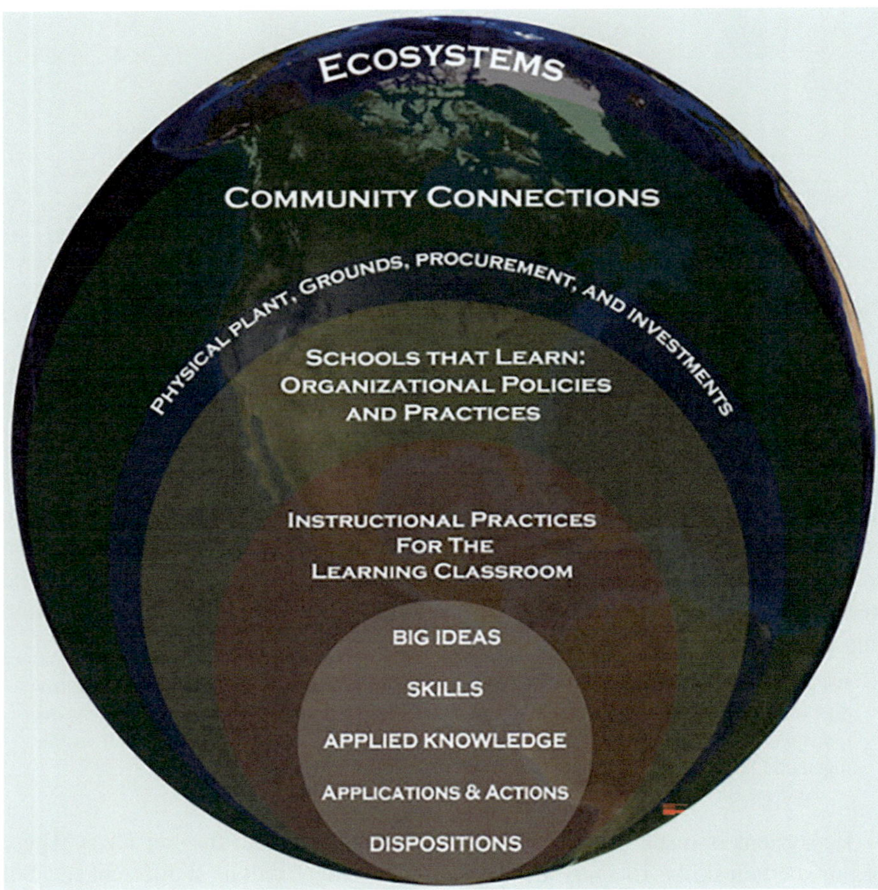

Fig. 6 Core understanding of ecosystem. (Source: (Bernier, n.d.)

Fig. 7 Student interpretation story map. (Source: Jennifer MacDonald, n.d.)

The perception of the youth, those young minds and their interpretation to nature, environment and sustainability is so fascinating. Who else could have thought that a water drop could hold so much importance and value. Another interesting expression of a student's thought process concerning the importance of water is depicted in the Fig. 8. I find it intricate and beautiful with so many layers of understanding.

When we try to map the factors of sustainability, it gives us a chance to understand in detail the functionalities and measurables of sustainability to a deeper extent. Imagine this, right from childhood how much memories do we hold related to water. All the ponds, oceans, seas, lakes, trips taken on boats and ships, birds flying, fishing, and the list goes on. Same applies to other natural sources around us. How do we learn to preserve and take care of the environment? How do we know about the dooming climate changes and the serious modifications that are happening in the global climate? The global warming is a very serious issue. We are all accountable and we all need to self-reflect to our contribution for the better and for the worse.

It is so surprising to me that a single drop of water can hold so many memories to someone. If we can relate ourselves to every aspect of nature as we are surrounded by it. By not taking it lightly or taking it for granted we can do so much for this society. It is roughly said that we have less than 16 years to put an end to the global warming

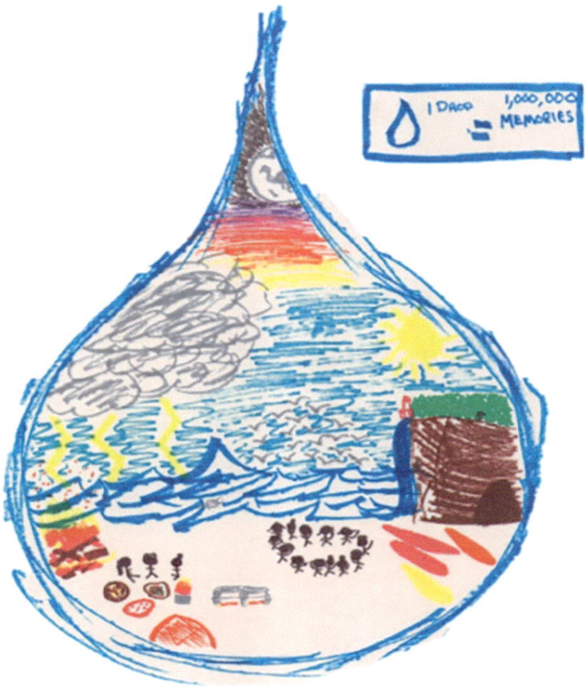

Fig. 8 Memories of water. (Source: Jennifer MacDonald, n.d.)

that is happening. Are we living a better society for our future? Are we agreeing that summers are getting hotter than usual, and winters are also extreme. The tectonic plates are shifting at a greater level. This is something that we really have to ponder upon. Massive environmental changed could prove to be disruptive (Fig. 9).

It is understandable that self-reflecting is the primary aim for communities that learn. When we are able to self-reflect, we can collectively learn from the classroom which will allow us to reflect through the schools that we learn. This in turn will keep us informed about the investments and physical plans around us. As a community we can grow to be educated about sustainability. Knowing ourselves, our identities, and understand more and more about sustainability-related factors from schools and colleges will help in forming a community that cares responsibly about the environment. When this happens, we can all benefit, grow, and learn together as a community. How powerful could that impact be? Let us all aim to live in a better world and leave a better world for our future generations to come.

Let us analyze and understand what makes the core when it comes to the measurable standards for education for sustainability. In Fig. 10, if you notice the bottom most foundation is implying to place-based awareness. In other words, firstly be

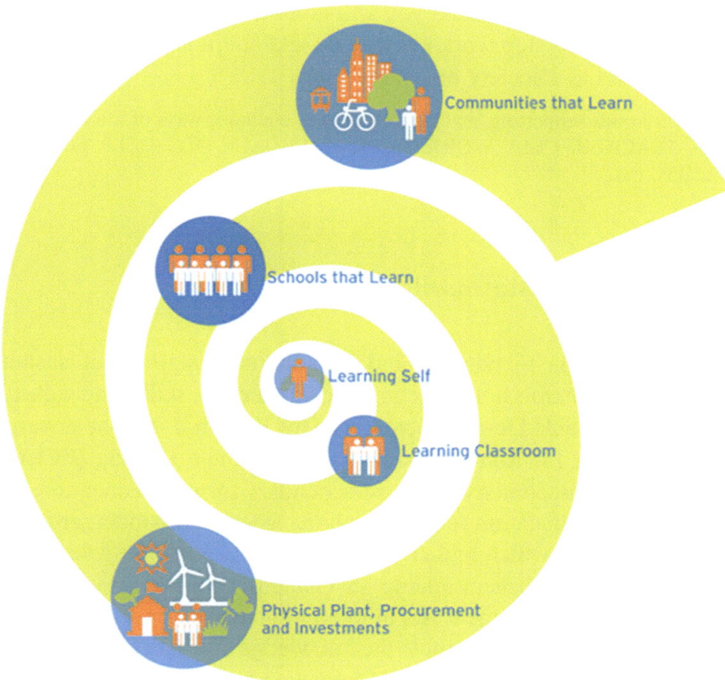

Fig. 9 ESD spirality. (Source: Education for Sustainability, 2012)

Fig. 10 Efs Standards. (Source: (Education for Sustainability, 2012)

aware of what is happening around you and around where you live related to sustainability. Then step by step work the way up and attain an imminent transformation as a culture and as a society on a large scale.

> Sustainability means supporting healthy, harmonious, equitable ways of life for future generations.—TEACHER, CELF WORKSHOP, SCARSDALE, NY (CELF Education for Sustainability, n.d.)

6 Education for Sustainability in Canada

By 2030, Canada aims to educate and impart knowledge about sustainability through education (Commission, Teaching the knowledge, skills, and values needed to achieve the Sustainable Development Goals by 2030, n.d.-a). The government is working closely with Environment and Climate Change Canada (ECCC). They seem to have primary, secondary, and post-secondary forms of education. They call this the K–12 as this education for sustainability goes from kindergarten to Grade 12 (Fig. 11). Following which higher education also plays a key role. An example could be the various publishers adding sustainability-related articles to the textbooks. The Canadian government has allotted almost three million dollars for the education for sustainability to promote the knowledge and care for our environment (Canada, G. o., 2024). One of the notable education for sustainability organizations includes Sustainability and Education Policy Network (SEPN). They are doing a tremendous job in doing the research and analysis focusing on education for sustainability (Kristen Hargis, n.d.) (Fig. 12).

> *In this back-to-school season, I am delighted to announce our support for these organizations and their environmental education projects that are working for future generations. People of all generations need to understand climate issues, if they are to make a difference. Likewise, they need to know what they can do, how they can do it, and how it will affect them and their community.*
>
> *—The Honourable Steven Guilbeault, Minister of Environment and Climate Change*

Whatever the goals being set to be achieved, it can only be achieved if there exists a means to verify, assess, and correct the course of actions when needed. Figure 13 depicts a typical assessment style that helps us in understanding how effective education in sustainability is perceived through various educational institutions. There must be a deeper engagement instead of shallow. The deeper engagement happens when one can relate to their own life and how sustainability acts as a core to make one's life better. It is only ideal if there is a measuring scale or system to analyze and assess the standards when it comes to the Education for Sustainability. This is where STARS which is a program under AASHE plays a crucial role. STARS stands for the Sustainability Tracking, Assessment and Rating System (Education, A. f., n.d.) and AASHE stands for the Association for the Advancement of Sustainability in Higher Education (The Association for the Advancement of Sustainability in Higher Education, n.d.) (Fig. 14).

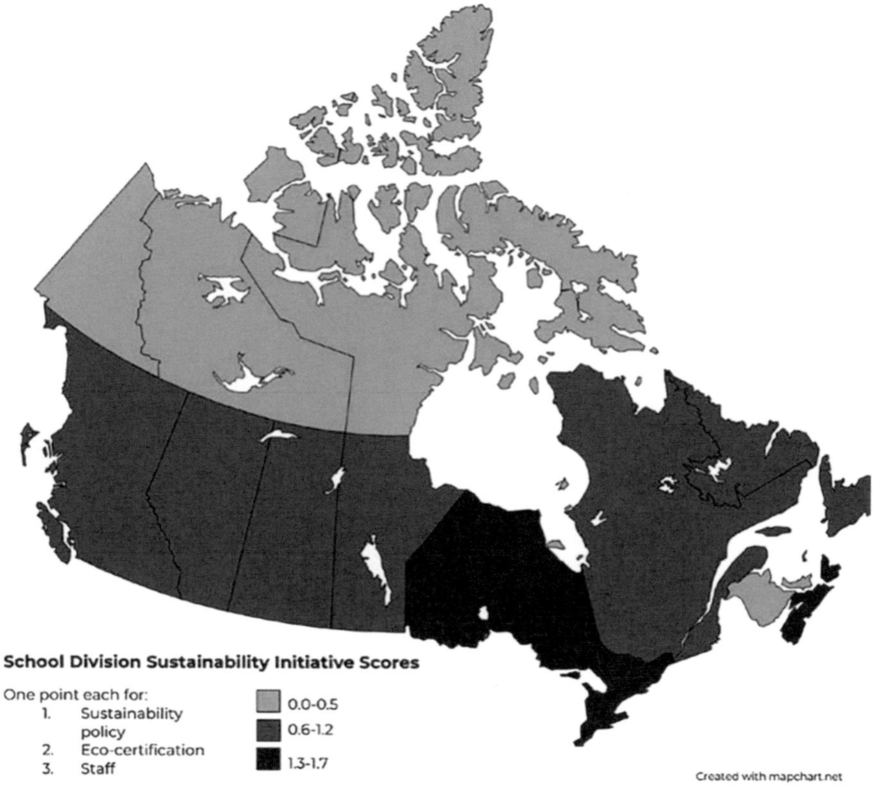

Fig. 11 Canadian map showing the ten provinces with corresponding points for sustainability (Beveridge, 2015)

Education for Sustainable Development Working Group (ESDWG) ensures that the goals of Councils of Ministers of Education Canada (CMEC) (Canada, C. o., n.d.) (Council of Ministers Education, n.d.) are met. They aim to centralize the metrics throughout the federal alignment and not just provincially. They ensure that at various levels of education, the required knowledge is recieved about sustainability (Fig. 15).

We live and grow in the world with complex relationships, and it is important for us to think how to deal with difficulties to do that we need to be able to differentiate between various levels of relationships in our lives. When we can do that, the next step does not come as a surprise; we have various possibilities for our futures as there are a lot of uncertainties and change is the only constant in our lives. Furthermore, the way we intercept all these rules and regulations matters. These make us liable to our own lives and for us to understand the values of sustainability. We need to be able to come to terms with all our future goals and what kind of steps we need to take or follow in order for us to achieve them. After that we need to be

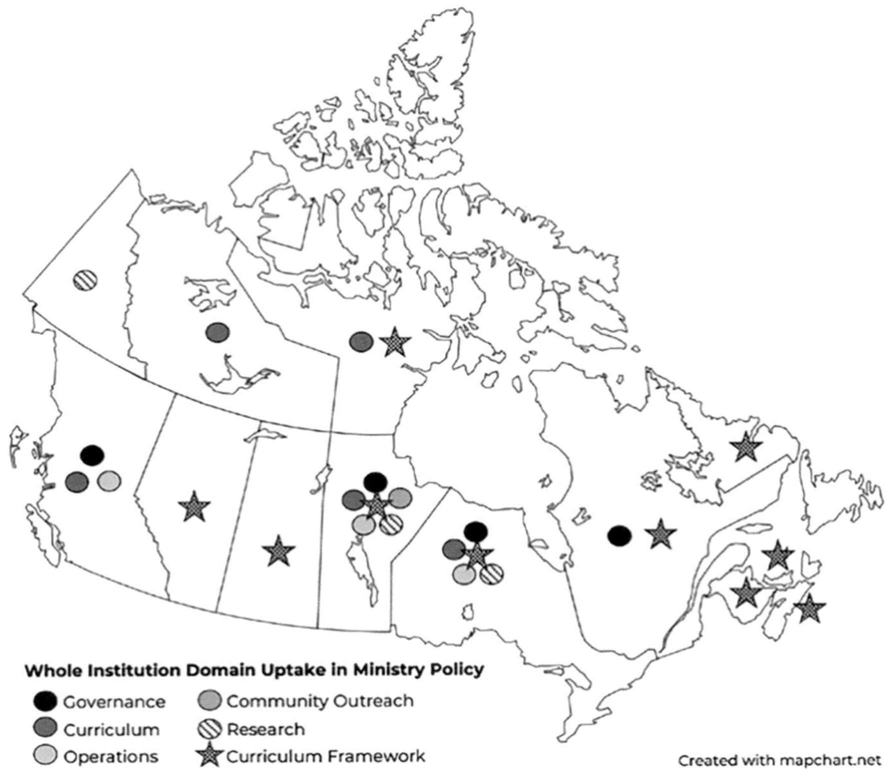

Fig. 12 Province wise adaptation to Education for Sustainability in Canada (Beveridge, 2019)

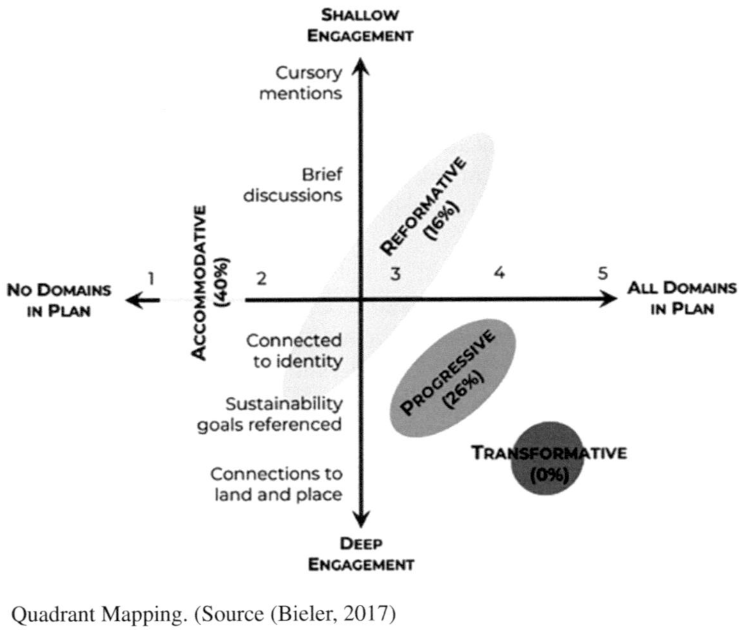

Fig. 13 Quadrant Mapping. (Source (Bieler, 2017)

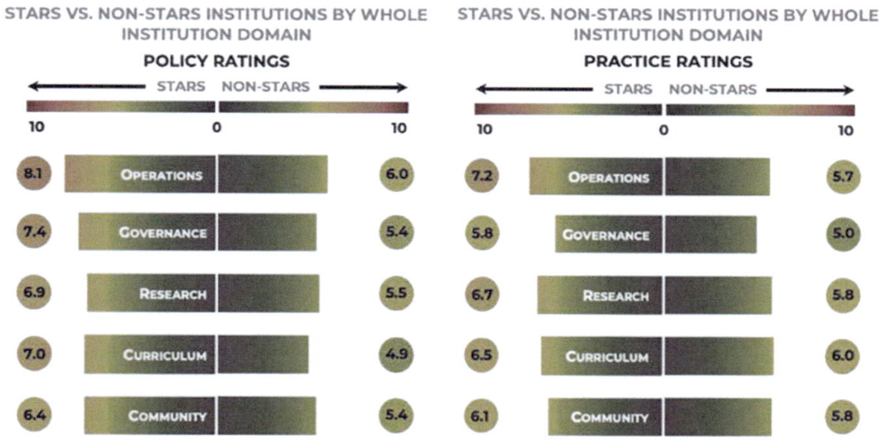

Fig. 14 This figure represents two types of rating based on policy and practice, respectively, by educational institutions. (Source: McKenzie, 2023)

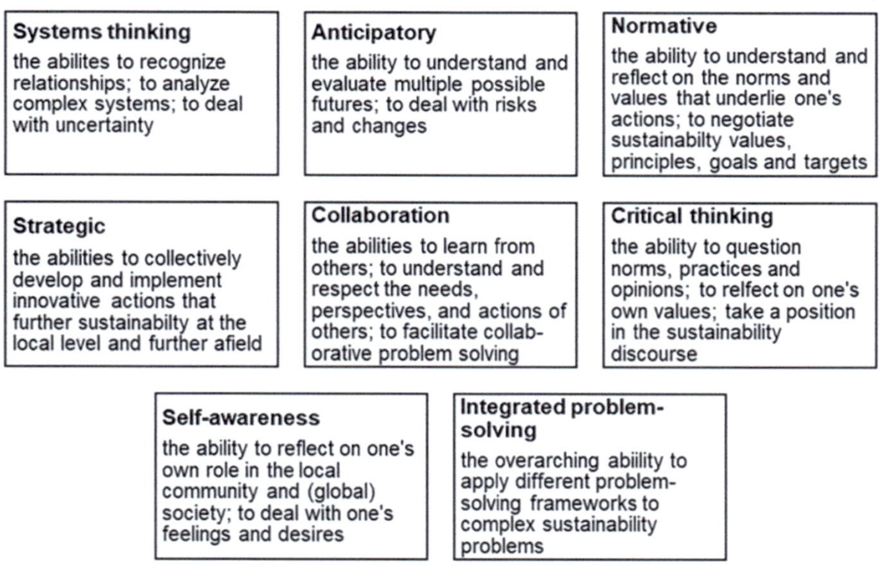

Fig. 15 Key factors of Sustainability. (Source: M, 2017)

able to put together whatever we can do with our capability to understand sustainability to how it can be done from the place that we originate from. As we live in a society and coexist with others, it is important for us to understand the differences and agree to disagree respectfully. This will help us to focus on solutions rather than problems. Additionally this would facilitate the understanding of the cost effect of relationships. Whether we agree to everything that the society puts on us or do we have our own opinions and our own interpretation to situations. We must know how to think and analyze everything critically. We know how to understand about

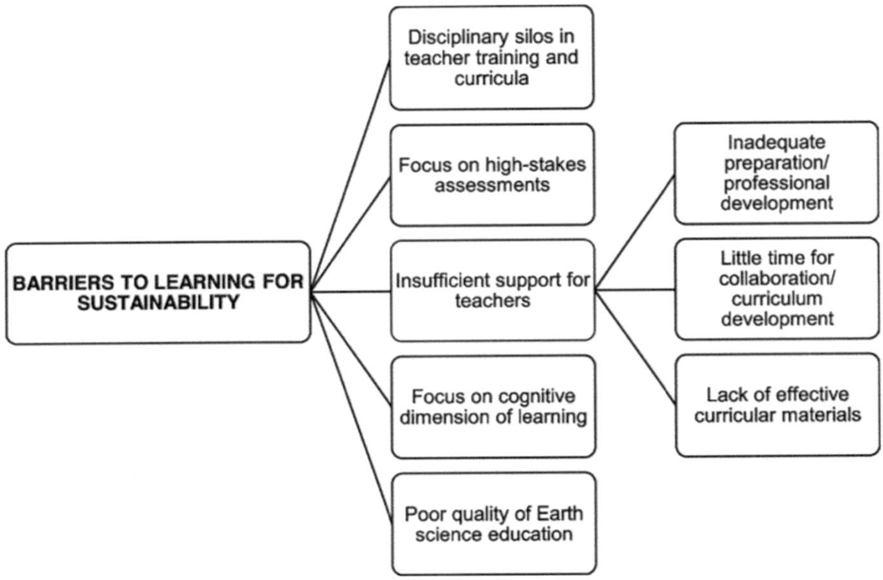

Fig. 16 Understanding barriers for sustainability. (Source: Metzger, 2023)

sustainability and take a stand in our own strongest suits. Connecting with oneself or being aware of oneself is the key strategy to sustainability. Each one of us have a role when it comes to having a sustainable environment. We all have different wishes, wants and needs. If we all can work together, we can achieve a better environment with lesser global climatic challenges.

When it comes to education for sustainability, there are a lot of challenges, especially for teachers as there are not enough resources, technology and funds in most of the cases. The Fig. 16 definitely depicts the challenges involved primitively. There is not enough time and the materials are neither fully ready nor prepared in the desired manner to be given to teachers. The teachers would need these in order for them to teach at various levels of educations. The assessments do not focus much on sustainability-based questions therefore the students are not really focusing on such topics. Education and sustainability must be made part of the curriculum in order for the students to understand and apply the concepts (Fig. 17).

Set service can help each of the schools preferably known as green schools to assess where they stand when it comes to sustainability and the scores can be taken every year to see how they have improved from the past; such means of metrics and measures can help us to attain the required results in a preferable time frame.

Green Schools: Quick Survey

Please answer the following questions based on a scale from 0 to 4. As a guideline, '0' is *no* or *not at all*, 1 is *rarely*, 2 and 3 are *sometimes*, and 4 is *yes* or *always*.

Green Team

Does your school have a green team? _____

Does your green team include students, teachers, and staff? _____

Energy

Does your school have an energy conservation program? _____

Does your school make an effort to reduce heating costs? _____

Do people turn off the lights when rooms are not in use? _____

Do people turn off computers and monitors when not in use? _____

Does your school use fluorescent and compact fluorescent bulbs? _____

Waste

Does your school have a recycling program that collects more than paper? _____

Does your school have a composting program? _____

Has your school ever done a waste audit? _____

Does your school have waste-free lunches? _____

Are school printers set to double side as a default? _____

Does your school use recycled paper? _____

Water

Does your school have a water conservation program? _____

Does your school have aerators or auto on/off devices on faucets? _____

Does your school have low-flow toilets? _____

Do people drink tap water instead of bottled water? _____

Does your school use environmentally friendly cleaning products? _____

Fig. 17 Understanding Green Schools Survey. (Source: Education, M. o., n.d.)

School Grounds

Does your school have a garden and/or a naturalized green space (more than just a grass field)? ____

Does your school avoid using pesticides on the grass and plants? ____

Do students spend time outside playing, studying, or socializing? ____

Transportation

Does your school have an anti-idling program? ____

Do people walk, bike, or skateboard to school? ____

Do people carpool or use public transportation to get to school? ____

Does your school have secure bicycle storage? ____

Scoring for 25 questions

0-25

There are many different things you can do to make your school greener. Start off by addressing waste and energy and continue building from there.

26-50

Your school has made some changes. Keep up the momentum.

51-75

Your school is making changes to become greener – continue working on expanding programs in all areas.

76-100

Your school has made significant changes to become sustainable and is a great role model for other schools!

Fig. 7.17 (continued)

7 Conclusion

This study concludes that in order for us to have a strong positive relationship and to promote Education for Sustainability, we need to start from the base as smaller communities and work our way toward the bigger communities. Starting from schools with different age groups and providing different content as per the different needs and adaptability. This will enhance and ensure the effectiveness for education for for sustainability as long as the changes to be decided ought to be made within, henceforth it will not only be successful but effective. Right from early childhood to advanced educational stages, in each and every step, wherein forms of materials and support guidance groups must be validated to go through this process.

The Environment and Climate Change Canada (ECCC) is trying to make a centralized curriculum when it comes to the sustainability-based education; the Children's Environment Literacy Foundation or self is having a sustainability-based approach when it comes to education and they even have a zero waste classroom to begin with (Commission, Teaching the knowledge, skills, and values needed to achieve the Sustainable Development Goals by 2030, n.d.-a).

Although the awareness about education for sustainability has predominantly increased but still there requires a lot of improvement. Organizations like UNESCO are helping and playing their part in this, but unless and otherwise the initiatives are taken and adapted by every institution and the education for sustainability goal it is not going to be successful. Collaboration as a society, in favor of the educational institutions at various levels are the prime need of the hour for us to bridge this gap between what is expected and what is existing in this society.

References

Berlin Declaration on Education for Sustainable Development. (2021). *UNESCO World Conference on Education for Sustainability, online* (pp. 2–15). UNESCO. Retrieved from https://unesdoc.unesco.org/ark:/48223/pf0000381229

Bernier, A. (n.d.). The EFS collective. Retrieved from Education for a Sustainable Future: https://www.efscollective.org/benchmarks

Beveridge, D. M. (2015). Sustainability in Canadian post-secondary institutions: The interrelationships among sustainability initiatives and geographic and institutional characteristics. *International Journal of Sustainability in Higher Education, 16*, 611–638.

Beveridge, D. M. (2019). A national census of sustainability in K-12 education policy: Implications for International Monitoring, Evaluation, and Research. *Canadian Journal of Educational Administration and Policy, 188*, 36–52.

Bieler, A. (2017). Strategic planning for sustainability in Canadian Higher Education. *Sustainability, 9*, 161.

Canada, C. o. (n.d.). *Council of Ministers of Education Canada*. Retrieved from https://cmec.ca/en/

Canada, G. o. (2024, September 13). *Environment and Climate Change Canada*. Retrieved from Government of Canada: https://www.canada.ca/en/environment-climate-change/news/2024/09/government-of-canada-advances-environmental-education-across-canada-with-29-million-investment.html

CELF Education for Sustainability. (n.d.). Retrieved from https://celfeducation.org/wp-content/uploads/2021/11/CELF-EfS-Principles-and-Practices-2020.pdf

Clarke, P. (n.d.). *Education for sustainability-becoming naturally smart*. Routledge.

Commission, U. C. (n.d.-a). *Teaching the knowledge, skills, and values needed to achieve the Sustainable Development Goals by 2030*. Retrieved from Education for Sustainable Development: https://en.ccunesco.ca/initiatives/education-for-sustainable-development#:~:text=With%20support%20from%20Environment%20and,ESD%20and%20Global%20Citizenship%20Education

Commission, U. C. (n.d.-b). *UNESCO Inititatives-Education for sustainable development*. Retrieved from UNESCO Canadian Commission: https://en.ccunesco.ca/initiatives/education-for-sustainable-development

Council of Ministers Education, C. (n.d.). *Education for Sustainable Development Working Group*. Retrieved from https://www.cmec.ca/Publications/Lists/Publications/Attachments/222/ESD-collaboration-action.pdf

Deborah Nusche, M. F. (2024). Rethinking education in the context of climate change: Leverage points for transformative change. In *OECD Education Working Paper; Organisation for Economic Co-operation and Development* (pp. 5–71).

Education, A. f. (n.d.). *The sustainability tracking, Assessment & Rating System*. Retrieved from https://stars.aashe.org/

Education for Sustainability. (2012). Retrieved from EFS Standards & Performance Indicators: https://www.truevaluemetrics.org/DBpdfs/Initiatives/CloudInstitute/CloudInstitute-EfS-Standards-Performance-Indicators-2012-K-12.pdf

Education, M. o. (n.d.). *Sustainable Schools Best Practices Guide*. Retrieved from https://www2.gov.bc.ca/assets/gov/education/kindergarten-to-grade-12/teach/teaching-tools/environmental-learning/sustbestpractices.pdf

European Commission: Directorate-General for Education, Y. S. (2022). *Publications Office of the European Union*. Retrieved from Learning for the green transition and sustainable development: staff working document accompanying the proposal for a Council recommendation on learning for environmental sustainability. https://doi.org/10.2766/02392

Jennifer MacDonald, U. O. (n.d.). *Getting There from Here: Mapping as a Process for Relationship Renewal*. Retrieved from file:///C:/Users/Rubina%20Ashmi%20Nabin/Downloads/CJEE+26+Macdonald.pdf

Kristen Hargis, M. M. (n.d.). Chapter 15. The State of Environmental and Sustainability Education in Canada—A review of past, current and future directions.

Lausselet, N. Z. (2022). *Bonding with the world: A pedagogical approach*. Retrieved from https://doi.org/10.1007/978-3-031-04108-2_15.

M, R. (2017). *Education for sustainable development goals: Learning objectives*. UNESCO.

McKenzie, M. A. (2023). Sustainability in the Canadian post-secondary education system. In *Sustainability and Education Policy Network, University of Saskatchewan, Saskatoon, Canada*.

Merriam, S. B., & Tisdell, E. J. (n.d.). *Qualitative research. A guide to design and implementation*. Jossey-Bass.

Metzger, S. P. (2023). *Barriers to learning for sustainability: a teacher reviews*. Retrieved from Sustainable Earth Reviews. https://doi.org/10.1186/s42055-022-00050-3

Miles, M. B. (1994). *Qualitative data analysis: An expanded sourcebook*. Sage.

Scott, W. (2003). *Sustainable Development and Learning: Framing the issues*. Taylor & Francis.

Siddhartha Bhattacharyya, N. K. (2021). *Cognitive data models for sustainable environment*. Academic Press.

The Association for the Advancement of Sustainability in Higher Education. (n.d.). Retrieved from https://www.aashe.org/

UNESCO. (2019). Framework for the implementation of education for sustainable development (ESD) beyond 2019. In *General Conference* (pp. 15–19). UNESCO.

UNESCO. (2020). *Education for sustainable development: A roadmap*. UNESCO.

UNESCO. (n.d.). Education for sustainable development. Retrieved from UNESCO: https://www.unesco.org/en/sustainable-development/education

Vinod Chandra, A. H. (n.d.). *Research methodology*. Pearson Education India.

Wheeldon, J. (2009). Framing experience: Concept maps, mind maps, and data collection in qualitative research. *International Journal of Qualitative Methods, 8*, 68–83.

Urbanization and Sustainable Cities

Lamyaa Moulad and Othmane Ait Lmoudden

1 Introduction

There has been a notable urban move in the global and cities are expanding at an unprecedented speed due to increasing urbanization, its level (denoted as URB), quantified as a percentage of the total population that lives in the urban areas. Over half of the population of the globe already resides in cities, and this proportion is expected to rise considerably over the next several decades. However, a number of problems follow this expansion, which have profound effects on urban resilience (Chen et al., 2024), ranging from social injustice and housing shortages to pollution and resource depletion. Despite the fact that cities have been built to meet human demands and necessities, it is becoming challenging to create sustainable communities due to planners' general interest in creating intelligent cities or giving the ecological aspect of sustainability top priority. The human perspective on urbanization is considered a future issue in achieving the Sustainable Development Goals (SDGs), this absolutely means that the key aspect of urban development is sustainable urbanization, hence the important significance of integrating sustainable practices into urban design. In this context, effective environmental sustainability initiatives propose how to solve environmental issues and raise the standard of the built environment as a whole, now the goal is try to create inclusive, secure, resilient, and sustainable communities (Newman, 2006; Chen et al., 2024) (Figs. 1 and 2).

Human well-being and the environment both benefit from sustainable cities and communities. The significance of creating inclusive, resilient, and

L. Moulad (✉)
Economics and Management Dépertement, Hassan II University, FSJES AC, Casablanca, Morocco

O. A. Lmoudden
Computer Science Départment, Hassan II University, FSBM, Casablanca, Morocco

A. Chakir et al. (eds.), *Green Society, Environmental Strategies and Sustainable Development*, https://doi.org/10.1007/978-3-031-94086-6_8

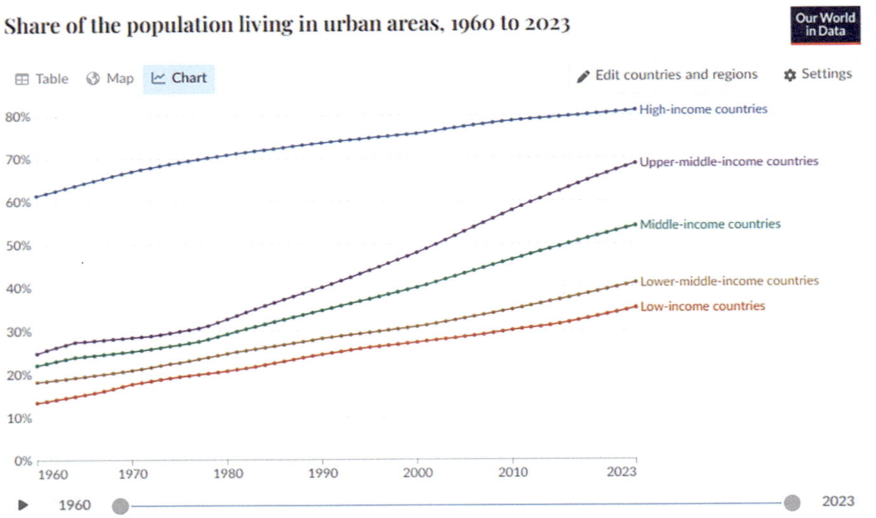

Fig. 1 Share of the population living in urban areas, 1960–2023. (Data Source: UN Population Division (via World Bank) (2025))

Fig. 2 Population of the world's largest cities, 1960–2023. (Data Source: United Nations, Department of Economic and Social Affairs, Population Division (2018))

sustainable cities is becoming more and more evident as the world's population continues to move into urban areas, so for future generations, we may build urban environments which help people and the environment by prioritizing sustainable energy, affordable housing, green areas, waste management, and community inclusivity.

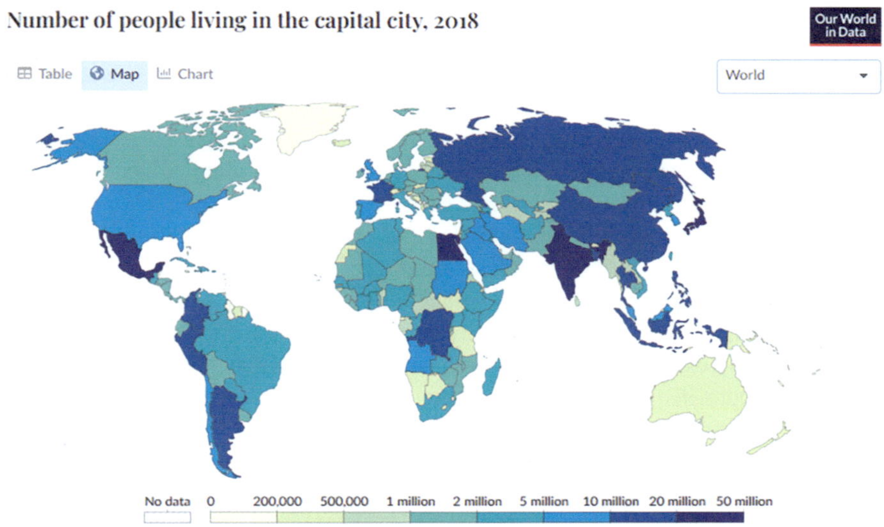

Fig. 3 Number of people in the capital city, 2018. (Data source: United Nations, Department of Economic and Social Affairs, Population Division (2018))

With over half of the world's population currently living in cities and that number predicted to rise to 70% by 2050, urbanization is accelerating globally in the twenty-first century. While there are social and economic benefits to this fast urbanization, there are also significant sustainability issues. Although they promote development, cities also use a lot of resources and degrade the environment. The world is facing a lot of stress due to the environmental imprint of cities, which includes trash creation and energy use. Rapid urbanization can also exacerbate inequality by causing social issues like overcrowding, subpar housing, and reduced access to basic amenities. Cities are particularly susceptible to the effects of climate change, which further complicates matters and maintains a sustainable future for cities and their residents by striking a balance between social justice, economic growth, and environmental preservation, in order to guarantee cities and their residents a bright future (Newman, 2006) (Fig. 3).

This chapter aims to examine how urban transformation affects sustainable development, with an emphasis on urban regeneration and sustainable urbanization and also main elements and projects involved in creating sustainable cities and communities will be discussed.

2 Concepts and Strategies to Build Sustainable Cities

A comprehensive strategy that establishes a balance between social justice, economic prosperity, and environmental responsibility is needed to build sustainable cities, by reducing its ecological impact and sustaining its natural resources, a sustainable city places a high priority on environmental integrity. Through a strong diversified economy that serves all citizens, it maintains economic vitality, hence social equality is crucial because it guarantees that everyone has equitable access to opportunities and resources. In order for the city to endure and bounce back from shocks like the effects of climate change, resilience (Lu and Teng 2023) is essential. Furthermore, livability emphasizes developing lively, appealing urban areas that provide a good standard of living (Surya et al., 2020) (Fig. 4).

A variety of related solutions are essential, like compact, multi-use developments, accessible neighborhoods that are produced by sustainable land (Wardekker, 2018) use and urban planning, which reduces expansion and reliance on transportation, also green roofs and urban trees are examples of green infrastructure that offers ecosystem services, reducing heat and enhancing air quality so public transportation and active modes like walking and bicycling are given priority in sustainable transportation systems; moreover, the concepts of the circular economy are implemented through resource efficiency and management of waste, which reduce waste and save resources, without forgetting green building techniques and investments in robust networks that are used in sustainable infrastructure and buildings (Nukala & Mutz, 2015).

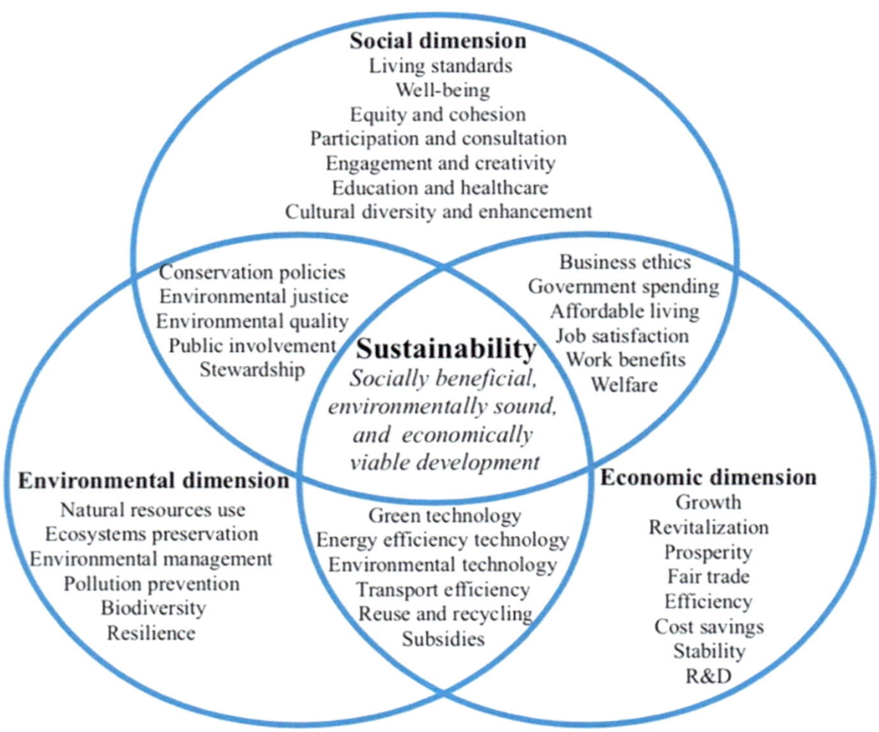

Fig. 4 A framework for balancing the three dimensions of sustainability

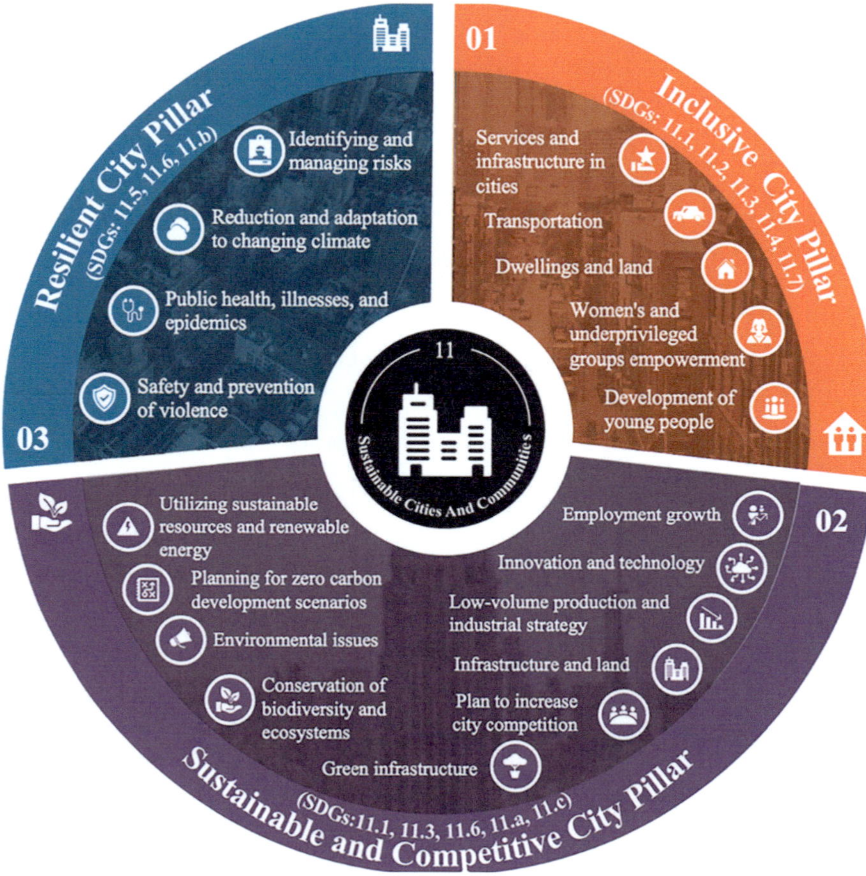

Fig. 5 Sustainable cities and communities pillars

For these strategies to have a synergistic effect, they must be combined, and for instance, green infrastructure improves livability, while compact design facilitates public transportation. Although financial limitations and community acceptance are two obstacles for adopting these techniques into practice, yet stakeholder participation, creative finance, and meticulous planning are crucial, with techniques suited to the specific needs of each community (Hameed, 2021) (Fig. 5).

3 Best Practices for Sustainable Urbanization

The development of sustainable urbanization depends on learning from successful models. Through practical examples and lessons learned, case studies provide insightful information about how cities have efficiently adopted sustainable

Fig. 6 Smart sustainable cities and sustainable urban development

projects. One of these illustrations highlights how flexible sustainable solutions can be by showcasing a range of strategies in different metropolitan settings. In order to achieve a synergistic effect, cases that succeed frequently showcase integrated techniques that combine many different approaches. For example, a city may integrate green infrastructure development, compact urban planning, and transport expenses (Surya et al., 2020; Bibri et al., 2020).

By illustrating how cities adapt solutions to meet their unique requirements and constraints, these examples also highlight the significance of local context, and by examining these situations, important success characteristics including good policy frameworks, community involvement, creative finance methods, and strong political leadership can be found, so examining the difficulties faced and the lessons discovered by less successful projects is just as crucial, though. This makes it possible to comprehend the intricacies of sustainable urban development in a more sophisticated manner. Then, researching best practices serves as a practical basis for building more resilient and sustainable cities by motivating and directing communities, urban planners, and legislators which accelerate the worldwide shift to a sustainable urban future by encouraging information exchange and cross-learning (Fig. 6).

4 Inclusive Urban Development: The Role of Community Engagement

Building fully sustainable cities requires inclusive urban development, and in order to guarantee that urban development meets the demands of all inhabitants, particularly marginalized and vulnerable groups, community engagement is not just a token gesture. Then by impacting planning procedures and insuring that their perspectives are heard, significant engagement gives communities the capacity to shape

Fig. 7 Adapted from Pretty (1995), Arnstein (1969), International Association for Public Participation (IAP2), and White (1996)

their areas, and also given that development choices take into account the many goals and perceptions of those who inhabit and surround the city, final results are more just and equal (Nukala & Mutz, 2015) (Fig. 7).

Also by encouraging a sense of collaborative responsibility and ownership, approaches that are inclusive improve social value and cohesiveness in the community because people are more inclined to encourage and make investments in the future prosperity of their urban environment when they actively participate in its design. Likewise community involvement can contribute insightful local knowledge to the planning process, resulting in better informed and contextually relevant solutions (Fawaz & Marques, 2025; OECD, 2025) (Figs. 8 and 9).

Hence in order to reduce risks and guarantee that the advantages of urbanization are distributed fairly, it can also assist in identifying and addressing possible adverse effects of growth on certain populations. Finally, inclusive urban development which is motivated by actual community involvement is not only morally required but also practically necessary to create resilient, sustainable, and prosperous communities for individuals (Fawaz & Marques, 2025).

5 Governance and Policy for Sustainable Urbanization

A key factor in promoting sustainable urbanization is governance and policy. A supportive environment that encourages revolutionary transformation requires strong governance mechanisms. Effective management, open decision-making techniques,

Fig. 8 Inclusive urban planning initiatives: engaging communities

Fig. 9 Role of community engagement in creating livable cities

and cooperative alliances between the public and private sectors are all necessary for this. In order to guarantee that sustainability issues are integrated into all departments and sectors, integrated planning is essential (Hameed, 2021). This involves avoiding segmented strategies and supporting a comprehensive vision that takes into account how urban systems are interlinked. Moreover, in order to stop unsustainable behaviors and encourage sustainable ones, new policy tools are required (Balk et al., 2022; Hoffmann, 2021) (Fig. 10).

This can include regulation, taxation, funding, and systems based on markets that encourage the use of renewable energy, sustainable transportation, and resource effectiveness. Robust monitoring and evaluation systems are also necessary for effective governance in order to assess developments, spot problems, and modify policies as necessary. In the race for sustainable urban development, this guarantees responsibility and permits ongoing progress. In the end, the basis for building cities that are socially just, economically successful, robust to future challenges, and

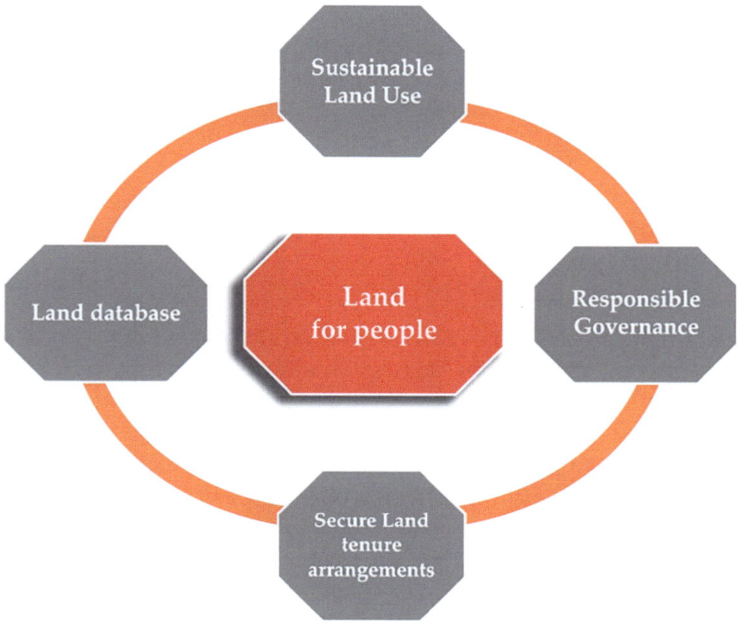

Fig. 10 Aspects of land use source and the position of governance

environmentally sustainable consists of solid governance and smart policies (Nukala & Mutz, 2015; Kalfas, 2023).

6 Reframing Urbanization: Cities as Dynamic Ecosystems

It is crucial to rethink urbanization by considering cities as dynamic and complex ecosystems rather than just as hubs of economic activity. This ecological perspective emphasizes how urban systems which include not only individuals but also infrastructure, the environment, and all living things are linked and connected. Comparable to natural ecosystems, cities display information, resource, and energy flows. In order to manage urban systems sustainably, it is essential to comprehend these flows. The significance of ecological concepts in urban planning and design is emphasized by the recognition of cities as ecosystems (Surya et al., 2020; Hameed, 2021).

This requires making green infrastructure a priority, preserving biodiversity, reducing resource usage and generate waste, and pushing the ideas of the circular economy. The general well-being of urban inhabitants is improved, resilience (Zhou & Qi, 2023), is promoted, and ecosystem services are improved when an ecosystem-based approach to urban development is used. It calls for a comprehensive and integrated strategy that takes into account how social, economic, and environmental elements interact. In the end, redefining urbanization as an ecological undertaking enables us to build cities that are socially just, environmentally sustainable, and economically thriving, guaranteeing a peaceful coexistence of urban growth and the natural environment (Lu and Teng 2023).

7 The Future of Sustainable Urbanization: Challenges and Opportunities

There are exciting prospects as well as significant challenges in the way of sustainable urbanization. Urban systems face significant pressure from a growing population, the impacts of climate change, limited resources, and enduring inequality. Cities will have to deal with the economic and social gaps that urbanization may aggravate, manage the growing demands for resources like electricity and water, and face increasingly frequent catastrophic weather events. However, there are a lot of possibilities for enhancement with technological advancements. Cities may become more effective, resilient, and sustainable with the use of technologies for smart cities, data analytics, and creative solutions in fields like waste management, transportation, and renewable energy (Hameed, 2021) (Fig. 11).

Reducing waste and reusing resources will be key components of the switch to a circular economy. Moreover, building fair and equitable cities will need supporting inclusive approaches to governance as well as offering people the resources they need to take part in urban planning and decision-making. The global shift to sustainable urban development will be accelerated in large part by international cooperation and knowledge exchange. In the end, our cities' future relies on our capacity to welcome inventiveness, give sustainability first priority, and create inclusive, resilient urban settings that can satisfy the demands of both the current and coming generations.

7.1 Examples of Technologies for Urban Sustainability

Numerous inspiring examples of the use of technologies to improve quality of life and urban sustainability are actually cited in the literature and have proven all that has been mentioned above about this concept, above is a sharp view in this sense after a literature search (PBL Netherlands Environmental Assessment Agency, 2019; Lee et al., 2021).

Copenhagen (Andersen et al., 2019) is often cited as the typical example of a smart city focused on sustainable mobility. The city has set up a dense network of cycle paths, bike sharing stations, and infrastructure encouraging the use of cycling in daily life. This approach has significantly reduced traffic congestion, improved air quality, and promoted a healthier lifestyle.

Fig. 11 Sustainable urban development

Singapore (Global Green Growth Institute, 2018) is a pioneer in smart cities. The city-state has deployed a multitude of technologies to optimize the management of its infrastructure, improve public services, and foster innovation. Some of the most remarkable initiatives include installing devices throughout the city to monitor air quality, noise levels, and other environmental parameters. Singapore has also implemented a highly efficient public transport system, a network of smart taxis, and smart waste management using connected bins to optimize collections and reduce costs.

Barcelona (Martinez-Fernandez et al., 2021) has adopted a participatory approach to the smart city, involving citizens in the co-creation of solutions. The city has developed a digital platform that allows residents to report problems, propose ideas, and monitor the progress of projects. Some of Barcelona's flagship initiatives include the CityOS digital platform that connects the city's different systems (transport, energy, waste, etc.) and allow for the collection and analysis of large amounts of data; Barcelona has implemented superblocks, neighborhoods where car traffic is limited, and public spaces are prioritized.

Also other notable examples are included in this context such as:

- Amsterdam: The Dutch city is known for its ambitious sustainability policies, particularly in water and energy management (City of Amsterdam, 2020).
- Seoul: The South Korean capital is a highly connected city, using technology to improve safety, transport management, and public services (Joo & Kim, 2016).
- San Francisco: This American city is a pioneer in the field of clean technologies and the circular economy.
- China by deploying ICT, technological diffusion and economic growth in Chinese cities (Chen et al., 2024; Li & Wu, 2023), in general, ICT allows the generation and distribution of decentralized information and ideas in production processes that are increasingly rely on information as an input.

Despite their many opportunities, smart cities pose many challenges, and data governance in smart cities is a crucial issue that raises many questions. Indeed, the massive collection of data, essential to the functioning of these cities, requires rigorous and ethical management. Among these issues we find (City of Portland, 2021) Protection of personal data because the collection of sensitive data (location, consumption habits, etc.) raises ethical and legal questions. It is essential to put in place rigorous protection mechanisms to guarantee data confidentiality, as well as transparency, since citizens must be informed of how their data is collected, used, and stored, and of course the consent of citizens must be obtained clearly and explicitly before any use of their data, then data security: so, smart cities are prime targets for cybercriminals. It is therefore necessary to put in place robust security systems to protect data against attacks. Also, the loss or corruption of data can have significant consequences on the functioning of the city. It is essential to put in place effective backup and restoration systems and also data quality, so data collected must be reliable and accurate to enable informed decision-making; they must be protected against any alteration or falsification and accessible to the various stakeholders of the city (administrations, businesses, and citizens) in compliance with security rules. It is necessary to clearly define the ownership of the data collected, particularly with regard to public data and private data, even more so the valorization of data can generate revenue for communities. It is important to find a balance between the general interest and private interests, Furthermore, data governance in smart cities requires close collaboration between the various stakeholders (administrations, businesses, citizens, and researchers), then it is essential to define common rules to ensure fair and transparent use of data.

In order to meet these challenges, it is necessary to establish a solid data governance, based on transparency with citizens who must be informed of how their data is used and each actor involved in data management must be responsible for their actions. Also close collaboration between the different actors is necessary to define common rules and implement effective solutions, then protection of personal data and respect for privacy must be at the heart of concerns, and finally knowing that technologies are evolving rapidly. It is important to stay at the forefront to develop new data governance solutions.

This implies that data governance in smart cities is a complex issue that requires in-depth reflection and collaboration between all the actors involved. By putting in place rigorous protection mechanisms and promoting transparency, it is possible to build smart cities that respect the rights and freedoms of their citizens.

7.2 *Smart Cities and New Technologies for Sustainability*

Smart cities represent a new approach to urban planning, where information and communication technologies (ICT) are used to improve the quality of life of citizens and preserve the environment. By integrating intelligent systems in various areas such as transport, energy, waste management, or security, these cities aim to optimize resources, reduce the ecological footprint, and promote sustainable development. This is generally based mainly on the integration of several fundamental aspects, such as data collection and analysis, the Internet of Things, artificial intelligence, and citizen participation.

For data collection and analysis, sensors and monitoring systems can collect huge amounts of data of the city (traffic, energy consumption, air quality, etc.). This data is then analyzed to identify trends, problems, and optimization opportunities.

Moreover by using the IoT, which connects a large number of physical objects to the Internet, allowing them to be controlled remotely and data to be collected in real time. In smart cities, the IoT is used to optimize the management of infrastructure, public services, and resources.

Also, the integration of artificial intelligence (AI) makes it possible to analyze the data collected in more depth and to automate certain tasks. It is used to optimize transport, predict energy demand, improve security, or even personalize services to citizens.

And this without forgetting the factor of citizen participation, currently citizens are increasingly involved in the management of their city, thanks to online platforms and mobile applications. They can report problems, propose solutions, and follow the progress of projects (Fig. 12).

These smart cities present many opportunities to promote sustainability by reducing energy consumption, because thanks to smart energy management systems, cities can optimize their consumption and reduce their greenhouse gas emissions, it allows the improvement of air quality: Sensors allow to monitor air quality in real time, and the data collected can be used to implement corrective measures, without forgetting the reduction of waste: Smart waste management allows to optimize collections and promote recycling and especially the aspect of improving mobility: Intelligent transport systems, such as connected public transport and self-service bicycles, encourage the use of sustainable modes of transport and reduce congestion. These smart cities can also better prepare for the impacts of climate change, thanks to monitoring and early warning systems.

Fig. 12 Smart sustainable city scheme

This shows that smart cities represent a unique opportunity to build more sustainable, resilient, and livable cities; now by combining the latest technologies with a human-centered approach, it is possible to create urban environments where quality of life and environmental preservation go hand in hand.

As every new vision, smart cities also pose challenges such as the protection of personal data, knowing that the collection and analysis of large amounts of data raise important questions regarding the protection of privacy, also the cost of investments since the implementation of smart systems requires significant investments, which can be an obstacle for some cities, even more it is essential to ensure that all citizens can benefit from the advantages of smart cities and to reduce the digital divide, without forgetting the dependence on technologies because just a system failure or a cyberattack could have significant consequences on the functioning of the city.

By effectively integrating information and communication technologies into their management, smart cities offer considerable potential in terms of social and economic impacts.

Positive social impacts: Smart cities can offer other efficient and personalized public services, such as waste management, smart public lighting, or public transport management, also video surveillance, alert systems and smart patrols can help reduce crime and strengthen citizen safety, with the deployment of digital platforms this will allow citizens to actively participate in the life of their city, by proposing ideas, reporting problems or monitoring the progress of projects, they can help reduce inequalities by providing equitable access to services and infrastructure, also assisted technologies can make life easier for people with reduced mobility or the elderly, also help to reduce their ecological footprint by optimizing resource management and promoting sustainable modes of transport, having better air quality, efficient waste management and promoting physical activities contribute to improving the health of citizens.

Positive economic impacts: These smart cities are more attractive to businesses, which can stimulate job creation and promote innovation, then the technologies used in smart cities can improve the productivity of businesses and administrations. Speaking of tourism, smart cities can become more attractive tourist destinations, thanks to their modern infrastructure and innovative services, without forgetting innovation side and that smart cities promote the emergence of new businesses and new economic models within an innovative ecosystem (Fig. 13).

These smart cities offer considerable potential to improve citizens' quality of life and boost economic growth. However, their development must be accompanied by

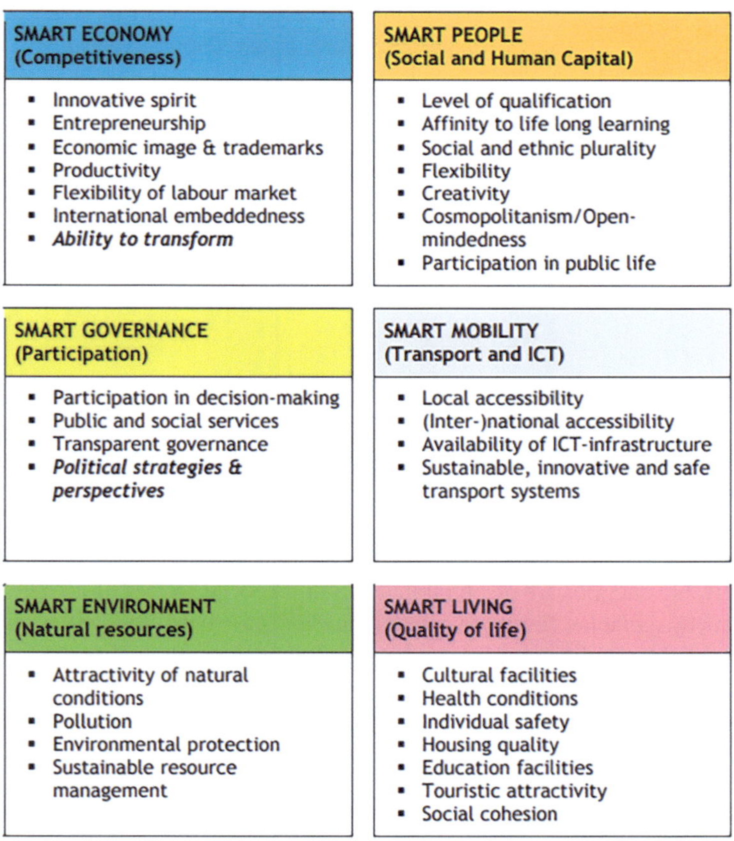

Fig. 13 Characteristics and factors of smart city

a thorough reflection on social, economic, and ethical issues. It is essential to put in place appropriate public policies to ensure a fair and equitable transition to smart cities.

Smart cities promise many benefits in terms of sustainable development. However, their environmental impacts are complex and deserve a nuanced analysis.

Among these environmental benefits, we can first mention the reduction of energy consumption, so thanks to optimized energy management systems, buildings can significantly reduce their energy consumption, the use of sensors, and automation systems allows lighting to be adapted to real needs, thus reducing energy consumption. Also, the improvement of air quality, so sensors allow continuous monitoring of air quality and identification of sources of pollution and even more the optimization of transport by promoting public transport, cycling and walking, smart cities contribute to reducing greenhouse gas emissions.

Moreover, and regarding waste, smart collection systems allow to optimize routes and reduce emissions related to waste transport, then mobile applications can encourage selective sorting and facilitate access to collection points.

Lastly, and for rational use of resources, smart management systems allow to detect leaks and optimize water distribution, without forgetting that smart cities can promote the development of urban agriculture, thus reducing emissions related to food transport.

What can currently be said is that smart cities offer considerable potential to improve the environmental performance of cities. However, their development must be accompanied by in-depth consideration of the environmental impacts of these technologies. It is essential to prioritize sustainable solutions, optimize energy consumption, and minimize waste production. In addition, a transition to renewable energy sources is essential to ensure a sustainable future for smart cities.

7.3 *Renewable Energy and Smart Cities*

Smart cities, a true field of innovation, are increasingly integrating renewable energy into their energy model. This transition is essential to reduce the carbon footprint of cities (Liu et al., 2022), improve air quality, and ensure a more sustainable energy supply.

Renewable energies are essential in smart cities for reduction of greenhouse gas emissions by replacing fossil fuels with renewable energy sources, cities contribute to the fight against climate change, then improvement of air quality, production of renewable energy does not generate polluting emissions, which significantly improves the air quality in cities, energy sources, cities reduce their dependence on fossil fuels and improve their resilience (Zhou & Qi, 2023; Tang et al., 2023) to fluctuations in energy prices, and development of renewable energies stimulates job creation in the installation, maintenance and research sectors. Hence the main sources of renewable energy used in smart cities are solar energy using solar panels, wind energy, geothermal energy, and biomass.

Smart cities integrate renewable energy by various techniques such as smart buildings are equipped with energy management systems that optimize energy consumption and integrate renewable energy such as solar photovoltaic, then electricity grids are modernized to integrate renewable energy in a flexible and efficient way, also cities develop charging infrastructure for electric vehicles and encourage the use of soft modes of transport, furthermore citizens are encouraged to produce their own energy and share it with the grid.

Numerous challenges can be met such as the cost of investments because the transition to renewable energy requires significant investments, then energy storage is necessary to develop efficient storage solutions to manage fluctuations in renewable energy production also integrating renewable energy into existing electricity grids can pose technical challenges.

Smart cities are constantly evolving, and new trends are emerging regularly. Here we can mention some of the most promising perspectives for the future such as hyper-personalization of services, accelerated sustainable development, ubiquitous artificial intelligence, chatbots, and virtual assistants will be increasingly present to answer citizens' questions and guide them, digital twins of cities, virtual 3D models of cities will simulate different scenarios and optimize decisions, Increased collaboration between stakeholders through partnerships between public authorities, businesses and citizens that will be strengthened to co-build the cities of tomorrow, and the opening of data will promote innovation and the development of new services.

The future of smart cities looks promising, with prospects for sustainable development, improved quality of life, and increased citizen participation. However, it is essential to address the challenges of security, ethics, and social inclusion if these cities are to serve the common good.

8 Conclusion

The phenomena of rapid urbanization is intricate and multidimensional. Although there are many obstacles to overcome, there are also chances to create cities that are more resilient, equitable, and sustainable. All societal parties, including enterprises, people, and governmental authorities, must be involved in the implementation of long-term initiatives in order to overcome these issues.

This chapter explored the intricacies of rapid urbanization and how it affects sustainability. Attention was drawn to the unprecedented rate of urban growth in the twenty-first century, which has been fueled by a number of causes. In order to set the stage for sustainable development, this chapter presented urbanization as both a challenge and an opportunity, emphasizing the need for integrated planning that balances ecological integrity and urban resilience. It also presented creative ideas for converting urban environments into sustainable cities, emphasizing governance structures and overall design that encourage environmental stewardship. They explore how to maximize the reduction of the ecological footprint while maximizing the quality of urban life through the use of green infrastructure, sustainable mobility, and smart urban planning. Case studies of cities that have effectively integrated sustainable practices are examined. These case studies illustrate a variety of strategies, from the inclusion of biodiversity-enhancing green spaces to decentralized waste management systems. Policymakers and urban planners can draw inspiration from these examples and move toward a more sustainable urban future. This chapter also highlights the importance of community involvement in the transition to sustainable urbanization. It argues that inclusion, or ensuring that disadvantaged communities are included in the planning process, is essential for cities to thrive in a sustainable manner. Cities can act as accelerators of equitable growth and environmental resilience by supporting participatory governance models in the fabric of urban development. Ultimately, this chapter urges parties to reconsider their conception of urbanization and to view cities as dynamic ecosystems that must coexist peacefully with the environment rather than simply as hubs of economic activity.

References

Andersen, T. M., et al. (2019). Green infrastructure in Copenhagen. *Environmental Research Communications, 1*(11), 111001.

Balk, D., Tagtachian, D., Jiang, L., Marcotullio, P., Cook, E. M., Jones, B., Mustafa, A., & McPhearson, T. (2022). Frameworks to envision equitable urban futures in a changing climate: A multi-level, multidisciplinary case study of New York City. *Frontiers in Built Environment, 8*, 949433.

Bibri, S. E., Krogstie, J., & Kärrholm, M. (2020). Compact city planning and development: Emerging practices and strategies for achieving the goals of sustainability. *Developments in the Built Environment, 4*, 100021.

Chen, M., et al. (2024). The impact of land marketization on urban resilience: Empirical evidence from Chinese cities. *Land, 13*(9), 1385. https://doi.org/10.3390/land13091385

City of Amsterdam. (2020). Amsterdam: Sustainable.

City of Portland. (2021). *Portland's Climate Action Plan. Masdar. About Masdar City*.

Fawaz, M., & Marques, E. (2025). Urban research publications limited, inclusive urban development: The role of community engagement. *IJURR, 49*(1).

Global Green Growth Institute. (2018). *Songdo International Business District: Case study*. Global Green Growth Institute.

Hameed, A. A. S. (2021). Urban and regional planning strategies to achieve sustainable urban development: (subject review). *International Journal of Advances in Scientific Research and Engineering, 7*, 22–27.

Hoffmann, J. (2021). Demographic change and land use. In T. Weith, T. Barkmann, N. Gaasch, S. Rogga, C. Strauß, & J. Zscheischler (Eds.), *Sustainable land management in a European context: A co-design approach* (pp. 63–74). Springer International.

Joo, J., & Kim, J. (2016). Sustainable urban development and green building: A case study of Songdo IBD, South Korea. *Sustainability, 8*(11), 1128.

Kalfas, D. (2023). Urbanization and land use planning for achieving the sustainable development goals (SDGs): A case study of Greece. *Urban Science, 7*(2), 43. https://doi.org/10.3390/urbansci.7020043

Lee, K., Heng, C. K., & Singapore Ministry of National Development. (2021). Sustainable development in singapore. sustainable urban development: The Case of Singapore. *Cities, 95*, 102382.

Li, Q., & Wu, Y. (2023). ICT, technological diffusion and economic growth in Chinese cities. *Empirical Economics, 64*, 1737–1768.
Liu, X., Xu, H., & Zhang, M. (2022). Impact and transmission mechanism of land leasing marketization on carbon emissions: Based on the mediating effect of industrial structure. *China Population, Resources and Environment, 32*, 12–21.
Lu, X., & Teng, Y. (2023). How innovation-driven policies enhance urban economic resilience: Analysis of the mechanism based on efficient market and effective government. *China Soft Science, 7*, 102–113.
Martinez-Fernandez, C., et al. (2021). City of Barcelona. Sustainable urban development strategy. In *Urban planning for city leaders: Barcelona's strategy for ecological urbanism*. Springer.
Naturvårdsverket. (2019). *Environmental objectives and strategies*. Swedish Environmental Protection Agency.
Newman, P. (2006). The environmental impact of cities. *Environment and Urbanization, 18*(2), 275–295. https://doi.org/10.1177/0956247806069599
Nukala, R. B., & Mutz, D. (2015). Strategic approach for sustainable land use in an emerging country: Case of India. In *Proceedings of the Annual World Bank Conference on Land and Poverty 2015, Washington, DC, USA, 23–27 March 2015*. Deutsche Gesellschaft für Internationale Zusammenarbeit (GIZ) GmbH.
OECD. (2025). *Zero-carbon buildings in cities, a whole life-cycle approach OECD urban studies*. OECD.
PBL Netherlands Environmental Assessment Agency. (2019). Amsterdam urban development and environmental quality.
Surya, B., Ahmad, D. N., Sakti, H. H., & Sahban, H. (2020). Land use change, spatial interaction, and sustainable development in the metropolitan urban areas, South Sulawesi Province, Indonesia. *Land, 9*, 95.
Tang, D., Li, J., Zhao, Z., Boamah, V., & Lansana, D. D. (2023). The influence of industrial structure transformation on urban resilience based on 110 prefecture-level cities in the Yangtze River. *Sustainable Cities and Society, 96*, 104621.
Wardekker, J. (2018). Resilience principles as a tool for exploring options for urban resilience. *Solutions, 9*(1). Retrieved August 27, 2024.
Zhou, Q., & Qi, Z. (2023). Urban economic resilience and human capital: An exploration of heterogeneity and mechanism in the context of spatial population mobility. *Sustainable Cities and Society, 99*, 104983.

Biodiversity and Ecosystem Conservation

Anitha Subbarayan

1 Introduction to Biodiversity and Its Dimensions

Biodiversity is fundamental to maintaining ecological balance and supporting essential facilities (World Wildlife Fund, 2024). Despite its significance, biodiversity faces severe threats due to human activities, necessitating urgent conservation measures (Mongabay, 2024). Biodiversity supports ecosystem services, underpins human well-being, and plays a vital responsibility in nourishing existence on globe (Wu et al., 2022). However, it is multifaceted, requiring systematic study under three primary dimensions as shown in Fig. 1.

1.1 Genetic Diversity

Genetic diversity states to the difference in inherent measureable inside populations and resist diseases (University of Oxford, 2024). For example, diverse wheat varieties exhibit increased resistance to stem rust (*Puccinia graminis*), ensuring stable crop yields (Beillouin et al., 2021). Maintaining genetic diversity is crucial for endangered species such as the Asiatic lion (*Panthera leo persica*), where a limited gene pool poses significant risks. Conservation strategies such as ex situ (e.g., gene banks and captive breeding) and in situ efforts (e.g., protected areas and wildlife corridors) play a pivotal role in preserving genetic resources (González-García et al., 2022). Genetic diversity encompasses the heritable genetic information stored in DNA. Genetic variation ensures species' adaptability and resilience to environmental changes, diseases, and other pressures. Genetic variation allows populations

A. Subbarayan (✉)
Computer Science and Engineering (Cyber Security), Sri Shakthi Institute of Engineering and Technology, Coimbatore, Tamil Nadu, India

A. Chakir et al. (eds.), *Green Society, Environmental Strategies and Sustainable Development*, https://doi.org/10.1007/978-3-031-94086-6_9

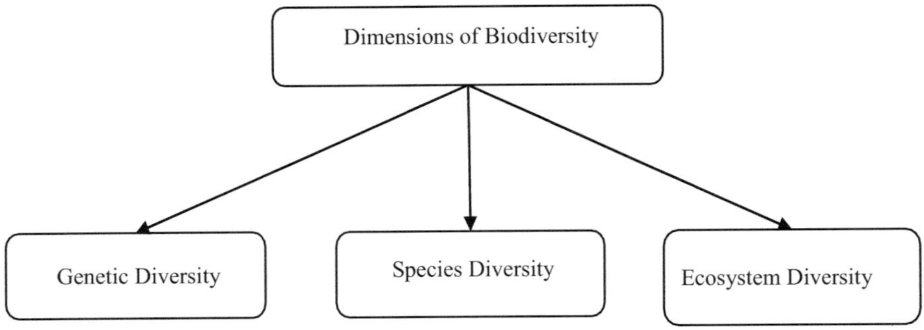

Fig. 1 Dimensions of biodiversity

to develop in reaction to environmental stressors like habitat modification. Tools like microsatellites and single nucleotide polymorphisms (SNPs) assess genetic variation. Whole-genome sequencing enables the study of genetic diversity at the molecular level, aiding in species conservation.

1.2 Species Diversity

A species diversity enhances ecosystem resilience, enabling recovery from habitat destruction and climate change (University of Oxford, 2024). Keystone species, such as sea otters (*Enhydra lutris*), regulate prey populations and maintain ecological stability (World Wildlife Fund, 2024). However, biodiversity hotspots, such as the Amazon Rainforest, face severe threats due to deforestation and illegal wildlife trade (Wired, 2025).

Species diversity comprises two main components: (1) Species richness, which is the total quantity of variety within a habitat. (2) Species evenness, which is the relative profusion of every variety within the community. The high level of species diversity increases ecosystem stability and its capacity to recover from disturbances. Key concepts in species diversity are: (1) Keystone Species: Species like sea otters (*Enhydra lutris*) play a disproportionately large role in ecosystem functionality by regulating prey populations. (2) Indicator Species: Certain species, such as amphibians, signal changes in environmental health due to their sensitivity to pollutants or habitat alteration. (3) Endemic Species: Geographically restricted species, like *Madagascar lemurs*, require specialized conservation strategies.

1.3 Ecosystem Diversity

Ecosystem diversity encompasses both biotic (living organisms) and abiotic (physical and chemical) components and their interactions, influencing ecological processes and ecosystem services. Terrestrial ecosystems, such as tropical rainforests with high biodiversity due to stable temperatures and abundant rainfall, and Arctic tundras with unique adaptations like antifreeze proteins in fish, demonstrate diverse ecological functions (Li et al., 2022a). Aquatic ecosystems, including oceans, rivers, lakes, and wetlands, also play crucial roles; coral reefs, known as the "rainforests of the sea," support vast marine life (Gotama et al., 2024), while mangroves serve as nurseries for fish and protect coastlines from storm surges. Ecosystem diversity drives essential processes like energy flow through food webs, nutrient cycling (e.g., nitrogen fixation by Rhizobium bacteria), and ecosystem.

The productivity is measured through net primary productivity (NPP), determining biomass availability for food webs, while greater biodiversity enhances ecosystem resilience and resistance to disturbances like deforestation and invasive species. Terrestrial ecosystems, particularly tropical rainforests, host nearly 50% of the world's species, while aquatic ecosystems face growing threats from climate change and pollution, underscoring the urgency of conservation efforts.

1.4 Interdependence Between Biodiversity Dimensions

The interdependence between biodiversity dimensions is essential for maintaining ecological stability, as genetic diversity drives species survival and evolution, species diversity enhances ecosystem resilience, and ecosystem diversity provides the structural foundation for sustaining biodiversity at all levels (Mongabay, 2024). Genetic diversity ensures adaptability to environmental changes, while species diversity supports ecosystem function and stability (World Wildlife Fund, 2024). In turn, ecosystem diversity creates the necessary habitats for various species, reinforcing the interconnectedness of all biodiversity components. Conservation efforts must integrate these dimensions to address the ongoing biodiversity crisis effectively (University of Oxford, 2024).

The loss of biodiversity in any of these dimensions can have cascading effects on ecosystems, ultimately threatening global ecological balance. Similarly, a decrease in species diversity can disrupt food webs and ecological interactions, leading to instability in ecosystems (World Wildlife Fund, 2024). Addressing major environmental threats such as habitat destruction, climate change, and pollution requires comprehensive and holistic strategies (Mongabay, 2024). Sustainable land-use practices, habitat restoration, and global cooperation are crucial for preserving biodiversity across genetic, species, and ecosystem levels (University of Oxford, 2024).

Preserving biodiversity requires coordinated efforts across scientific, policy, and community levels. Genetic diversity conservation can be promoted through seed banks, gene editing, and breeding programs for endangered species (World Wildlife Fund, 2024). Ecosystem conservation involves sustainable land and resource management, afforestation, and mitigating climate change impacts through carbon sequestration and rewilding initiatives (Zhang et al., 2021). Additionally, education and awareness campaigns can foster public engagement, inspiring local and global action to protect biodiversity. Governments, conservation organizations, and individuals must work together to implement sustainable solutions that promote ecological balance and safeguard biodiversity for future generations (University of Oxford, 2024).

Recent studies underscore the alarming trajectory of biodiversity attrition. The World Wildlife Fund (2024) reports a catastrophic 94% reduction, while global freshwater species have suffered an 83% decline. However, targeted conservation interventions in Europe have facilitated population resurgence among apex predators, such as brown bears (*Ursus arctos*) and gray wolves (*Canis lupus*), demonstrating the efficacy of well-structured biodiversity management frameworks.

1.5 Case Studies

Case Study: Amazon Rainforest
The Amazon, a critical global biodiversity hotspot, exhibits high levels of genetic, species, and ecosystem diversity, which contribute to its ecological robustness:

- **Genetic Diversity**: Encompasses a vast reservoir of genetic material among endemic flora and fauna, fostering adaptive potential.
- **Species Diversity**: Supports approximately 390 billion individual trees spanning 16,000 species, sustaining intricate trophic interactions.
- **Ecosystem Diversity**: Comprises heterogeneous ecological niches, including *várzea* (floodplains), terra firme rainforests, and montane cloud forests, each supporting distinct biotic assemblages.

Case Study: Community-Led Conservation in Kenya's Maasai Mara
Community-driven conservation initiatives in Kenya's Maasai Mara illustrate a paradigm shift toward participatory biodiversity governance. Over 80% of the region's wildlife inhabits conservancies operated by indigenous Maasai communities rather than government-managed national parks (Wired, 2025). These conservancies function as ecological corridors, enhancing species dispersal and genetic flow while simultaneously generating economic incentives through sustainable ecotourism. The Maasai people play an instrumental role in habitat rehabilitation and keystone species conservation, reinforcing the synergetic relationship between cultural heritage and ecological stewardship (Mongabay, 2024).

The erosion of any single biodiversity dimension—genetic, species, or ecosystem—instigates cascading failures, precipitating species extinctions and ecosystem destabilization. Biodiversity, in its entirety, underpins planetary homeostasis and human viability, influencing ecological resilience, adaptive evolution, and biogeochemical cycles. Addressing biodiversity attrition necessitates an integrative conservation strategy that holistically considers the interdependence of genetic variability, species richness, and ecosystem heterogeneity. Cutting-edge advancements in genomic conservation, real-time ecological surveillance, and computational ecosystem modeling present novel methodologies for biodiversity preservation. The systemic integration of these biodiversity dimensions is imperative for maintaining ecological equilibrium (Oguh et al., 2021). While anthropogenic pressures accelerate biodiversity loss, empirical evidence from conservation success stories, such as those in Europe and Kenya, reaffirms that strategic interventions can mitigate and even reverse biodiversity decline. To safeguard global biodiversity, conservation policies must be reinforced, land-use paradigms restructured, and transnational collaborations strengthened (University of Oxford, 2024).

2 Ecosystem Services: A Quantitative Framework

Recent advancements in computations have enabled precise quantification and mapping of these services, enhancing decision-making processes in environmental management. Figure 2 explores the various types of ecosystem services.

2.1 Provisioning Services

Provisioning services encompass the material outputs derived from ecosystems, which are essential for sustaining human life and economic development. These include food, freshwater, raw materials, and medicinal resources. Ecosystems provide a diverse array of food sources, including crops, livestock, fisheries, and forest products. Fisheries contribute 20% of animal protein to the diets of over 3 billion people globally. However, over 33% of global fish stocks are overexploited (Hua et al., 2022). Traditional agricultural systems leverage crop diversity to enhance resilience against climate fluctuations. Practices such as intercropping and agroforestry increase yield stability, with studies indicating a 20–40% increase in productivity. Simulation tools like DSSAT (decision support system for agrotechnology transfer) predict productivity across different ecosystems under varying climatic scenarios. Platforms like *Sea Around Us* analyze species-specific catch data to estimate fisheries' contribution to global food security.

Aspirin, derived from the willow tree (*Salix* spp.), revolutionized pain relief and anti-inflammatory treatment. Taxol, extracted from the Pacific yew tree (*Taxus*

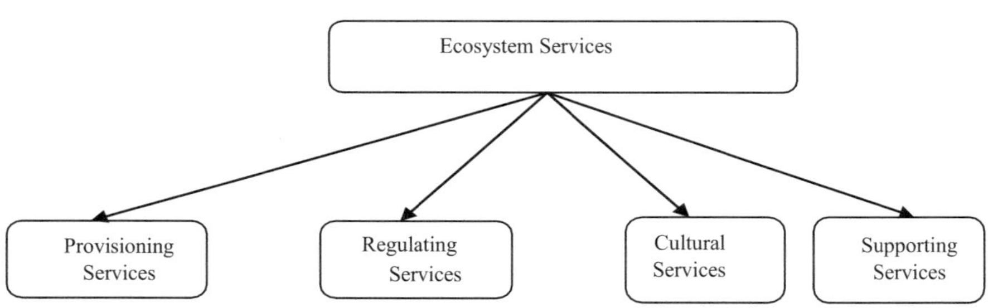

Fig. 2 Ecosystem services

brevifolia), is a critical chemotherapy agent for treating cancer, with an annual market value exceeding $1 billion. Statistical models assess the likelihood of discovering commercially viable medicinal compounds in different ecosystems. Economic valuation models estimate the pharmaceutical industry's reliance on biodiversity, informing conservation policies. Water yield model tools like *InVEST* (integrated valuation of ecosystem services and tradeoffs) assess freshwater availability by integrating precipitation, land cover, and evapotranspiration data. Systems such as SWAT (soil and water assessment tool) predict water flow and quality across different ecosystems.

2.2 Regulating Services

Regulating services help maintain environmental balance and mitigate the impact of both natural disasters and anthropogenic activities. These include climate regulation, pollination, and disaster risk reduction (Pereira et al., 2024). Forests and oceans sequester approximately 60% of anthropogenic CO_2 emissions annually. Amazonian Rainforests alone sequester an estimated 1.5 billion metric tons of CO_2 per year. Carbon sequestration models use tools like the *Carbon Benefits Project* estimate carbon stocks and fluxes in forests, soils, and other ecosystems. Remote Sensing: NASA's OCO-2 (Orbiting Carbon Observatory) tracks global CO_2 levels to assess ecosystem contributions to carbon sequestration. CO_2/year for various ecosystem type is provided in Table 1.

Pollinators are crucial for agricultural productivity, with nearly 75% of global food crops depending on pollination. Pollination services contribute an estimated $235–$577 billion annually to global crop output. Pollinator habitat quantification tool (PHQT) predicts pollinator abundance and their impact on crop yield. Ecological niche models assess habitat suitability for pollinators under different environmental conditions.

2.3 Cultural Services

Cultural services enhance human well-being through recreational, aesthetic, spiritual, and educational values. Natural landscapes attract millions of tourists, generating significant economic revenue. US national parks contribute over $40 billion

Table 1 Ecosystem type

Ecosystem type	Carbon sequestration potential (Mt CO_2/year)
Amazon rainforest	1.5 billion
Boreal forests	0.8 billion
Seagrass meadows	83 million

annually through tourism and recreation. Visitor Use Models: The recreation opportunity spectrum (ROS) quantifies the recreational value of ecosystems. Satellite imagery and GIS tools analyze land features that enhance aesthetic appeal. Many cultures derive spiritual significance from ecosystems, such as sacred groves in India. Quantification approaches such as surveys and contingent valuation methods estimate willingness-to-pay (WTP) for preserving culturally significant sites.

2.4 Supporting Services

Supporting services form the foundation for all other ecosystem services by maintaining fundamental ecological processes. Healthy soils support plant growth, agriculture, and forestry. Mycorrhizal fungi enhance phosphorus uptake, boosting crop productivity. The contribution to an ecosystem for the various systems is depicted in Table 2.

The integration of advanced computational models, GIS, and AI has significantly enhanced our ability to quantify and manage ecosystem services. This quantitative framework not only deepens our understanding of the ecological, economic, and societal roles of these services but also informs more effective conservation and policy strategies (Schmid & Schöb, 2022). As we face escalating environmental challenges, it is imperative to refine our valuation methodologies, improve large-scale monitoring, and develop adaptive management strategies. Aligning policy-driven conservation efforts with technological advancements will maximize the benefits of ecosystem services for future generations.

3 Anthropogenic Pressures on Biodiversity

Biodiversity, the cornerstone of ecological stability, is increasingly under threat due to anthropogenic activities. These human-induced pressures have accelerated the extinction rates of species and disrupted ecosystem functions, resulting in ecological imbalances. This section examines the various anthropogenic pressures on biodiversity, their mechanisms, and their far-reaching impacts as depicted in Fig. 3.

Table 2 Contribution to ecosystem

Process	Contribution to ecosystem
Nitrogen fixation	Supports agricultural productivity
Soil formation	Enhances plant growth
Decomposition	Recycles nutrients

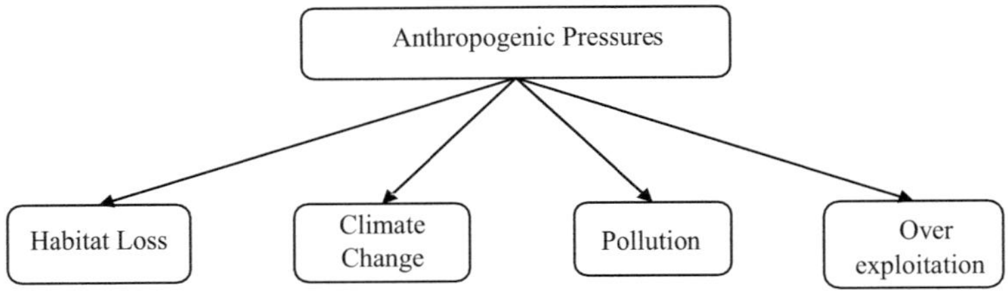

Fig. 3 Anthropogenic pressures

3.1 Habitat Loss and Fragmentation

Habitat loss occurs when natural habitats are converted or divided due to actions of human beings (Belaire et al., 2022). This phenomenon significantly threatens biodiversity by reducing the availability of essential resources and disrupting ecological balance. Mechanisms of habitat loss include deforestation: Large-scale logging and land clearing for agriculture and infrastructure development reduce forest cover, which is critical for species, such as orangutans, jaguars, and tigers. The global deforestation rate is approximately ten million hectares annually. Urban sprawl is the expansion of urban areas that encroaches on natural ecosystems, replacing them with impermeable surfaces like concrete and asphalt. The increasing demand for food leads to mono-cropping and industrial farming practices that eliminate native vegetation and reduce habitat diversity. For example, palm oil plantations in Southeast Asia have directed to a substantial deterioration in orangutan residents. About 77% of global agricultural land is used for livestock production, leading to significant habitat conversion (Bredemeier et al., 2022).

Consequences of habitat fragmentation are edge effects, population isolation, and reduced echo system services. In edge effects, fragmented habitats create more "edges," where microclimatic changes occur, making them unsuitable for interior-dwelling species. Forest fragmentation increases light penetration and wind exposure, altering temperature and humidity levels, which can disrupt species adapted to stable conditions. In population isolation, fragmentation restricts gene flow between populations, increasing the risk of inbreeding depression. The Florida panther (*Puma concolor coryi*) suffers from reduced genetic diversity due to habitat fragmentation. Only about 120–230 Florida panthers remain in the wild, making conservation efforts crucial.

Tools like remote sensing, GIS, habitat models, and biodiversity monitoring enable quantitative assessment of ecosystems. Tools like Google Earth Engine and Landsat imagery help map and quantify habitat loss. For example, NASA's MODIS satellite data has been used to track deforestation trends worldwide. Habitat suitability models include software like MaxEnt (maximum entropy modeling) that predicts species distributions under different habitat scenarios. For example, researchers use MaxEnt to assess potential habitats for endangered species under climate change scenarios. Biodiversity monitoring program initiatives such as the

Global Biodiversity Information Facility (GBIF) compile species occurrence data to analyze habitat changes.

3.2 Climate Change

Scientists utilize quantitative models to predict the long-term effects of climate change and inform conservation strategies. Rising global temperatures are causing species to migrate to difficult altitudes. The American pika (*Ochotona princeps*), a small mammal adapted to cold mountainous environments, has shifted its range upwards in response to rising temperatures (Brown et al., 2022).This trend has been observed in many species across different ecosystems, leading to changes in predator–prey dynamics and ecosystem composition. Changes in precipitation regimes significantly affect freshwater availability and vegetation patterns. Regions that once received stable rainfall are experiencing prolonged droughts, while others face increased flooding. A notable case study is the Amazon Rainforest, where prolonged drought conditions have led to higher tree mortality rates, reducing carbon sequestration and accelerating climate change. Studies show that between 2005 and 2015, drought-related tree mortality in the Amazon increased by approximately 30%, weakening the forest's ability to absorb carbon dioxide (CO_2) from the atmosphere. The changing hydrological cycle also affects agricultural yield.

The 2020 Australian bushfires were one of the most devastating environmental disasters, affecting over three billion animals, including endemic species such as koalas and kangaroos (World Wildlife Fund, 2020). In the USA, hurricanes like Katrina (2005) and Harvey (2017) have demonstrated the impact of extreme weather on coastal ecosystems, leading to loss of wetland habitats and increased salinity in freshwater systems. Wildfires in California and Siberia have also destroyed vast forested areas, reducing biodiversity and releasing massive amounts of carbon into the atmosphere. Scientists employ various quantitative models to predict climate-induced habitat changes and assess species vulnerability.

Global circulation models (GCMs) simulate climate patterns based on greenhouse gas emissions, providing insights into future temperature and precipitation changes. Species distribution models (SDMs) utilize ecological data to estimate potential shifts in species ranges under different climate scenarios. For example, SDMs predict that Arctic species, such as the polar bear (*Ursus maritimus*), will experience significant habitat loss as sea ice continues to decline. These models help conservationists develop mitigation strategies, such as establishing climate corridors to facilitate species migration. Table 3 depicts the impacts of climate changes.

Climate change is a hazard to universal biodiversity by altering temperature regimes, disrupting precipitation patterns, and intensifying extreme weather events. The ecological consequences of these changes include habitat loss, population declines, and increased vulnerability to extinction. Quantitative models provide valuable tools for predicting future impacts and guiding conservation efforts. Urgent

Table 3 Climate change impact

Climate factor	Impact	Statistical data
Global temperature increase	Species migration	1.1 °C increase since pre-industrial levels (IPCC, 2023)
Coral bleaching	Habitat loss	50% loss of Great Barrier Reef coral cover since 1995 (GBRMPA, 2021)
Drought in Amazon	Increased tree mortality	30% rise in mortality between 2005 and 2015 (Phillips et al., 2018)
Australian bushfires	Biodiversity loss	Three billion animals affected (WWF, 2020)
Arctic ice decline	Habitat loss for polar bears	40% reduction in sea ice extent since 1980 (NASA, 2020)

action is required to mitigate climate change through policy measures, habitat restoration, and sustainable practices to preserve ecosystems for future generations.

3.3 *Overexploitation of Resources*

Overexploitation of natural resources encompasses the unsustainable extraction of wildlife, forests, and marine life, leading to significant ecological and economic repercussions. Despite existing regulations, illegal activities such as poaching, unregulated logging, and overfishing continue to pose substantial threats to biodiversity and ecosystem health. Wildlife exploitation remains a critical concern, with poaching and illegal trade driving numerous species toward extinction. The illegal wildlife trade is estimated to generate up to $20 billion annually. This black market demand has led to alarming declines in species such as elephants, rhinos, and pangolins. For instance, African elephant populations have suffered a 30% decline between 2007 and 2014, primarily due to ivory poaching. Approximately one in five fish is caught illegally, contributing to the decline of species such as the Atlantic cod (*Gadus morhua*), whose populations collapsed in the early 1990s due to overfishing. Despite moratoriums and conservation efforts, many fish populations have struggled to recover, underscoring the need for sustainable fishing practices. Forest exploitation, particularly through illegal logging, significantly contributes to deforestation and habitat loss. This rampant illegal logging has led to the destruction of critical habitats, such as those in the Amazon rainforest, which has experienced increased deforestation rates in recent years. In 2023, global deforestation rates rose to 6.4 million hectares, moving further away from international commitments to end deforestation by 2030.

To combat overexploitation, various monitoring and enforcement tools have been developed. Catch documentation schemes (CDS) are implemented to trace the origin of seafood products, ensuring that fish are harvested sustainably and legally. Satellite monitoring technologies are also employed to detect illegal logging activities in real-time, enabling authorities to respond promptly. Addressing this crisis requires robust legal frameworks, international collaboration, and the implementation of advanced monitoring technologies. Prioritizing sustainable resource management and conservation efforts is essential for ecological balance.

3.4 Pollution

Pollution affects ecosystems and species through contamination of air, water, and soil. It disrupts ecological balance, reduces biodiversity, and poses significant risks to human health (Cardinale et al., 2012). Various pollutants originate from industrial activities, agriculture, urbanization, and improper waste disposal. Anthropogenic emissions and volatile organic compounds (VOCs) contribute to environmental degradation. Air pollution results from the emission of harmful substances, such as gaseous pollutants, particulate matter, and biological molecules into the Earth's atmosphere. It has adverse effects on human health, wildlife, and vegetation. Acid rain, a consequence of emissions from manufacturing activities, alters soil pH, leading to nutrient depletion and increased toxicity (Qu et al., 2024). It significantly damages forest ecosystems, particularly coniferous species at high altitudes, and contributes to aquatic acidification, impacting biodiversity. Statistical data shows that rainwater pH levels range between 4.2 and 4.8 in the Eastern United States, 4.0 and 4.5 in Central Europe, and 4.1 and 4.6 in China, affecting millions of square kilometers of arable land. The rainwater pH ranges of various regions are provided in Table 4.

Particulate matter (PM2.5), comprising ultrafine particles increases respiratory ailments, cardiac disorders, and lung function impairment. It also affects ecological processes by depositing toxic substances on vegetation and aquatic systems. Case studies reveal that Delhi, India, has an annual PM2.5 concentration of approximately 98 $\mu g/m^3$, while Beijing, China, has recorded peak levels of 500 $\mu g/m^3$ during severe smog episodes. In contrast, Los Angeles, USA, maintains an average PM2.5 level of around 12 $\mu g/m^3$, primarily due to vehicular emissions and industrial output. Water pollution arises from anthropogenic activities, including industrial effluents, agricultural runoff, and untreated sewage discharge, leading to extensive ecological damage. Eutrophication, driven by excessive nitrogen (N) and phosphorus (P) loadings, results in algal blooms that deplete dissolved oxygen levels, creating hypoxic zones that disrupt aquatic life. A study on the Gulf of Mexico's dead zone reveals that it spans approximately 17,000 square kilometers annually due to nutrient influx from the Mississippi River basin, leading to mass fish mortality and economic setbacks in the fisheries sector.

Table 4 Rainwater pH range

Region	Rainwater pH range
Eastern United States	4.2–4.8
Central Europe	4.0–4.5
China	4.1–4.6

Plastic pollution is a pervasive issue, with microplastics (particles smaller than 5 mm) detected in marine organisms and food chains. Statistical data highlights that the Pacific Ocean contains between 80 and 100 million tons of plastic debris, while the Atlantic Ocean and the Indian Ocean hold approximately 50–70 million tons and 30 to 50 million tons, respectively. Studies indicate that over 90% of seabirds have ingested plastic debris, resulting in gastrointestinal obstructions, malnutrition, and increased mortality rates. Soil contamination poses significant risks to agricultural productivity, soil microbiota, and human health. Heavy metal contamination from industrial discharges, including lead (Pb), cadmium (Cd), and mercury (Hg), disrupts soil microbial communities and reduces soil fertility. The bioaccumulation of these metals in crops leads to long-term health consequences. A case study on cadmium contamination in rice from China reveals that Cd concentrations exceeded 0.2 mg/kg in samples from Hunan province, with prolonged exposure linked to nephrotoxicity, osteoporosis, and metabolic disorders. The contamination of affected region is provided in Table 5.

Several advanced monitoring systems are employed to quantify pollution levels and assess environmental impact. The air quality index (AQI) serves as a crucial indicator of air pollution, integrating concentrations of PM2.5, NO_2, SO_2, CO, and ground-level ozone (O_3). In water pollution studies, predict pollutant transport and simulate watershed responses. Pollution exerts profound ecological and socio-economic consequences, necessitating immediate policy actions, technological advancements, and sustainable management practices. The escalating levels of air, water, and soil contamination demand stringent regulatory frameworks, enhanced waste treatment technologies, and widespread adoption of green infrastructure solutions. Implementing stricter emission controls, promoting circular economy models, and fostering community-based environmental stewardship will be pivotal in mitigating pollution impacts. Continuous environmental monitoring and interdisciplinary collaboration will be essential in safeguarding ecological integrity for future generations.

3.5 *Invasive Alien Species (IAS)*

Invasive alien species (IAS) are non-native organisms that disrupt local ecosystems by outcompeting indigenous species, altering habitats, and impacting ecological processes. IAS pose a significant threat to biodiversity, economic stability, and human livelihoods, making their study and management crucial for environmental

Table 5 Contamination of affected region

Contaminant	Affected region	Concentration (mg/kg)
Cadmium (Cd)	Hunan, China	>0.2
Lead (Pb)	Industrial zones	Varies
Mercury (Hg)	Mining areas	Varies

sustainability. These species are introduced through various pathways, including global trade, accidental transportation, and intentional releases for economic or aesthetic purposes. IAS spread through multiple mechanisms that enable them to establish and dominate new environments. For example, water hyacinth (*Eichhornia crassipes*) spreads rapidly across water bodies, forming dense mats that clog waterways, reduce oxygen levels, and severely impact aquatic biodiversity. Some IAS act as novel predators, disrupting food chains and leading to population declines or extinctions of native species. IAS can also serve as vectors for diseases that affect native species and human populations. The chytrid fungus (*Batrachochytrium dendrobatidis*), introduced by invasive amphibians, has decimated global frog populations. The hybridization between the North American mallard (*Anas platyrhynchos*) and native duck species has altered gene pools in several regions. Other mechanisms, such as allelopathy, enable IAS to release biochemicals that inhibit the growth of surrounding native flora, further reinforcing their dominance in new environments.

IAS significantly impact economies and ecosystems through various direct and indirect consequences. Invasive species cause over $1.4 trillion in damages globally each year, affecting agriculture, fisheries, forestry, and human health. They lead to increased costs in pest control, reduced agricultural yields, and economic losses in tourism and trade. The European gypsy moth (*Lymantria dispar*) defoliates millions of acres of forests annually in North America, leading to substantial losses in the timber industry. Similarly, the red imported fire ant (*Solenopsis invicta*), an aggressive invader, causes billions of dollars in damages by affecting agriculture, human health, and electrical infrastructure. IAS alter ecological balances by reducing native biodiversity, modifying soil chemistry, disrupting food webs, and increasing fire risks. The introduction of cheatgrass (*Bromus tectorum*) in the Western United States has increased fire frequency, altering native plant communities. Zebra mussels (*Dreissena polymorpha*), introduced via ballast water in the Great Lakes, have clogged water intake pipes, affected native mollusk populations, and altered aquatic food chains. The estimated economic impact is $500 million annually in control and damages. The African giant land snail (*Achatina fulica*) is a highly destructive agricultural pest, consuming a vast range of crops and spreading human pathogens such as *Angiostrongylus cantonensis*, causing meningitis. In Hawaii, miconia (*Miconia calvescens*), known as the "green cancer," displaces native vegetation, promotes soil erosion, and reduces biodiversity in Hawaiian forests. In Australia, the introduction of the cane toad (*Rhinella marina*) to control sugarcane pests led to widespread ecological disruption, as the toad itself became an invasive threat due to its toxic secretions, which are lethal to native predators. The introduction of rabbits (*Oryctolagus cuniculus*) in Australia also serves as a classic example of an IAS with severe consequences, as they have caused extensive land degradation and competition with native species, leading to large-scale ecosystem changes.

To access and manage IAS, scientists employ several advanced quantitative tools. Invasive species distribution models (ISDMs) predict the potential spread of IAS under various climate surroundings. For example, MaxEnt modeling is used to forecast the spread of *Parthenium hysterophorus* in Australia. Economic impact assessments evaluate the cost–benefit analysis of IAS management strategies, helping policymakers allocate resources effectively. Cost estimates for eradicating *Rhododendron ponticum* from the UK exceed $50 million. Genetic and genomic tools, including DNA barcoding and whole-genome sequencing, help track invasion pathways and identify genetic variations in IAS. Remote sensing and geographic information system (GIS) technologies have also emerged as essential tools for mapping and monitoring IAS spread. The IAS management enables researchers to develop predictive models that help optimize control measures and resource allocation. Citizen science initiatives, such as smartphone apps, allow the public to report IAS sightings.

Invasive alien species (IAS) is a significant global challenge, affecting biodiversity, economies, and ecosystem functions. Effective IAS management requires a combination of preventive measures, early detection, and rapid response strategies. Advancements in ecological modeling, genetic analysis, and economic assessment support conservation planning, inform policy decisions, and promote sustainable resource usage. Strengthening international collaboration, public awareness, and policy enforcement is essential. Governments and environmental organizations must implement stricter biosecurity measures, invest in sustainable eradication programs, and promote habitat restoration initiatives to curb the detrimental effects of IAS. Through continuous research and adaptive management strategies, global efforts can reduce the long-term ecological and economic costs of IAS, ensuring the protection of native ecosystems for future generations.

3.6 Agricultural Practices

Modern agricultural practices have significantly increased food production to sustain growing populations. However, these methods exert considerable pressure on biodiversity, soil health, and ecological balance. The lack of genetic variation makes crops highly susceptible to widespread disease outbreaks. For instance, the Irish potato famine in the 1840s was exacerbated by dependence on a single crop variety (*Solanum tuberosum*), leading to mass starvation when potato blight (*Phytophthora infestans*) spread unchecked. The excessive use of pesticides has adverse ecological consequences (Ige et al., 2025). Chemicals such as neonicotinoids have been linked to declines in pollinators, particularly bees, threatening pollination-dependent crops. Studies indicate that global bee populations have declined by nearly 40% due to habitat loss and pesticide exposure. Pesticide contamination also affects soil microbiota and water quality, leading to long-term ecosystem imbalances.

To assess the impact of modern agricultural practices, scientists use several quantitative tools: Agro-biodiversity indices: These measures help policymakers to assess sustainability levels (Rosenfield et al., 2022). Pesticide Risk Assessment Models: These evaluate the ecological risks of chemical usage by analyzing toxicity levels, persistence, and bioaccumulation in food chains. By integrating sustainable agricultural techniques, biodiversity loss can be mitigated while maintaining food security.

3.7 Impacts of Anthropogenic Pressures

Anthropogenic pressures on biodiversity are multifaceted and deeply interconnected. Social events such as deforestation, pollution, overfishing, climate change, and urbanization have significantly contributed to biodiversity loss worldwide. Addressing these challenges requires global cooperation, innovative conservation strategies, and sustainable resource management. This section explores various anthropogenic pressures, their consequences, and strategies for sustainable biodiversity management. Deforestation is one of the leading causes of biodiversity loss. The Amazon Rainforest, which hosts about 10% of the world's known species, is severely threatened by agricultural expansion, logging, and infrastructure development. Illegal logging and land conversion for cattle ranching and soy farming are major contributors, leading to species extinction, climate change acceleration, and loss of indigenous communities' homes.

The long-term consequences included reduced fish populations and habitat degradation. Excessive hunting, fishing, and poaching threaten species survival. Overfishing has led to 33% of global fish stocks being overfished, with the Atlantic Bluefin Tuna population declining by 85% since the 1970s. A critical case study is African Elephant poaching, where due to high demand for ivory, elephant populations have declined by 30% in seven years. Anti-poaching efforts, such as increased surveillance and bans on ivory trade, have helped curb illegal hunting. Expanding cities, roads, and industrial zones reduce natural habitats and fragment ecosystems. Urban expansion in Asia has had a significant impact, with China's urban population increasing from 19% in 1980 to over 60% in 2020. Habitat loss has led to declining populations of species such as the Giant Panda and the Siberian Tiger. Technological advancements offer promising solutions to mitigate anthropogenic pressures. Remote sensing and GIS are used for tracking deforestation and habitat changes. Artificial intelligence (AI) helps in species identification and poaching detection. Drones are employed to monitor wildlife populations and illegal activities in conservation areas. The impact of anthropogenic pressures is depicted in Table 6.

The impacts of anthropogenic pressures on biodiversity are severe and wide-ranging. Human-induced environmental changes have led to habitat destruction, climate shifts, pollution, overexploitation, and urban expansion, threatening countless species and ecosystems. While the challenges are immense, solutions exist through sustainable practices, technological advancements, and global cooperation. Incorporating ecological, social, and economic factors into policies and practices fosters balanced decision-making that supports sustainable development, social well-being, and economic resilience. Administrations, preservation societies, industries, and individuals need to work together to mitigate these threats and ensure a balanced coexistence with nature.

Table 6 Impact of anthropogenic pressures

Pressure type	Impact on biodiversity	Key statistics
Deforestation	Habitat loss, species extinction	420 million ha lost (1990–2020)
Climate change	Disrupted ecosystems, coral bleaching	1.2 °C temperature rise
Pollution	Marine and terrestrial species affected	Eight million tons of plastic/year
Overexploitation	Declining fish and wildlife populations	33% of fish stocks overfished
Urbanization	Habitat fragmentation	60% urbanization in China (2020)

4 Advanced Strategies for Conservation

The conservation of biodiversity has evolved significantly from traditional approaches to advanced, multidisciplinary strategies. These strategies leverage cutting-edge technology, innovative methodologies, and interdisciplinary collaboration to address complex environmental challenges. Advanced conservation techniques integrate ecological, social, and economic dimensions to ensure the sustainable management of resources while minimizing anthropogenic pressures. This section discusses the most advanced strategies for biodiversity conservation, emphasizing their scientific, technological, and policy-based foundations.

4.1 Ecosystem-Based Conservation

The long-term sustainability of ecological processes and services promotes biodiversity while maintaining ecosystem functions. Landscape-scale conservation focuses on conserving biodiversity across large landscapes, considering ecological connectivity and habitat heterogeneity. This approach is essential in maintaining genetic diversity and ensuring species survival across fragmented habitats. Wildlife corridors play a crucial role in connecting fragmented habitats, allowing gene flow and species migration. Buffer zones are another critical component of landscape-scale conservation, acting as transitional areas around protected regions to minimize human impact. For example, buffer zones around India's Kaziranga National Park help protect endangered species like the Indian rhinoceros by regulating human encroachment and limiting activities such as agriculture and logging.

Marine protected areas (MPAs) safeguard marine biodiversity by restricting human activities in designated zones. These areas are vital for preserving marine habitats, maintaining fish stocks, and ensuring ecological balance. As of 2023, approximately 8% of the global ocean is protected under MPAs, with a target of 30% by 2030 under the Global Biodiversity Framework. The impact of ecosystem-based conversion is provided in Table 7.

Ecosystem-based conservation strategies are essential for maintaining biodiversity, ensuring ecosystem stability, and mitigating anthropogenic pressures. These approaches integrate science, policy, and community involvement to protect landscapes, marine ecosystems, and degraded habitats. Wildlife corridors facilitate

Table 7 Impact of ecosystem-based conservation

Conservation strategy	Key impact	Notable statistics
Wildlife corridors	Enhances genetic diversity and species migration	80% reduction in wildlife–vehicle collisions (Banff, Canada)
Marine protected areas (MPAs)	Protects marine biodiversity and fish stocks	8% of global ocean protected, 30% target by 2030
Reforestation initiatives	Increases forest cover and carbon sequestration	Costa Rica's forest cover increased from 21% to 50%
Wetland restoration	Improves water filtration and flood control	Everglades project restoring natural water flow

species migration, MPAs safeguard marine biodiversity, and restoration projects enhance ecological functions. With continued investment and global cooperation, these advanced conservation strategies will play a pivotal role in securing a sustainable future for biodiversity. Administrations, protection societies, and local communities should combine to expand protected areas, restore degraded ecosystems, and implement policies that balance development with environmental protection.

4.2 *Technological Innovations in Conservation*

By integrating cutting-edge information and communication technology (ICT) solutions, conservation efforts have become more effective, scalable, and data-driven. This section explores various ICT-based technologies, their applications, case studies, and statistical insights into their impact on conservation. Remote sensing and GIS provide high-resolution satellite imagery, geospatial analysis, and real-time environmental monitoring. These technologies are instrumental in tracking deforestation, habitat fragmentation, climate changes, and illegal land-use modifications. One of the most well-known satellite monitoring programs is NASA's Landsat program, which provides detailed imagery for ecological monitoring. Research indicates that deforestation in the Amazon has been reduced by 44% in areas monitored through satellite imaging and predictive mapping. The European Space Agency's (ESA) Copernicus suite also offers extensive Earth observation data, which has helped in tracking illegal logging and land encroachments. GIS applications in conservation have enabled predictive modeling of species migration patterns, aiding in establishing wildlife corridors and protected areas.

Artificial intelligence and machine learning have transformed conservation by investigating huge datasets to identify trends, predict future environmental changes, and automate species identification (Takacs & O'Brien, 2023). Similarly, Google's DeepMind has been deployed to analyze poaching patterns, allowing conservationists to proactively deploy anti-poaching measures. AI-driven drone surveillance has significantly increased the detection of illegal fishing activities, leading to a 60% reduction in unauthorized fishing in protected marine areas. Recent advancements in genetic technologies have revolutionized biodiversity conservation through species identification, genetic rescue, and disease prevention. DNA barcoding allows scientists to identify species from small genetic samples, helping in biodiversity assessments and illegal wildlife trade monitoring. CRISPR gene-editing technology has been explored to address genetic diseases in threatened species, such as the black-footed ferret and Tasmanian devil, improving their survival rates.

IoT-enabled devices have significantly improved real-time environmental monitoring, enabling conservationists to collect continuous data on ecosystem health. Sensor networks measure temperature, humidity, pollution levels, and water quality, helping to detect environmental threats before they become catastrophic. A remarkable example is the use of SMART collars equipped with GPS and biometric sensors, which track the movement and health of endangered species, such as elephants, tigers, and snow leopards. In Africa, IoT-based anti-poaching systems combining motion sensors and real-time alerts have reduced poaching incidents by 50%. Underwater IoT devices have also been deployed to monitor coral reef health, with recent findings showing a 25% improvement in reef restoration success rates due to early threat detection. Cloud computing has enabled conservationists to store and process vast amounts of ecological data efficiently. Large-scale datasets from remote sensors, drones, and environmental monitoring stations can be examined, consenting for quick decision-making. Platforms like Google Earth Engine provide cloud-based processing for large geospatial datasets, enabling researchers to assess deforestation rates, climate trends, and habitat loss dynamically. Mobile computing has facilitated field-based data collection and analysis through smartphone applications and mobile sensors. Conservationists can now access GIS maps, species tracking data, and environmental reports from any location. Mobile applications like iNaturalist and eBird allow citizen scientists to contribute biodiversity observations, enhancing global conservation efforts.

Generative AI is being increasingly used in conservation to create realistic simulations of ecosystem changes under different environmental conditions. AI models can generate synthetic satellite images to predict future deforestation, climate impact, and species migration patterns. These predictive capabilities allow conservationists to take preventive action before environmental degradation occurs. Advanced data analytics techniques, including big data processing and predictive modeling, have strengthened conservation decision-making. By analyzing massive datasets from satellites, drones, and IoT sensors, conservation organizations can detect trends and allocate resources more efficiently. For example, predictive analytics have helped reduce wildlife–human conflicts by identifying high-risk zones for elephant migration and implementing preventive measures. Blockchain is emerging as a transformative tool in conservation by ensuring transparency and traceability in wildlife trade, carbon offset programs, and protected area funding. Through decentralized ledgers, conservation organizations can track donations and ensure funds reach their intended projects. For example, the World Wildlife Fund (WWF) has implemented blockchain-based tracking systems to authenticate sustainable fishing practices, significantly reducing fraud in seafood labeling. Similarly, smart contracts are being used to verify land ownership in indigenous communities, preventing land grabs and illegal deforestation. The impact of technological innovations during the year 2024 is provided in Table 8.

Table 8 Impact of technological innovations

Technology	Application	Notable impact (2024)
Remote sensing and GIS	Tracks deforestation and illegal land use	44% reduction in deforestation (Amazon)
AI and machine learning	Predicts species distribution and automates monitoring	90% reduction in manual data processing time (wildlife insights)
Genetic tools	Enhances genetic diversity and species identification	30% increase in Florida panther population viability
IoT and sensor networks	Monitors environmental parameters in real-time	50% decrease in poaching incidents in Africa
Cloud computing	Processes and stores large-scale ecological data	Google earth engine analysis of habitat loss
Mobile computing	Enables real-time conservation fieldwork	Increased citizen science contributions through mobile apps
Generative AI	Simulates future ecological changes	Improved climate resilience planning through predictive AI models
Blockchain technology	Enhances transparency in conservation funding	25% improvement in sustainable fisheries tracking

Technological advancements have significantly enhanced biodiversity conservation by providing efficient and data-driven solutions. Remote sensing enables large-scale habitat monitoring, AI and machine learning automate species tracking, genetic tools improve species survival, and IoT allows for real-time ecosystem monitoring. Cloud computing facilitates large-scale data storage and processing, while mobile computing enhances fieldwork efficiency. Generative AI and predictive analytics offer foresight into environmental changes, enabling proactive conservation efforts. Blockchain technology ensures transparency in conservation efforts, making funding and trade more accountable. By integrating these technologies into conservation policies and practices, stakeholders can develop more effective strategies to mitigate biodiversity loss and promote environmental sustainability.

4.3 *Community-Based Conservation*

Community-based conservation (CBC) is an approach that integrates local communities in biodiversity conservation efforts, ensuring ecological sustainability and socio-economic benefits. By empowering communities, conservation initiatives can achieve greater success in protecting natural resources while improving livelihoods. This approach combines participatory decision-making, incentive mechanisms, and indigenous knowledge systems to enhance conservation outcomes. It recognizes that conservation is most effective when local people are actively involved in managing and protecting their surrounding environment. Participatory conservation involves engaging local stakeholders in decision-making and management processes. When communities have a voice in conservation planning, they are more likely to support and sustain these efforts. Co-management agreements between governments and local groups have proven effective in many conservation projects. A case study from Namibia demonstrates the success of participatory conservation. The establishment of communal conservancies has led to a 50% increase in wildlife populations, including endangered species such as black rhinos and elephants. These conservancies allow local communities to manage and benefit from natural resources, reducing conflicts between humans and wildlife.

Two major incentive mechanisms include: Payment for ecosystem services (PES) programs provide financial compensation to individuals or communities for their contributions to ecosystem protection. These payments help offset opportunity costs associated with conservation. Costa Rica's PES program, which started in 1997, has led to a 40% increase in forest cover by compensating landowners for reforestation, sustainable agriculture, and watershed protection efforts. Studies show that communities participating in PES programs report higher incomes and greater environmental awareness. Ecotourism generates revenue by attracting visitors to conservation areas, creating economic opportunities for local communities while promoting biodiversity conservation. In Rwanda, gorilla trekking has generated over $400 million in revenue over the past decade. A quota of these funds is assigned to local communities, improving setup and healthcare while reinforcing conservation efforts. As a result, the mountain gorilla population has increased by 23% in the last 20 years.

Indigenous communities possess extensive ecological knowledge, which can enhance conservation strategies. Traditional resource management practices contribute to sustainable land use, species conservation, and climate adaptation. The Kayapo people of the Amazon manage over ten million hectares of rainforest using traditional ecological knowledge. Their conservation efforts have successfully protected large areas from deforestation and illegal logging. Research indicates that indigenous-managed lands have lower deforestation rates compared to government-protected areas. Another case study from Australia highlights how aboriginal fire management techniques, such as controlled burning, reduce the risk of large-scale wildfires and enhance biodiversity. The impact of community-based conversion is provided in Table 9.

Community-based conservation is a highly effective approach that integrates environmental sustainability with socio-economic benefits. By involving local communities, providing financial incentives, and leveraging indigenous knowledge, conservation efforts can achieve long-term success. Countries that have adopted participatory conservation, PES programs, and ecotourism have demonstrated significant improvements in biodiversity protection and community well-being. Moving forward, strengthening policy frameworks, increasing community engagement, and expanding funding opportunities will further enhance the impact of community-based conservation initiatives globally.

Table 9 Impact of community-based conservation

Conservation strategy	Key impact	Notable statistics (2024)
Participatory conservation	Enhances wildlife populations and habitat protection	50% increase in wildlife in Namibia conservancies
PES programs	Increases reforestation and community income	40% increase in forest cover in Costa Rica
Ecotourism	Supports conservation funding and local economies	$400 million generated from gorilla tourism in Rwanda
Indigenous knowledge	Protects large forest areas and prevents deforestation	10 million hectares protected by the Kayapo people

4.4 Policy and Legal Frameworks

Robust policy frameworks and legal instruments are essential for implementing conservation strategies effectively. These frameworks provide the necessary structure for protecting biodiversity, regulating land use, and ensuring sustainable resource management. Effective policy mechanisms operate at multiple levels, from international agreements to national policies and local enforcement measures. This section explores key policies, legal tools, and enforcement strategies that drive conservation efforts worldwide. Global biodiversity conservation efforts are guided by international treaties and agreements that establish shared goals and collaborative mechanisms. As of 2024, over 195 countries have ratified the convention, with commitments to restore at least 30% of degraded ecosystems by 2030. Over 38,000 species are currently protected under CITES, leading to a 70% reduction in illegal ivory trade. The Paris Contract includes commitments to protect forests and other ecosystems as carbon sinks, contributing to biodiversity conservation efforts.

Countries implement national policies that align with international conservation goals while addressing local ecological challenges. India's Project Tiger, launched in 1973, has successfully increased tiger populations from fewer than 2000 to over 3900 in 2023 by creating a network of 53 tiger reserves. Environmental Impact Assessments (EIA) is mandated in many countries to evaluate potential ecological impacts of development projects and propose mitigation measures. Brazil's NBSAP has contributed to a 45% reduction in deforestation in the Amazon over the past decade. Legal instruments ensure compliance with conservation policies and create incentives for sustainable development. Biodiversity offsets require developers to compensate for habitat destruction by restoring or protecting other areas. Australia's biodiversity offset policies mandate habitat restoration projects for mining activities, resulting in the renovation of over 500,000 hectares of degraded property. The Polluter Pays Principle holds industries accountable for environmental damage, requiring them to fund restoration efforts. The European Union's environmental liability directive has generated over €12 billion in restoration projects since its implementation. Conservation easements enable landowners to voluntarily restrict land use for biodiversity protection. In the USA, conservation easements have permanently protected over 30 million acres of private land. Effective enforcement of conservation laws relies on technological advancements and community participation. Real-time monitoring using drones and AI-assisted surveillance has improved law enforcement in wildlife reserves. In Kenya, drone surveillance has led to a 62% decrease in poaching incidents within protected areas. Satellite monitoring platforms, such as Global Forest Watch, track deforestation and habitat changes, helping reduce illegal deforestation in Indonesia by 25%. Indigenous groups and local communities are actively involved in environmental protection. In Bolivia, community rangers have been instrumental in safeguarding 1.5 million hectares of rainforest from illegal activities.

International and national conservation policies have had measurable impacts on biodiversity protection. CITES has contributed to a 70% reduction in illegal wildlife trade. India's Project Tiger has increased the number of tigers from fewer than 2000 in 1973 to over 3900 in 2023. Australia's biodiversity offset policies have restored over 500,000 hectares of land. Drone surveillance in Kenya has resulted in a 62% decline in poaching incidents. Community rangers in Bolivia have helped protect 1.5 million hectares of rainforest from deforestation and illegal activities. Strong policy and legal frameworks are the foundation of effective biodiversity conservation. International agreements set global conservation targets, while national policies implement localized strategies to protect ecosystems. Legal tools such as biodiversity offsets, conservation easements, and the Polluter Pays Principle provide financial incentives for sustainable practices. Meanwhile, technological advancements and community-based enforcement ensure compliance and real-time monitoring. Moving forward, enhancing cross-border cooperation, increasing investment in conservation law enforcement, and integrating emerging technologies will be critical in achieving long-term biodiversity protection.

4.5 *Sustainable Resource Management*

It integrates ecological conservation, economic feasibility, and social responsibility to maintain biodiversity and ecosystem services. The rapid depletion of natural resources due to industrialization and population growth necessitates strategic interventions. Sustainable agriculture employs methods that enhance soil fertility, soil stability, enhances carbon sequestration, and provides habitats for various species. Crop rotation, which involves alternating different crops, reduces soil depletion, pest infestations, and enhances nutrient recycling. Organic farming avoids synthetic fertilizers and pesticides, preserving soil biodiversity and preventing water contamination. Precision farming utilizes IoT and AI-based monitoring systems to optimize resource utilization and minimize waste. A real-world example of sustainable agriculture is the shade-grown coffee plantations in Latin America. Studies indicate that such farms harbor up to 30% more bird species compared to sun-exposed monoculture farms (Levis et al., 2024). Sustainable fisheries aim to regulate fishing activities to prevent overexploitation and ensure long-term fish stock health. Quota systems are implemented by governments to set limits on the allowable catch, preventing overfishing. Community-managed fishing zones allow local communities to oversee fishing regulations and maintain stock balance. A case study of sustainable fisheries is the Marine Stewardship Council (MSC). Fisheries certified by the MSC adhere to sustainable practices, reducing overfishing by 15–20% annually. In 2021, over 420 fisheries worldwide were MSC-certified, covering 15% of global seafood production.

Sustainable forestry ensures that timber harvesting meets present needs without depleting forests or harming biodiversity. Reduced impact logging (RIL) minimizes soil erosion, maintains canopy structure, and preserves wildlife habitats. Selective logging, which involves harvesting only mature trees, prevents deforestation. Reforestation and afforestation efforts contribute to counterbalancing deforestation and improving carbon sequestration. An example of sustainable forestry is the Forest Stewardship Council (FSC)-certified logging operations, which balance ecological and economic needs, reducing deforestation rates by 30% compared to non-certified logging (WWF, 2021). Water is a critical resource that must be managed efficiently to prevent shortages and maintain ecosystem balance. Water recycling involves treating wastewater for reuse in industrial and agricultural applications. Drip irrigation reduces water wastage by directly delivering water to plant roots. Desalination technologies convert seawater into potable water using energy-efficient methods. Singapore's water management system is a leading example, as the country has developed a comprehensive water recycling and desalination strategy, meeting up to 40% of its water demand through advanced purification techniques.

Solar and wind power harness renewable energy to reduce carbon emissions. Smart grids enhance electricity distribution efficiency with AI-driven optimization. Energy-efficient appliances reduce energy consumption through advanced technology. Biofuels and hydrogen fuel cells provide cleaner alternatives to traditional fossil fuels. Germany's renewable energy transition exemplifies sustainable energy management, as the country aims to generate 80% of its electricity from renewables by 2030, reducing dependency on coal and nuclear power. The impact of various resource management is provided in Table 10.

Sustainable resource management is essential for maintaining ecosystem integrity. Governments, industries, and local communities must collaborate to implement eco-friendly policies and technological advancements to optimize resource use.

4.6 Climate Transformation

Addressing climate transformation is integral to biodiversity conservation. Climate change has led to shifts in ecosystems, affecting species distribution, habitat viability, and survival rates. Implementing climate-smart conservation strategies helps mitigate these impacts and supports long-term ecological resilience. Climate-smart conservation integrates climate adaptation and mitigation strategies into conservation planning. One effective approach is protecting climate refugia—areas less affected by climate change, such as shaded valleys, high-altitude forests, and wetlands. These regions serve as safe havens for species adapting to shifting environmental conditions. For instance, the Great Barrier Reef Marine Park Authority has employed a climate-resilient zoning plot to protect coral reefs from the adverse

Table 10 Sustainable resource management

Resource type	Key practice	Environmental impact reduction (%)	Economic benefit increase (%)
Agriculture	Crop rotation	25	15
Agriculture	Organic farming	30	20
Fisheries	Quota systems	20	18
Fisheries	Community fishing zones	22	16
Forestry	Reduced impact logging	35	25
Forestry	Selective logging	28	22
Water	Rainwater harvesting	40	30
Water	Drip irrigation	35	28
Energy	Solar and wind power	50	40
Energy	Smart grids	45	35

effects of ocean acidification and temperature rise. Another climate-smart initiative is genetic conservation, which involves preserving genetic diversity to enhance species' adaptability. Seed banks, such as the Svalbard Global Seed Vault, store genetic material of plant species to prevent biodiversity loss due to climate change. Reforestation and afforestation initiatives absorb atmospheric carbon, mitigating climate change effects. Increasing forest cover enhances carbon capture, reduces greenhouse gas emissions, and strengthens global climate resilience.

Assisted migration involves relocating species to areas with favorable climate conditions to ensure their survival. This proactive strategy mitigates habitat loss and prevents species extinction due to shifting climate zones. A distinguished work is the relocation of the *Torreya taxifolia* tree, an endangered species, to cooler regions in the USA to counteract warming temperatures. Similarly, the coral reef restoration project in the Maldives transplants heat-resistant coral species to ensure reef survival amid rising ocean temperatures (Li et al., 2022b). Moreover, conservation programs for endangered species, such as the Pine Marten Recovery Project in the UK, utilize assisted migration to reintroduce populations to areas with declining threats. The various climate transformation strategies are tabulated in Table 11.

Climate transformation is a critical challenge that requires targeted interventions to sustain biodiversity and ecosystem health. Climate-smart conservation, carbon sequestration, and assisted migration serve as pivotal strategies. Case studies such as The Great Green Wall, Svalbard Global Seed Vault, and coral reef restoration projects demonstrate the effectiveness of proactive measures in combating biodiversity loss. Statistical data reinforces the necessity of large-scale global implementation to ensure the survival of diverse ecosystems. Through continued research, policy-making, and collaborative global efforts, ecosystems can be preserved, ensuring resilience against future climate fluctuations. Governments, conservationists, and local communities must work collectively to implement sustainable solutions, ensuring a balanced coexistence between human progress and environmental preservation.

Table 11 Climate transformation strategies

Strategy	Key practice	Carbon reduction ($MTCO_2$/yr)	Biodiversity impact (%)	Global implementation (%)
Climate-smart conservation	Climate refugia protection	N/A	40	55
Climate-smart conservation	Genetic conservation	N/A	35	50
Carbon sequestration	Reforestation	2.6 billion	50	60
Carbon sequestration	Afforestation	3.2 billion	55	58
Assisted migration	Species relocation	N/A	45	30
Assisted migration	Coral reef transplantation	N/A	38	25

4.7 Advanced Metrics and Monitoring

Effective conservation efforts rely on precise measurement tools to assess biodiversity trends, habitat conditions, and ecological stability (Weiskopf et al., 2022). Biodiversity indicators provide quantitative assessments of species population trends, extinction risks, and ecosystem health. The Living Planet Index (LPI), a globally recognized metric, compiles data from thousands of monitored species to measure biodiversity changes over time. The 2022 WWF Living Planet Report revealed a 69% decline in global vertebrate populations since 1970, underscoring the urgency of conservation measures. The IUCN Red List Index is another crucial tool that evaluates species' risk of extinction. Currently, 28% of assessed species face extinction threats, with amphibians experiencing the highest risk. Conservation success stories demonstrate the impact of targeted biodiversity interventions. This species, once classified as Critically Endangered, saw its population grow from 94 individuals in 2002 to over 1300 in 2022, due to habitat restoration and strict anti-poaching regulations.

Satellite-based Earth observation has revolutionized environmental monitoring by providing real-time, large-scale data on deforestation, land degradation, and habitat shifts (Ekka et al., 2023). The Copernicus program, a key player in this field, supplies high-resolution imagery to support biodiversity conservation efforts. NASA's Global Ecosystem Dynamics Investigation (GEDI) project employs LiDAR technology to generate 3D maps of forest structure and biomass, aiding in the assessment of carbon sequestration and illegal logging activities. Additionally, drone surveillance has emerged as a crucial tool in conservation. Organizations like Rainforest Connection deploy acoustic sensor-equipped drones to detect deforestation in tropical forests, facilitating rapid responses to illegal activities. Public participation in scientific research significantly enhances data collection and conservation efforts. Platforms such as eBird and iNaturalist engage global communities in documenting species distributions, yielding extensive datasets that

Table 12 Effectiveness of conservation monitoring technologies

Monitoring approach	Technology utilized	Accuracy (%)	Data scope	Adoption rate (%)
Biodiversity metrics	Genetic analysis, surveys	85	Global	70
Satellite monitoring	Copernicus, GEDI	90	Large scale	65
Drone surveillance	LiDAR, acoustic sensors	92	Targeted	50
Crowdsourced data	Mobile apps, citizen science	80	Broad reach	75

support ecological studies. For instance, eBird data has been instrumental in identifying migratory stopover sites for species such as the Blackpoll Warbler (*Setophaga striata*). This information has guided the establishment of protected areas along critical migration routes. Similarly, the Great Sunflower Project has mobilized thousands of participants to track pollinator activity, generating valuable insights into declining bee and butterfly populations and informing habitat restoration initiatives. The effectiveness of Conservation Monitoring Technologies is provided in Table 12.

Innovative conservation monitoring strategies integrate cutting-edge technology, public engagement, and robust policy frameworks to address biodiversity loss effectively. Success stories such as the Iberian lynx recovery, GEDI's forest mapping, and eBird-powered avian conservation highlight the potential of data-driven interventions in preserving ecosystems. Empirical data underscores the impact of these monitoring methods, with satellite technology achieving 90% accuracy in tracking environmental changes and biodiversity indicators informing 70% of global conservation strategies. Future conservation efforts must emphasize interdisciplinary collaboration, investment in emerging technologies, and adaptive management approaches to ensure long-term ecological resilience. Harnessing real-time data, predictive analytics, and community-driven initiatives will be pivotal in sustaining biodiversity and mitigating environmental threats.

5 Biodiversity and Sustainable Development: Integration into Policy

5.1 The Nexus Between Biodiversity and Sustainable Development

Biodiversity is fundamental to sustainable development, providing social well-being, economic stability, and environmental resilience. The integration of biodiversity into policy frameworks is critical for maintaining these services. This section explores the complex relationship between biodiversity and sustainable development, analyzing policy mechanisms, implementation challenges, and integration strategies at local, national, and global levels. Biodiversity contributes significantly to multiple aspects of sustainable development, influencing food security, water quality, public health, and climate resilience. Protecting biodiversity ensures long-term socio-economic stability by sustaining essential resources and ecosystem functions.

Contributions to Sustainable Development Goals (SDGs): SDG 2 (Zero Hunger): Biodiversity in agriculture enhances food security by ensuring genetic diversity in crops and livestock, improving resilience to pests, diseases, and climate variability. SDG 6 (Clean Water and Sanitation): Healthy ecosystems, such as wetlands and forests, regulate water cycles and maintain water quality, reducing contamination and water scarcity. SDG 13 (Climate Action): Biodiversity plays a role in stabilizing the climate by sequestering carbon, reducing extreme weather events, and providing habitats that support ecological resilience. The decline in biodiversity increases the risk of zoonotic diseases, as observed during the COVID-19 pandemic, where habitat destruction heightened human–wildlife interactions and facilitated virus transmission. Vulnerable communities, especially in developing countries, suffer the most from biodiversity loss due to their reliance on natural resources for sustenance and economic activities. The depletion of fisheries, forests, and arable land exacerbates poverty and food insecurity. Costa Rica has integrated biodiversity into its national policy, implementing payment for ecosystem services (PES) plans for preserving forests, leading to a 50% increase in forest cover since the 1980s. The European Green Deal: This policy framework prioritizes biodiversity conservation, committing to protect 30% of land and marine areas in Europe by 2030. The economic impact of biodiversity integration is provided in Table 13.

Biodiversity and sustainable development are intrinsically linked and integrating biodiversity into policy frameworks. Case studies from Costa Rica, the European Union, and India demonstrate that proactive policies can yield significant environmental and economic benefits. To achieve sustainable development, interdisciplinary collaboration, scientific research, and robust governance mechanisms must be prioritized. Global efforts should focus on strengthening biodiversity conservation policies to ensure a resilient and sustainable future.

5.2 *Policy Instruments for Biodiversity Integration*

Governments and policymakers employ a variety of regulatory tools and strategic frameworks to incorporate biodiversity conservation into sustainable development. By aligning environmental stewardship with economic policies, these mechanisms help to ensure ecological balance while fostering long-term economic and social stability. This section explores key policy instruments, their practical applications, and their role in global biodiversity conservation efforts. Under the Convention on Biological Diversity (CBD), NBSAPs serve as comprehensive policy roadmaps guiding national biodiversity conservation. A significant example is India's NBSAP, which effectively merges localized biodiversity conservation initiatives with national policy frameworks.

By fostering community-driven ecological restoration projects, this strategy promotes habitat conservation while accommodating economic development. These initiatives encourage sustainable land use and environmental stewardship. A notable success story is Costa Rica's PES initiative, which has significantly contributed to forest regeneration by providing financial rewards for activities such as afforestation and watershed preservation. Since its inception, Costa Rica has successfully restored

Table 13 Economic impact of biodiversity integration

Region	Biodiversity policy implemented	GDP growth impact (%)	Environmental benefit (%)
Costa Rica	Payment for ecosystem services	2.5	50
European Union	Green deal conservation plan	1.8	30
India	Agroforestry initiatives	2.2	40

60% of its forest cover, leading to enhanced carbon sequestration and greater ecosystem resilience. Environmental Impact Assessments (EIAs) serve as essential regulatory tools that assess the potential biodiversity impacts of development projects. By ensuring rigorous environmental evaluation prior to project approval, EIAs help mitigate habitat destruction and promote sustainable infrastructure growth. For example, Brazil enforces EIAs for large-scale infrastructure projects within the Amazon Rainforest. This regulatory requirement ensures that roads, dams, and agricultural expansions incorporate biodiversity-conscious strategies, thereby minimizing deforestation and habitat fragmentation (Young & Castro, 2021).

Over the years, global conservation efforts have significantly increased the share of land and marine regions under protection. As of 2023, an estimated 17% of terrestrial land and 8% of marine areas have been designated as protected spaces. Initiatives like the African Union's Pan-African Protected Area Network emphasize transboundary conservation, bolstering regional biodiversity management efforts. The impact of diverse Biodiversity Policy Mechanisms is provided in Table 14.

The integration of biodiversity considerations into policy frameworks is vital for achieving global environmental and economic sustainability. Case studies from India, Costa Rica, Brazil, and Africa demonstrate the effectiveness of these policies in promoting ecological resilience and sustainable development. Moving forward, conservation policies must emphasize enhanced enforcement, international collaboration, and advanced technology-driven conservation methods. By fostering adaptive and data-driven policy approaches, governments can secure biodiversity conservation while ensuring long-term economic and societal prosperity.

5.3 *Challenges in Policy Integration*

The incorporation of biodiversity into sustainable development frameworks is fraught with challenges stemming from economic limitations, political opposition, technological deficiencies, and institutional barriers. These obstacles significantly impact conservation efforts, reducing their efficiency and effectiveness. This section examines the major hindrances to biodiversity policy integration, potential remedies, and case studies that highlight successful mitigation strategies. Economic growth and biodiversity conservation frequently come into conflict, as large-scale infrastructure expansion, urbanization, and industrial agriculture prioritize immediate economic benefits over long-term ecological sustainability. The destruction of

Table 14 Impact of biodiversity policy mechanisms

Policy instrument	Key implementation region	Ecological benefit (%)	Economic benefit (%)
NBSAPs	India's national strategy	45	30
PES	Costa Rica's Forest recovery	60	40
EIAs	Brazil's Amazon regulations	50	25
Protected areas	African union conservation	55	35

forests for industrialization and intensive farming disrupts ecosystems, leading to habitat degradation and biodiversity loss (Santos et al., 2021). The Amazon Rainforest faces continuous deforestation driven by infrastructure projects, such as the BR-163 Highway in Brazil. While the road has improved trade and economic connectivity, it has also escalated deforestation rates. According to the World Resources Institute (WRI), deforestation in the region increased by 34% between 2018 and 2022, highlighting the ongoing struggle between development and conservation. Reliable biodiversity data is crucial for shaping informed policies and evaluating their effectiveness. However, many regions lack advanced monitoring systems, creating significant gaps in conservation planning and implementation.

Emerging technologies such as satellite-based remote sensing, AI-powered data analytics, and IoT-integrated environmental sensors are improving biodiversity tracking and analysis. A significant challenge to biodiversity conservation is the funding gap. Global biodiversity protection requires an estimated $700 billion annually, while current financial allocations remain between $124 and $143 billion per year, leaving a considerable shortfall. The GEF, one of the leading environmental funding organizations, has provided over $21.5 billion for biodiversity conservation projects. However, the scale of global conservation needs surpasses the available financial resources, limiting the implementation of large-scale biodiversity initiatives, particularly in developing regions.

The enforcement of biodiversity policies is often undermined by weak governance structures, corruption, and limited institutional capacity. Despite well-established regulations, many countries struggle to implement and enforce conservation laws effectively. According to a 2022 UNODC report, illegal wildlife trafficking generates an estimated $23 billion annually, intensifying biodiversity loss across Southeast Asia and Africa. The biodiversity policy mechanisms are provided in Table 15. The integration of biodiversity into sustainable development strategies remains a formidable challenge due to conflicting economic interests, inadequate funding, insufficient data systems, and weak governance structures. Case studies such as Amazon deforestation, GBIF's data initiatives, and GEF's funding programs illustrate both the barriers and opportunities in biodiversity conservation. To overcome these challenges, a comprehensive approach is needed, incorporating robust governance mechanisms, advanced technological solutions, increased financial investment, and cross-border collaboration. Policymakers must balance ecological preservation with economic development to guarantee a supportable and robust forthcoming for biodiversity conservation.

Table 15 Biodiversity policy mechanisms

Key challenge	Impact on biodiversity	Estimated financial loss ($)	Suggested solution
Economic vs. ecological priorities	Habitat destruction, species loss	$1.6 trillion annually (WRI)	Sustainable land-use planning
Lack of comprehensive data	Ineffective conservation strategies	N/A	AI and IoT-enhanced monitoring
Financial shortfalls	Limited conservation initiatives	$700 billion annual gap (UNEP)	Green finance and investments
Poor policy enforcement	Illegal logging, poaching	$23 billion lost (UNODC)	Strengthened governance frameworks

5.4 *Strategies for Effective Policy Integration*

Overcoming challenges in biodiversity policy integration requires a comprehensive approach that combines scientific innovation, governance frameworks, and community participation (Schmeller & Bridgewater, 2025). By embedding biodiversity considerations into various sectors, nations can achieve ecological resilience while maintaining economic development. Incorporating biodiversity into different industries enhances sustainability and minimizes environmental degradation. In agriculture, promoting agroforestry, permaculture, and organic farming helps maintain genetic diversity in crops and livestock, improving ecosystem stability. In urban planning, developing green infrastructure, biodiversity-friendly cities, and ecological corridors supports urban biodiversity. Singapore integrates biodiversity conservation into urban development through initiatives such as green roofs, vertical gardens, and extensive urban parks. The City in Nature Vision 2030 aims to increase green cover by 50%, improving air quality and urban ecosystem resilience. Assigning economic value to ecosystem services strengthens biodiversity conservation in policy frameworks. This transnational effort, supported by governments and private sectors, safeguards marine biodiversity in Southeast Asia's Coral Triangle, which sustains 120 million people through fisheries, coastal protection, and tourism. The initiative highlights the economic benefits of marine conservation, estimated at $2.3 billion annually in ecosystem services. Adaptive Policy Frameworks Dynamic policies that evolve based on scientific research, technological advancements, and socio-economic shifts ensure long-term conservation effectiveness.

For example: Australia's Adaptive Management of the Great Barrier Reef employs an approach to mitigate threats such as coral bleaching, pollution, and climate change. Through continuous monitoring and policy adjustments, conservation efforts remain responsive to emerging challenges, ensuring reef resilience. The impact of biodiversity integration strategies is provided in Table 16.

Effective biodiversity policy integration requires sectoral mainstreaming, economic valuation of ecosystem services, and adaptable governance models. Case studies from Singapore, Southeast Asia, and Australia demonstrate successful policy applications. Future conservation efforts should prioritize evidence-based policy

Table 16 Impact of biodiversity integration strategies

Strategy	Implementation region	Environmental benefit (%)	Economic benefit (%)
Urban biodiversity planning	Singapore's green strategy	50	30
Marine conservation	Coral triangle initiative	60	40
Adaptive policy frameworks	Great barrier reef management	55	35
Agricultural biodiversity	Agroforestry and organic farming	45	25

adaptation, cross-sectoral collaboration, and increased financial investments to ensure sustainable biodiversity management worldwide.

5.5 *Case Studies: Effective Policy Implementation*

The successful incorporation of biodiversity conservation into policy frameworks has demonstrated significant positive impacts in multiple regions worldwide. By harmonizing economic progress with environmental stewardship, various governments and institutions have implemented strategic policies that promote sustainability. This section examines key case studies that illustrate how effective biodiversity policies can drive economic, ecological, and societal benefits.

South Africa has successfully merged biodiversity conservation with economic advancement through its Biodiversity Economy Strategy (BES). This initiative encourages ecotourism, wildlife management, and bioprospecting, fostering employment opportunities while preserving ecosystems. More than 10,000 jobs have been created within biodiversity-dependent sectors. Sustainable resource utilization has strengthened conservation efforts, particularly in national parks and wildlife reserves. Ecotourism generates an estimated $10 billion annually, reinforcing biodiversity as an economic asset. Bhutan integrates biodiversity preservation into its national policy through its Gross National Happiness (GNH) Index, emphasizing environmental protection as a key pillar of national well-being. Key policies are the nation's constitution mandates at least 70% forest cover, Stringent environmental laws ensure that infrastructure projects do not adversely impact biodiversity and community forestry programs. Bhutan remains carbon negative, absorbing more CO_2 than it emits. Extensive forest coverage improves water security and mitigates the effects of climate change. The nation's commitment to sustainable agriculture and ecotourism significantly contributes to economic stability. The European Green Deal is a holistic policy initiative designed to integrate biodiversity restoration into climate adaptation and economic development. Key measures are 30% of terrestrial and marine areas in the EU are protected under conservation policies. The 2030 Biodiversity Strategy prioritizes restoring degraded ecosystems and boosting afforestation. The EU Taxonomy for Sustainable Activities encourages financial investments in biodiversity-friendly initiatives. The impacts are the EU plans to plant 3 billion trees by 2030 to enhance carbon sequestration and restore ecosystems. Over €100 billion has been allocated for green investments aimed at biodiversity conservation. Agricultural reforms target 25% of EU farmland to be converted to organic farming by 2030.

These case studies illustrate how biodiversity policy integration can successfully balance economic development with environmental preservation. South Africa's Biodiversity Economy Strategy highlights the financial viability of ecotourism and sustainable wildlife management. Bhutan's Gross National Happiness Index exemplifies how prioritizing environmental well-being can enhance national sustainability, while the European Green Deal underscores large-scale efforts in ecological restoration. For the future, global policymakers should adopt adaptive, data-driven, and community-inclusive conservation strategies to ensure sustainable biodiversity management. Increased financial commitments, international cooperation, and embedding biodiversity in national economic frameworks will be essential to maintaining ecological resilience worldwide.

5.6 Future Directions

As biodiversity continues to be impacted by climate change, technological progress, and geopolitical shifts, sustainable development policies must evolve to address these emerging challenges. Enhancing international collaboration, utilizing innovative technologies, and implementing adaptive governance frameworks will be crucial to ensuring long-term biodiversity conservation and ecosystem resilience. Biodiversity conservation often requires cross-border efforts, particularly in regions where ecosystems extend across multiple nations. Strengthened global partnerships facilitate shared resources, joint conservation strategies, and policy alignment to protect critical habitats. The CBFP unites Central African administrations, non-governmental organizations, and international agencies to endorse sustainable forestry, anti-deforestation policies, and climate adaptation measures. The impacts are that it preserves 500 million acres of tropical forest spanning six nations, provides livelihoods for 75 million people who depend on forest resources, and contributes to climate change mitigation through significant carbon absorption. Technological advancements are transforming biodiversity conservation through artificial intelligence (AI), machine learning (ML), blockchain, and remote sensing, optimizing data collection, analysis, and managing. AI and ML can be used in species identification, ecological forecasting, and automated biodiversity monitoring. Blockchain ensures transparency in conservation finance and the ethical sourcing of biodiversity-based products (Seidl et al., 2024). Remote sensing and drones provide precise data on deforestation, marine ecosystem changes, and species migration patterns.

In the Amazon Rainforest, AI-based acoustic sensors from Rainforest Connection detect illegal logging and alert authorities in real time. This system has contributed to a 30% reduction in unauthorized deforestation in protected regions. Recognizing biodiversity as a fundamental economic resource is essential for sustainable development.

Quantifying the financial value of ecosystem services encourages investment in conservation initiatives. Key Economic Strategies: Biodiversity financing and green bonds facilitate investments in conservation projects and sustainable land use. Natural capital accounting integrates biodiversity into national economic planning and financial assessments. Corporate biodiversity initiatives encourage businesses to adopt sustainable supply chain practices and reduce environmental footprints (Feger & Mermet, 2022). The various future biodiversity strategies are shown in Table 17.

The future of biodiversity policy depends on global collaboration, technological advancement, and economic transformation. Strengthening cross-border partnerships, embracing AI-driven monitoring, and embedding biodiversity into financial policies are essential for effective conservation. Case studies from the Congo Basin, the Amazon, and the Dasgupta Review highlight the practical benefits of these forward-thinking approaches. To ensure ecological sustainability, decision-makers must implement data-driven, adaptable, and internationally coordinated biodiversity strategies. Increasing conservation funding, fostering multi-sector engagement, and integrating biodiversity into global economic models will be key to securing a resilient and thriving natural world for future generations. Table 18 demonstrates the scale and impact of biodiversity dimensions, ecosystem services, and anthropogenic pressures.

Table 18 summarizes the biodiversity dimensions, ecosystem services, and anthropogenic pressures. Integrating biodiversity into sustainable development policies is both a necessity and an opportunity. By aligning biodiversity conservation with economic and social goals, policymakers can create a resilient future that benefits both people and the planet. Advanced strategies, coupled with international cooperation and innovative technologies, are essential to address the complexities of biodiversity loss in the Anthropocene era. Biodiversity and ecosystem conservation are cornerstones of a sustainable future. As humanity faces pressing challenges like climate change and resource depletion, maintaining ecological balance is paramount. Collaborative efforts involving governments, communities, researchers, and industries are essential to safeguard the planet's rich biological heritage. By adopting holistic and inclusive strategies, we can ensure that biodiversity thrives for generations to come, aligning with the broader vision of a green society and sustainable development. Future research must prioritize predictive models for biodiversity under climate change scenarios, while policy efforts must incentivize large-scale restoration and conservation projects globally.

Table 17 Future biodiversity strategies

Strategy	Focus area	Estimated impact (%)	Global implementation (%)
International partnerships	Cross-border conservation	50	40
AI and ML for biodiversity tracking	Species monitoring and forecasting	60	35
Blockchain for conservation	Ethical resource management	45	30
Natural capital valuation	Economic integration of biodiversity	55	25

Table 18 Impact of biodiversity dimensions

Category	Subcategory	Details
Dimensions of biodiversity	Genetic diversity	Genetic variation within species ensures adaptability to environmental changes and disease resistance. Example: Diverse wheat varieties have shown increased resistance to stem rust, averting losses in global wheat production worth billions annually
	Species diversity	Encompasses species richness (total species count) and evenness (relative abundance). Approximately 1.74 million species are documented globally, with estimates suggesting 8.7 million species exist on earth. Keystone species like sea otters regulate prey populations, maintaining ecosystem balance
	Ecosystem diversity	Variety of ecosystems includes tropical rainforests (covering 6% of earth's surface but harboring over 50% of all species) and coral reefs (supporting 25% of marine life despite covering less than 1% of the ocean floor)
Ecosystem services	Provisioning services	Provide food, water, and medicines. For example, fisheries contribute 20% of animal protein to the diets of over 3 billion people globally. The global market value of plant-derived medicines is estimated at over $150 billion annually
	Regulating services	Include climate regulation (forests absorb 2.6 billion metric tons of CO_2 annually), water purification (wetlands remove up to 70% of nitrogen in water), and pollination (75% of global crops depend on pollinators, contributing $235–$577 billion annually)
	Cultural services	Generate recreation and spiritual benefits. Example: US national parks contribute over $40 billion annually to the economy. Sacred groves in India are culturally significant and preserve biodiversity
	Supporting services	Services including soil formation (critical for agriculture, saving $3.5 trillion annually in avoided costs) and nutrient cycling (nitrogen fixation by bacteria supports agricultural productivity)
Anthropogenic pressures	Habitat loss	Global deforestation rate is ten million hectares annually (FAO, 2023), reducing habitats for species like orangutans. Urban expansion and agricultural monocultures fragment ecosystems, limiting gene flow and species survival
	Climate change	Rising temperatures have caused species like the American pika to migrate to higher altitudes. Coral bleaching events linked to climate change have led to a 50% decline in coral cover in the great barrier reef since 1995

(continued)

Table 18 (continued)

Category	Subcategory	Details
	Pollution	Plastic pollution affects over 90% of seabirds, with eight million metric tons of plastic entering oceans yearly. Eutrophication has created over 400 hypoxic dead zones globally, covering an area of 245,000 km^2, affecting marine biodiversity
	Overexploitation	Overfishing has caused the collapse of major fish stocks like Atlantic cod in the 1990s. Illegal wildlife trade, valued at \$7–\$23 billion annually, threatens species like pangolins and rhinos. Logging, both legal and illegal, accounts for significant biodiversity loss in tropical forests
	Invasive species	Invasive alien species cause global damages exceeding \$1.4 trillion annually. Example: Water hyacinth reduces oxygen levels in aquatic ecosystems, threatening native biodiversity. Brown tree snakes in Guam have driven multiple bird species to extinction

References

Beillouin, D., Ben-Ari, T., Malézieux, E., Seufert, V., & Makowski, D. (2021). Positive but variable effects of crop diversification on biodiversity and ecosystem services. *Global Change Biology, 27*(19), 4697–4710.

Belaire, J. A., Higgins, C., Zoll, D., Lieberknecht, K., Bixler, R. P., Neff, J. L., & Jha, S. (2022). Fine-scale monitoring and mapping of biodiversity and ecosystem services reveals multiple synergies and few tradeoffs in urban green space management. *Science of the Total Environment, 849*, 157801.

Bredemeier, B., Herrmann, S., Sattler, C., Prager, K., van Bussel, L. G., & Rex, J. (2022). Insights into innovative contract design to improve the integration of biodiversity and ecosystem services in agricultural management. *Ecosystem Services, 55*, 101430.

Brown, M. B., Morrison, J. C., Schulz, T. T., Cross, M. S., Püschel-Hoeneisen, N., Suresh, V., & Eguren, A. (2022). Using the conservation standards framework to address the effects of climate change on biodiversity and ecosystem services. *Climate, 10*(2), 13.

Cardinale, B. J., et al. (2012). Biodiversity loss and its impact on humanity. *Nature, 486*, 59–67. https://doi.org/10.1038/nature11148

Ekka, P., Patra, S., Upreti, M., Kumar, G., Kumar, A., & Saikia, P. (2023). Land degradation and its impacts on biodiversity and ecosystem services. In *Land and environmental management through forestry* (pp. 77–101).

FAO. (2023). *Loss and damage and agrifood systems—Addressing gaps and challenges*. FAO.

Feger, C., & Mermet, L. (2022). New business models for biodiversity and ecosystem management services: Action research with a large environmental sector company. *Organization & Environment, 35*(2), 252–281.

González-García, A., Palomo, I., González, J. A., García-Díez, V., García-Llorente, M., & Montes, C. (2022). Biodiversity and ecosystem services mapping: Can it reconcile urban and protected area planning? *Science of the Total Environment, 803*, 150048.

Gotama, R., Baker, D. M., Guibert, I., McIlroy, S. E., & Russell, B. D. (2024). How a coastal megacity affects marine biodiversity and ecosystem function: Impacts of reduced water quality and other anthropogenic stressors. *Ecological Indicators, 160*, 111683.

Great Barrier Reef Marine Park Authority (2021). Available at: https://elibrary.gbrmpa.gov.au/jspui/bitstream/11017/3838/2/Science-and-Knowledge-Needs-Snapshot.pdf [Accessed 05 March. 2025].

Hua, F., Bruijnzeel, L. A., Meli, P., Martin, P. A., Zhang, J., Nakagawa, S., et al. (2022). The biodiversity and ecosystem service contributions and trade-offs of forest restoration approaches. *Science, 376*(6595), 839–844.

Ige, O. E., Aliu, F. P., Omole, A. E., & Alabi, O. O. (2025). Ecotoxicological impacts of pesticide. In *The interplay of pesticides and climate change: Environmental dynamics and challenges* (pp. 231–252). Springer Nature Switzerland.

IPCC. (2023). Summary for policymakers. In H. Lee & J. Romero (Eds.), *Climate change 2023: Synthesis report. Contribution of working groups I, II and III to the sixth assessment report of the intergovernmental panel on climate change* (pp. 1–34). IPCC.

Mongabay (2024). Available at: https://news.mongabay.com/2024/11/help-farmers-adopt-agroecology-to-protect-biodiversity-and-climate-commentary/[Accessed 06 March. 2025

Levis, C., Flores, B. M., Campos-Silva, J. V., Peroni, N., Staal, A., Padgurschi, M. C., & Clement, C. R. (2024). Contributions of human cultures to biodiversity and ecosystem conservation. *Nature Ecology & Evolution, 8*(5), 866–879.

Li, P., Li, D., Sun, X., Chu, Z., Xia, T., & Zheng, B. (2022a). Application of ecological restoration technologies for the improvement of biodiversity and ecosystem in the river. *Water, 14*(9), 1402.

Li, S., Yu, D., Huang, T., & Hao, R. (2022b). Identifying priority conservation areas based on comprehensive consideration of biodiversity and ecosystem services in the Three-River headwaters region, China. *Journal of Cleaner Production, 359*, 132082.

NASA (2020). Earth's changing climate. NASA Earth Observatory. Available at: https://earthobservatory.nasa.gov [Accessed 05 March. 2025].

Oguh, C. E., Obiwulu, E. N. O., Umezinwa, O. J., Ameh, S. E., Ugwu, C. V., & Sheshi, I. M. (2021). Ecosystem and ecological services; need for biodiversity conservation-a critical review. *Asian Journal of Biology, 11*(4), 1–14.

Pereira, H. M., Martins, I. S., Rosa, I. M., Kim, H., Leadley, P., Popp, A., & Alkemade, R. (2024). Global trends and scenarios for terrestrial biodiversity and ecosystem services from 1900 to 2050. *Science, 384*(6694), 458–465.

Phillips, S. J., Anderson, R. P., & Schapire, R. E. (2018). Maximum entropy modeling of species geographic distributions. Ecological Modelling, 190(3–4), 231–259. https://doi.org/10.1016/j.ecolmodel.2005.03.026

Qu, X., Li, X., Bardgett, R. D., Kuzyakov, Y., Revillini, D., Sonne, C., & Delgado-Baquerizo, M. (2024). Deforestation impacts soil biodiversity and ecosystem services worldwide. *Proceedings of the National Academy of Sciences, 121*(13), e2318475121.

Rosenfield, M. F., Brown, L. M., & Anand, M. (2022). Increasing cover of natural areas at smaller scales can improve the provision of biodiversity and ecosystem services in agroecological mosaic landscapes. *Journal of Environmental Management, 303*, 114248.

Santos, J. L., Moreira, F., Ribeiro, P. F., Canadas, M. J., Novais, A., & Lomba, A. (2021). A farming systems approach to linking agricultural policies with biodiversity and ecosystem services. *Frontiers in Ecology and the Environment, 19*(3), 168–175.

Schmeller, D. S., & Bridgewater, P. (2025). Transformative change in a nexus world: Product and process in the eleventh plenary of the intergovernmental science-policy platform on biodiversity and ecosystem services (IPBES-11). *Biodiversity and Conservation, 34*, 1–7.

Schmid, B., & Schöb, C. (2022). Biodiversity and ecosystem services in managed ecosystems. In *The ecological and societal consequences of biodiversity loss* (pp. 211–231).

Seidl, A., Cumming, T., Arlaud, M., Crossett, C., & van den Heuvel, O. (2024). Investing in the wealth of nature through biodiversity and ecosystem service finance solutions. *Ecosystem Services, 66*, 101601.

Takacs, V., & O'Brien, C. D. (2023). Trends and gaps in biodiversity and ecosystem services research: A text mining approach. *Ambio, 52*(1), 81–94.

University of Oxford. (2024). Genetic diversity and adaptation. In J. Smith (Ed.), *Evolutionary biology and conservation* (pp. 150–175). Oxford University Press.

World Wildlife Fund, (2020). Available at: https://www.worldwildlife.org/threats/deforestation-and-forest-degradation [Accessed 07 March. 2025].

World Wildlife Fund, (2024). Available at https://www.worldwildlife.org/initiatives/wildlife-conservation [Accessed 08 March. 2025].

Weiskopf, S. R., Myers, B. J., Arce-Plata, M. I., Blanchard, J. L., Ferrier, S., Fulton, E. A., & Rosa, I. M. (2022). A conceptual framework to integrate biodiversity, ecosystem function, and ecosystem service models. *Bioscience, 72*(11), 1062–1073.

Wired. (2025). The impact of biodiversity loss on ecosystem resilience. In H. Carter (Ed.), *Wired Science* (pp. 250–275). Springer.

Wu, S., Chen, Y., Hao, C., Liu, K., Zhang, W., & Zhang, L. (2022). Promoting biodiversity conservation requires a better understanding of the relationships between ecosystem services and multiple biodiversity dimensions. *Frontiers in Ecology and Evolution, 10*, 891627.

WWF. (2020). Australian bushfires and biodiversity loss. In S. Patel (Ed.), *Wildfire impacts on ecosystems* (pp. 300–325). Springer.

Young, C. E. F., & Castro, B. S. (2021). Financing mechanisms to bridge the resource gap to conserve biodiversity and ecosystem services in Brazil. *Ecosystem Services, 50*, 101321.

Zhang, R., Wang, J., & Niu, S. (2021). Toward a sustainable grazing management based on biodiversity and ecosystem multifunctionality in drylands. *Current Opinion in Environmental Sustainability, 48*, 36–43.

Financing Sustainability: Economic Models and Investment

Abdolhossein Zameni Pejman

1 Introduction

Sustainability, as defined by the Brundtland Commission (1987), means meeting our current needs without preventing future generations from meeting theirs. In practice, sustainability policies aim to consider how present-day activities, whether they involve business operations, government initiatives, or individual choices, shape people's lives, the environment, and the economy over time. Awareness has grown in recent decades about how human behavior contributes to climate change, biodiversity loss, and pollution, prompting governments and industries around the world to develop more eco-friendly regulations and corporate policies. Many organizations have responded by investing in technologies and practices that reduce their environmental impact and protect natural resources for the years ahead.

Within financial research, the concept of sustainable finance has become both dynamic and fast-paced. Migliorelli (2021) points out that varying definitions, ideas, and industry standards have turned this field into a complex landscape. Such diversity can make it harder to establish a common theoretical footing, which is why some scholars have suggested reframing sustainable finance as "finance for sustainability." They argue that this would help focus the discussion on shared priorities and clearer guidelines.

A detailed study by Granier et al., (2021), which reviewed sustainability-related finance research published between 1981 and 2018, identified five major themes at the heart of scholarly debate. These themes include the performance of socially responsible investment (SRI) funds, the role of corporate social responsibility, the behavior of responsible companies and associated stock market indices, the strategies of investors in sustainable markets, and the part played by pension funds in

A. Zameni Pejman (✉)
Management Faculty, University of Niagara Falls Canada, Niagara Falls, ON, Canada
e-mail: pejman.zameni@unfc.ca

A. Chakir et al. (eds.), *Green Society, Environmental Strategies and Sustainable Development*, https://doi.org/10.1007/978-3-031-94086-6_10

sustainable development. This body of work highlights the many layers of sustainable finance and shows why it requires approaches that blend environmental and social outcomes with strong financial returns (Granier et al., 2021; Migliorelli, 2021).

Sustainability itself rests on three main pillars. One pillar is environmental sustainability, which seeks to conserve resources, protect ecosystems, and prevent irreversible damage to the planet. Another is economic sustainability, centered on establishing long-term economic stability in ways that do not drain essential resources or harm future prosperity. The third pillar is social sustainability, involving fair treatment, healthy living conditions, and social well-being for current populations and those yet to come (Purvis et al., 2019). In essence, sustainability calls for maintaining a balanced relationship between humans and nature for generations (EPA, 2024), taking care to use resources responsibly and keep social, economic, and environmental factors in harmony.

This broad vision aligns with the 2030 Agenda for Sustainable Development, adopted by the United Nations (UN) member states in 2015 (Department of Economic and Social Affairs Sustainable Development, 2025). The agenda sets out 17 Sustainable Development Goals (SDGs), inviting every country, no matter its stage of economic growth, to work together on eliminating poverty, improving healthcare and education, driving economic progress, mitigating climate change, and protecting ecosystems on land and at sea. Because these challenges connect closely with one another, the effort to end poverty must happen side by side with broader initiatives to strengthen society and safeguard the environment.

In the business world, sustainability does not stop at ecological concerns. Harvard Business School (Spiliakos, 2018) stresses two major indicators for measuring responsible business practices: one is environmental impact, covering how a firm's operations affect the planet, and the other is social impact, referring to the ways a company influences communities and society at large. Companies aim to be positive forces in at least one of these areas, acknowledging that sustainability can benefit both the environment and people. Over time, the notion of corporate sustainability has grown into an ethical stance, largely thanks to a growing public awareness about the long-term drawbacks of chasing quick profits at the cost of the planet's health and social equity. This outlook urges companies to balance short-term performance with enduring commitments that uphold ecological integrity and social welfare. Possible actions include lowering greenhouse gas emissions, boosting energy efficiency, practicing responsible sourcing, and managing waste in ways that lower pollution. Many organizations are now making explicit pledges to achieve sustainability targets, such as cutting packaging waste or reducing carbon emissions by a chosen percentage within a set time frame and have made these commitments publicly known. Walmart, for instance, plans to reach zero emissions in its operations by 2040 (Schneider Electric Blog, 2023). Morgan Stanley hopes to achieve net-zero "financed emissions" by 2050, and Google aims to run on carbon-free energy by 2030 (Jackson, 2024; Derrick, 2024). These moves reflect a larger trend of major companies weaving sustainability into their fundamental business strategies.

The push toward greener methods is particularly apparent in the energy sector. Instead of dedicating all resources to expanding fossil fuel extraction, many energy firms now set explicit goals for growing the share of power they derive from wind, solar, and hydroelectric sources. Yet the growth of eco-friendly policies can also lead to controversy, especially when companies appear to be "greenwashing." This term refers to giving the impression of environmental responsibility as a marketing tool rather than truly adopting meaningful changes. Businesses accused of greenwashing are often found moving some parts of their operations to locations with fewer regulations and lower labor costs, a practice that can mask the real environmental and social impacts of their production processes.

Research supports the view that sustainability strategies sometimes intertwine with decisions to shift operations across borders. In a study of data from 1080 multinational companies, Lartey et al. (2021) found that environmental sustainability efforts played a significant role in whether firms outsourced work to other countries. This relationship showed up in both developing and advanced markets, highlighting how global business strategies and sustainability goals can interact in complex ways.

1.1 Challenges Surrounding Business Sustainability

Transitioning to sustainability can be challenging. The Santa Fe Institute (2020) explains that companies often face three major problems when they try to improve their environmental performance. First, it can be hard to figure out a single company's exact impact on the environment because many factors shape this impact, and data can be scattered. Second, it is difficult to compare the environmental effects of different activities, since each type of process or product can create its own risks. Third, it is not easy to predict how people and businesses in the economy will react when rules or incentives change. Despite these problems, many investors see sustainability as a "fundamental" part of their strategies. Surveys show that around half, or even more, of investors care about sustainability when deciding where to put their money. However, there is no universal agreement on this topic. For example, in July 2021, Security Exchange Commission (SEC) Commissioner Hester Peirce (2021) raised concerns that making rules for environmental, social, and governance (ESG) disclosures might go beyond the SEC's authority and could upset financial systems (Deloitte, 2024). Peirce argued that sustainability metrics are "inherently political," designed to guide money toward certain businesses, and this could clash with the SEC's neutral role (ESGDIVE, 2024). She felt that adopting these measures would break with the agency's traditional stance of being fair and impartial. In another example, Eiji Hirano, who once led Japan's Government Pension Investment Fund, suggested in Bloomberg News interviews that there might be a bubble in ESG investing. He believed the fund should take a closer look at its own ESG strategies before continuing on the same path.

1.2 Benefits of Business Sustainability

When done well, sustainability can bring many rewards for companies, both ethically and financially. Using resources in a responsible way can help a business stay strong over time, and cutting waste or pollution can reduce costs. For example, putting in energy-saving lights or water systems can lower monthly bills and build a good reputation. Companies that focus on sustainability may also get tax breaks or other government benefits. Investors increasingly pay attention to such efforts. A study from HEC Paris (Bonnefon et al., 2019) found that shareholders attach high value to ethical behavior. In fact, they might pay $0.70 more per share if a company donates $1 or more per share to charity. At the same time, the study noted that firms seen as having a bad social impact can see their worth go down. Insights from interviews with leaders at 43 global investment firms led Harvard Business Review (Eccles & Klimenko, 2019) to say that the idea some business leaders hold, that ESG issues are not widely accepted by investors, is now outdated. This shift in investor views appears in the strong rise of the Principles for Responsible Investment, a United Nations-backed group that pushes for ESG factors in investment decisions. In 2006, it had 63 signatories managing $6.5 trillion in assets; by 2018, that grew to 1715 signatories managing $81.7 trillion (Signatory directory | PRI, 2025). This growth suggests that more investors are tying their decisions to ethical, social, and environmental considerations.

1.3 Creating a Sustainable Business Strategy

Many companies now try to weave sustainability into their core plans. They often do this the same way they handle other important strategies. The first step is to spot a clear problem, such as producing too much waste or having hiring practices that harm local communities. Then, the company sets clear goals and decides how to measure progress. For example, it might choose to cut its carbon footprint by a certain amount or aim for a target to hire more people from diverse backgrounds. These measurable goals show whether the plan is working. Once the plan is in place, the company carries it out and checks the results. As the business grows and changes, it may need to update its targets and methods. Even with a solid plan, companies can run into common obstacles. One is the gap between what they know and what they actually do: executives may talk about sustainability as a key value but fail to take real steps. Another challenge is mixing up what is required by law with what is a choice for better competitiveness. Although improving sustainability can boost how a firm competes in the marketplace, it is not the same as simply obeying rules (Haanaes & Olynec, 2022).

1.4 Real-World Example

A well-known success story is Unilever. This company, which owns brands like Dove, Axe, Ben & Jerry's, and Hellmann's, introduced the Unilever Sustainable Living Plan in 2010. This 10-year plan aimed to cut the environmental effects of its brands while also creating fairer job conditions. By the end of this plan, Unilever showed real gains in shrinking its environmental footprint and improving its bottom line. Between 2008 and 2018, it saved over 1 billion euros by saving water and energy. It also boosted opportunities for women, making Unilever one of the top workplaces for students in 50 countries (Unilever, 2010, 2020a, 2020b, 2021).

Different groups rank sustainable companies, each using its own criteria. Corporate Knights, a research firm in Canada, puts out a yearly list of the 100 most sustainable businesses. Top names have included Vestas Wind Systems and Chr. Hansen Holding from Denmark, Autodesk Inc. from the United States, Schneider Electric from France, and City Developments from Singapore (EricssonSCR, 2024; McCormick & Company, 2024; Mccarthy, 2024). Conversely, many firms depend on resources that cannot be replaced fast enough to match how quickly they are used, such as fossil fuels. That makes these products unsustainable. Some resources, timber from forests, fish stocks, coral reefs, and other wildlife, can be sustained if people harvest them at levels that allow them to grow back. As consumers learn more about the damage caused by unchecked growth, a rising number of businesses want to cut their harm on both the planet and local communities. Adopting sustainable methods also helps them meet rising customer interest in buying from responsible brands.

In the next sections, this chapter covers several important topics. It explains why sustainability and finance are closely linked and stresses the need to change the way we think about money and investments. It looks at theories that support sustainable financing and shows how to include environmental concerns in financial planning. It also discusses older economic views that often left out ecological impacts, leading to harmful practices. Different investment paths, like impact investing, green bonds, and public-private partnerships, are explored, with examples of projects that bring both returns and social or environmental benefits. This chapter underlines the need for teamwork among governments, businesses, and nonprofits to create a financial system that supports responsible growth. At the same time, it recognizes barriers such as tough regulations, low transparency, and tricky risk assessments in green finance. To handle these issues, the authors suggest finding more public support for eco-friendly ventures, setting clear ways to measure environmental impacts, and boosting everyone's knowledge about sustainable finances. Finally, this chapter ends by stressing that funding sustainability is urgent and calls on all groups to help shape a future where investments truly care for the planet.

2 Intersection of Sustainability and Finance

The financial industry plays a major role in promoting and funding sustainability projects. It can help pay for research in renewable energy and support companies that follow ethical and sustainable labor standards. Sustainable finance involves making investment decisions by considering environmental, social, and governance (ESG) factors in business activities or projects. The environmental part focuses on problems like climate change and the need to use natural resources wisely. The social side covers human and animal welfare, fair consumer policies, and diverse hiring practices. Meanwhile, governance deals with how organizations are managed, how they treat their workers, and how they set compensation in both public and private sectors. By looking at these ESG factors, sustainable finance not only pushes positive change but also aligns economic activities with society's broader goals. This helps create a global economy that is both responsible and sustainable.

The growth of sustainable business methods is changing the financial profession, pushing experts to gain new skills in areas like ESG reporting and analysis. Today, accounting and finance staff often have to keep track of things like greenhouse gas emissions, differences in pay between men and women, and compliance with ethical standards. They also need to measure the risks of renewable energy and figure out which sustainability investments make sense. While these tasks can be hard, they also open new doors. Professionals who learn these skills can increase their organization's value, grow their careers, and add to global sustainability efforts. By embracing these new abilities, financial specialists can turn traditional roles into powerful ways to make a difference, achieving both personal success and positive impacts on the environment and society.

At the same time, the finance world is shifting toward "stakeholder capitalism." This means that companies have to think about the needs of many different groups, not just shareholders. As a result, they must include environmental, social, and governance issues in their capital spending decisions. Employees, local communities, customers, regulators, and even the natural world are now part of the corporate decision process.

Sustainable finance has become important for two main reasons. First, there is a growing recognition that businesses should account for all types of capital, including the health of our ecosystems and wildlife. This is especially clear given the drastic drop in wildlife populations since 1970. Many believe companies need to understand how much they rely on nature and how their actions affect it. One example is the Climate Action 100+, in which investors managing $54 trillion are pressuring top polluters to make real changes. Meanwhile, regulators such as the SEC are strengthening ESG reporting requirements, showing that sustainable practices are a top concern.

Second, investors want more openness and responsibility from companies. The rapid growth of ESG-branded assets, which could reach $53 trillion by 2025, reflects this rising interest. Women and younger investors, in particular, often look for clear ethical and social benefits before putting their money into a firm. Large investment companies, which manage a big share of publicly listed stocks, have also become more active in improving the way corporations work. One famous example was the 2021 vote against some of Exxon Mobil's directors. In this new financial world, focusing on ESG issues is crucial for businesses that want to attract investors and remain strong over time.

2.1 *Sustainability and Green Financing*

Green financing is a major change in how money is directed within our economy. It aims to move capital, coming from banks, micro-credit, insurance, and investment, to projects that support sustainable development. This includes efforts by government agencies, private companies, and nonprofit groups to encourage an environmentally friendly type of economic growth. Three main principles guide green financing. The first is improved risk management, which means companies and investors should identify and reduce the environmental and social risks linked to their activities. The second is finding projects that can deliver both good financial returns and positive environmental outcomes at the same time. The third is greater accountability, where there is a stronger focus on transparent and responsible financial decisions.

Several methods can help boost green financing. One approach is to update national policies, so they motivate companies to invest in sustainable projects. Another is to align government incentives with green objectives, making it more appealing for businesses to adopt eco-friendly models. More involvement across different economic sectors, along with strong support from public agencies, can also speed up green financing. Investing in cleaner technology encourages the growth of green economies that use resources responsibly. Expanding green bonds is another way to raise funds specifically for environmental aims. By using these tools, green financing can build a global economy that mixes financial progress with the protection of nature.

The United Nations Environment Programme (UN Environment) leads efforts to guide worldwide financial systems toward the 2030 sustainable development agenda. It focuses on directing money to meet the Sustainable Development Goals (SDGs) and recognizes that financial markets have great power in shaping future consumption and the condition of our ecosystems. UN Environment's green financing work concentrates on three broad areas. The first is creating a supportive setting for sustainability by helping governments set policies that encourage green finance. The second is developing strong public–private partnerships, including the use of green bonds. The third is building skills at a local level, such as by offering small loans for community-led projects.

Through its resource efficiency program, UN Environment offers many services to different nations, such as policy reviews, help in forming sustainable finance strategies, and advice to central banks and regulators. It aims to reinforce local financial markets and promote policy actions across multiple countries. The organization also brings policymakers and business leaders together to create a green economy, hoping these policy measures will inspire both government bodies and private investors to participate. Its inclusive strategy involves partnerships with banks, investors, micro-credit providers, insurance companies, and public agencies, all working together to shape a more sustainable and fair financial system (Climate Bonds Team, 2018; ICMA, 2023; Janssens, 2024; UN Environment Programme, 2025a).

2.2 *Green Finance Mechanisms and Information Technology's (IT) Role in Sustainable Development*

Green finance relies on several tools to fund eco-friendly projects. One is the use of green bonds, which are forms of debt meant for environmental work. These bonds give investors returns on their money while directing funds to projects like renewable energy or wildlife conservation. Another tool is green loans, which offer more favorable terms to groups that focus on reducing carbon footprints or improving efficiency in places like offices and factories. There are also sustainable investment funds, where money is pooled to support businesses and projects that meet ESG standards. These funds give people and institutions a chance to earn financial returns while backing responsible activities.

In advancing the Sustainable Development Goals (SDGs) and promoting green finance, information technology plays a large role by offering solutions that improve clarity, efficiency, and overall impact. Through integrated reporting systems, IT can blend financial and sustainability data, providing more accurate and trustworthy results. Technology like artificial intelligence (AI) can examine huge amounts of data, helping to predict risks and track how well a project is doing. Strong data analytics also give a clearer picture of the true effects of green investments, making it simpler to measure both their successes and challenges. In addition, devices linked through the Internet of Things (IoT) can constantly monitor real-world conditions, like pollution levels or resource use, helping adjust strategies as needed.

Technology's contributions do not stop there. Advanced data management techniques help process large volumes of environmental and financial information, refining how we measure and report sustainability. Blockchain systems can provide permanent, transparent records of transactions such as carbon credit trades, ensuring accountability and building trust. AI can also create forward-looking models for climate risks or help optimize investment portfolios to meet certain sustainability targets. Digital platforms make it easier for individuals to invest in green projects, and crowdfunding sites can give smaller initiatives a way to find financial backing. In addition, IoT devices allow real-time checks on environmental changes, and cloud computing makes it possible for many partners to work together on the same sustainability project while cutting down on energy use.

By embracing these technologies, green finance can channel funds to projects that have the best potential to protect nature and foster long-term economic growth. This combination of financial innovation and digital tools helps create a greener financial landscape, showing how we can balance the need for economic development with the health of the planet.

3 Theoretical Frameworks

In this section, we look at the theories that help blend sustainability into financial processes. These theories have appeared to explain and guide the way we include environmental, social, and governance (ESG) factors in finance. Six key ideas stand out in this area: priority theory, peer emulation theory, life span theory, system disruption theory, positive signaling theory, and resource theory. These models aim to give academics, policymakers, economists, researchers, and students a useful set of tools for analyzing and improving sustainable finance. They have grown in response to the rising importance of sustainability in financial decision-making and the need for a stronger theoretical base (Ozili, 2022).

3.1 Priority Theory of Sustainable Finance

Priority theory suggests that the amount of work and resources that economic actors devote to sustainable finance depends on how important they see it compared to other goals in a country or region. This sense of importance can be judged by how much effort different groups coordinate, how quickly they come to an agreement, and how fast they take action. When sustainable finance tops the list of things to do, financial institutions are more likely to build it into their core business practices. Still, focusing on sustainable finance can create tough choices for some countries, especially those with urgent development needs. Priority theory admits that objectives can shift over time and that highlighting sustainability does not guarantee success. Even so, it helps us see how the value placed on sustainability influences the steps economic actors take (de Souza Cunha et al., 2021; Kuhn, 2022; Krüger et al., 2016; Setyowati, 2023).

3.2 The Peer Emulation Theory of Sustainable Finance

Peer emulation theory says that in the absence of uniform standards for sustainable finance, economic players tend to copy the actions and policies of groups they admire. This often happens when those groups share similar views on social, political, or economic issues, including beliefs about climate change. One benefit is that agents can quickly adopt tested policies without spending too much time and money creating their own. Over time, they can improve on what they have copied. However, there is a risk that imitation may limit creativity and might not work well in places where the markets, regulations, or political climate differ. Despite these weaknesses, peer emulation theory shows us how sustainable finance practices can spread through shared models, especially when agents look to successful peers for ideas (Ditlev-Simonsen & Midttun, 2011).

3.3 The Life Span Theory of Sustainable Finance

Life span theory is adapted from Vernon's product cycle hypothesis (Vernon, 1979). It views sustainable finance offerings, such as green bonds, as products that move through stages of introduction, growth, maturity, and decline. Economic actors gauge how long these products will stay relevant and base their commitment on whether they think these products will fade soon or bring long-lasting change to the market. If a company believes a sustainable finance tool will quickly disappear, it may only offer limited support. But if it sees a lasting shift, it may invest heavily in that tool. This theory highlights how investor expectations can shape the demand for sustainable finance and shows that personal views on product life cycles can be uncertain, sometimes resulting in mistakes about how long these products will remain in use.

3.4 System Disruption Theory of Sustainable Finance

System disruption theory proposes that pushing sustainable finance goals can shake up established financial systems and businesses that rely on older practices. The resulting turmoil might cause these groups to resist or refuse to join the shift toward sustainability if they see more costs than benefits. The theory points out that open sharing of information can help people make more informed choices about whether and how to embrace sustainable finance. Yet, this theory does not mean we always have to tear down the traditional system. In some cases, sustainable finance can work as a smaller part of the existing market or alongside standard tools, letting businesses pick the ESG-friendly options that suit them (Gutterman, 2024; Ozili, 2021; Krüger et al., 2016).

3.5 Positive Signaling Theory

Positive signaling theory suggests that economic actors often disclose their commitment to sustainability in order to draw support from outsiders. Such disclosures may appear as public statements or extra details in annual reports. Firms might announce a new line of green bonds to catch the attention of investors who care about sustainability, while governments might highlight national sustainability policies to draw foreign aid or technical help. Although this openness can lower the information gap between investors and organizations, talking about sustainable finance does not guarantee actual follow-through. Some might over-promise to distract from poor financial results linked to polluting activities. As a result, it is important for observers to ensure real actions back up the positive signals (Quatrini, 2021; Park, 2018).

3.6 Resource Theory of Sustainable Finance

Resource theory argues that countries move toward sustainable finance at different speeds based on the human-made resources they have. Such resources might include healthy foreign reserves, budget surpluses, stable financial systems, advanced technology, strong oversight, and solid climate monitoring. Countries with better resources can sometimes reach their sustainable finance goals more quickly than those without them. This outlook allows each nation to address its sustainability aims at a pace that fits its abilities, but it also has a downside. Some might use weak resource levels as an excuse to avoid trying, and it could lead to unfair treatment of nations that lag behind in building these assets. Even so, resource theory is useful in showing why progress in sustainable finance differs around the world and underscores the need for approaches that match each country's circumstances.

As sustainable finance continues to grow and change, these theories will likely expand to cover new developments and challenges. They help us understand how people act and make choices around green investments and encourage further research into this evolving field. This ongoing process ensures that scholars and practitioners have ample chances to refine current ideas and discover new ones, contributing to a well-rounded understanding of sustainable finance.

4 Traditional Economic Perspectives

Traditional economic perspectives often overlook how economic activities affect the natural world, and this can lead to harmful, short-sighted practices. One common problem is a narrow focus on economic flows while ignoring ecological elements. For instance, the circular flow diagram omits details about extracting raw materials, creating waste, and causing pollution (L'équipe Manutan, 2024). As a result, it offers an incomplete picture of how the economy interacts with the environment.

A second weakness is the tendency to overlook natural capital, treating resources and ecosystem services as if they were free or infinite. This oversight allows critical natural assets to be used faster than they can recover, threatening the long-term stability of the economy (Kumar, 2017). Because these resources are often not given any formal monetary value, there is little incentive to protect them.

A third concern involves market failures. Traditional models often fail to address external costs that harm the environment, such as pollution and resource depletion (Brown, 2025). When the real costs of these impacts are not reflected in prices, economic activity can become inefficient and environmentally damaging.

Another issue is the growth-centric approach embraced by standard economic thinking. Models that single-mindedly promote endless expansion seldom examine the toll on ecosystems or consider whether constant growth can be sustained (Davidson, 2000). This approach can drain resources and weaken nature's capacity to support future generations.

Valuing the environment accurately is also a challenge. There are limited valuation methods for environmental goods and services that are not traded in open markets (Grantham Research Institute on Climate Change and the Environment, 2014). Because many ecological assets lack a direct price tag, they tend to be overlooked or undervalued when making important economic decisions.

In addition, inadequate national accounting can hide the depletion of natural resources or the costs of environmental damage. Traditional measures like GDP do not show how economic gains might come at the expense of forests, water supplies, or other vital forms of natural capital. This can mislead policymakers and the public about the real health of an economy (Change Oracle, 2022; Steenmans & Lesniewska, 2023).

Because of these shortcomings, many experts now believe that conventional economic models need to evolve. They point out that the economy depends fundamentally on the planet's life-support systems and that many natural resources are finite. By failing to capture that reality, traditional models can encourage unsustainable behaviors. Consequently, there is growing interest in alternative approaches that more effectively incorporate environmental concerns and support sustainable development.

5 Investment Strategies

Investment strategies that advance sustainability are becoming increasingly popular. These approaches give investors a chance to align their portfolios with environmental and social goals while still pursuing financial returns. One strategy is impact investing, where investors seek measurable benefits for people or the planet along with profits. This often involves focusing on areas like affordable housing, access to healthcare, and environmental protection. Impact investing has grown from a niche idea to a sizable market, with worldwide assets under management reaching about $715 billion in 2020. Research shows that it can be especially appealing to younger investors who want to make a difference while also earning returns. During certain economic downturns, impact investing has performed as well as- or even better than- traditional investment approaches (Kerch, 2024).

Green bonds are another important method for financing environmental projects. These debt instruments raise money for initiatives such as clean energy, energy efficiency, or sustainable transportation. Since 2013, the global green bond market has expanded rapidly, going from $10.9 billion in issuance to over $500 billion in 2021. Many large institutional players, including pension funds and sovereign wealth funds, have contributed to this boom. Clear guidelines for reporting and verification have made green bonds more credible, and experts predict that the issuance of all types of sustainable bonds (green, social, sustainable, and sustainability-linked) will top $1 trillion by 2025. The European Union is also introducing the EU Green Bond Standard, a voluntary label that requires rigorous reporting and the commitment of at least 85% of bond proceeds to sustainable activities in line with EU Taxonomy guidelines (Kerch, 2024).

A related practice is ESG integration, where environmental, social, and governance factors are evaluated alongside a company's financial performance. The goal is to identify risks and opportunities linked to issues like climate change, labor practices, or board diversity, which can influence a firm's long-term success. Another form of sustainable investing is themed investing, where funds concentrate on specific areas like clean energy, sustainable agriculture, or water management. Studies increasingly show that companies with strong ESG track records can experience lower share-price swings and potentially better returns over time (World Economic Forum, 2024).

These different approaches provide investors with a wide range of ways to blend financial objectives with positive impacts on the environment and society. Going forward, as sustainability regulations become stricter and more investors look for eco-friendly choices, sustainable investing is expected to offer fresh opportunities for competitive returns. At the same time, investors will pay close attention to how well these approaches produce genuine progress on environmental and social challenges.

6 Case Studies

Across the world, various sustainable finance projects show how investors can achieve both economic gains and meaningful social or environmental improvements (Green Finance Platform, 2025). One standout example is Ørsted, a Danish energy company that shifted from fossil fuels to offshore wind. Once known as DONG Energy, Ørsted moved away from coal and focused on renewable energy infrastructure. By 2019, it was the world's top producer of offshore wind power, and it achieved 86% renewable energy generation far earlier than planned. At the same time, it greatly cuts its CO_2 emissions, with plans to be free of coal by 2023 and carbon neutral in power generation by 2025. This environmentally friendly shift also boosted Ørsted's market value, showing that a major green pivot can improve both sustainability and profitability (Haanaes & Chauvet, 2022; McKinsey, 2020; Science Based Targets, 2025).

Another example is the Zarafshan wind project in Uzbekistan. In 2022, First Abu Dhabi Bank (FAB) helped finance this 500-megawatt development, which is Central Asia's largest renewable energy project so far. Acting as green loan coordinator, FAB set up a $102 million loan with backing from Etihad Credit Insurance and worked with multiple development finance institutions that provided over $175 million in long-term loans. The project is expected to power 500,000 homes while cutting 1.1 million tons of CO_2 emissions each year. It should also help Uzbekistan reach its target of generating a quarter of its electricity from renewable sources by 2030. This partnership between public and private capital, known as blended finance, can be a powerful way to fund green projects in emerging markets.

A third illustration involves a debt-for-nature swap in Gabon, led by Bank of America in partnership with the Gabonese government and The Nature Conservancy. Completed in August 2023, the deal refinanced $500 million of Gabon's sovereign debt, freeing up $125 million for ocean conservation. It also carried political risk insurance from the US International Development Finance Corporation. Gabon aims to protect 30% of its lands, freshwater systems, and oceans by 2030, and annual payments from this deal will go to a conservation fund and endowment. Investors gained confidence through longer debt durations and improved credit conditions, making it a win–win for nature and the economy.

A final case is the HSBC and IFC Real Economy Green Investment Opportunity GEM Bond Fund (REGIO), which focuses on green bonds in emerging markets. By closing with $538 million in commitments, this fund is the first of its kind to target "real economy" issuers in developing countries. It aims to help them finance projects that limit climate change, while diversifying investment opportunities for its contributors (HSBC USA, 2020; UN Environment Programme, 2023; World Bank Group, 2024). This fund shows how collaboration between private financial institutions and international organizations can mobilize large sums of money for sustainable growth in regions that are most at risk from climate change.

These examples reveal a range of strategies that bridge profitability and positive change. From large corporate transformations to special financing setups, they prove that strong returns and environmental or social progress can go together. As more investors and companies realize that sustainable finance can open doors to growth and innovation, these success stories become templates for a more balanced and responsible way of doing business.

6.1 Conclusion

The case studies above show how sustainable finance can lead to both respectable returns and better outcomes for the planet and society. Increasingly, investors and businesses see that tackling environmental and social problems can give them a competitive edge and create lasting value. As rules change and market forces keep shifting, it is becoming more normal for companies to take sustainability seriously when making financial decisions. This trend will likely spur more creativity and have a bigger effect on how global finance evolves.

7 Collaboration Importance

The drive toward sustainability has shown that no single group can tackle big environmental or social problems alone. Governments, businesses, and nonprofit organizations each bring unique skills and resources to the table, but it is their combined effort that can truly push sustainable finance forward. In recent years, there has been a growing realization that these partnership, often called multi-stakeholder collaborations are essential for creating a strong financial system that supports responsible growth.

7.1 The Imperative for Collaborative Action

Global sustainability issues, such as climate change and resource depletion, are too big and too complex for any one sector to solve on its own. According to the United Nations Environment Programme Finance Initiative (UNEP FI), "Sustainable finance policy and regulation are crucial to align financial flows with the global sustainability agenda" (United Nations Environment, Finance Initiative, 2025). This kind of alignment requires input from both public and private sectors. Governments set regulations and offer policy incentives, while companies and financial institutions design practical market solutions to help meet environmental and social goals.

Nonprofit organizations play a key part as well by providing expertise on pressing issues like climate impacts, biodiversity loss, and human rights. They also lend credibility to sustainability initiatives by representing public interests and advocating for higher standards. As the problems become more urgent, these diverse groups have discovered that working together yields better, faster results than going alone.

7.2 Multi-stakeholder Approach

A multi-stakeholder approach unites different players, government agencies, corporations, and civil society, in the same conversation. This joint effort harnesses the influence of policymakers, the innovative drive of businesses, and the depth of knowledge nonprofits possess regarding social and environmental needs. The World Economic Forum notes that "Industry collaboration can break down sustainability barriers and drive systemic change" (Gomez, 2023). Such a model achieves several things:

Governments can pass laws and give financial incentives that encourage sustainable investment, such as tax breaks for green technology or stricter rules on carbon emissions.

Businesses develop new products and services that reduce environmental harm and respond to the growing consumer demand for ethical practices.

Nonprofits often advise on best practices, champion underrepresented causes, and ensure that social and environmental criteria remain at the forefront of decision-making.

This collective approach ensures no one group's views dominate. Instead, it weaves together regulatory frameworks, market forces, and grassroots advocacy to build a financial system that aims for both profit and positive impact.

7.3 Policy Alignment and Standardization

Collaboration also helps create coherent policies that guide everyone in the same direction. The UN Principles for Responsible Investment (PRI) underlines the importance of "harmonized and sound regulation that ensures clarity of direction" (PRI, 2022). When multiple sectors work together, they can streamline regulations, reduce conflicting rules, and develop common standards. Two important benefits of these joint efforts are:

Regulatory Coherence: Governments and industry bodies cooperate to avoid contradictory rules. The European Union's push for an EU Taxonomy is a good example. By agreeing on precise definitions and standards for sustainability, EU member states give investors a uniform set of rules to follow (PRI, 2025).

Global Standards: Many sustainability problems cross borders, so having agreed-upon frameworks and metrics helps align global investment toward common goals. The PRI and World Bank, for instance, have joined forces to create toolkits that policymakers can use to shape sustainable investment policies (PRI, 2025).

By clarifying the rules of engagement, policymakers and investors can better share a vision for sustainable finance, ensuring that money flows to projects with genuine, measurable impact.

7.4 Resource Pooling and Risk Mitigation

Large-scale environmental or social projects, such as building offshore wind farms or restoring entire forests, often require significant funding and come with high risks. Collaboration among governments, businesses, and nonprofits helps spread these risks and pool resources.

Shared Resources: Some sustainability projects operate like "staking pools," where multiple partners invest money, labor, or expertise in a shared goal, such as planting trees or installing solar panels. Because the costs and benefits are divided among all participants, large projects become more feasible.

Risk Distribution: In uncertain environments (like emerging markets), different types of organizations can cover different layers of risk. Development banks, for instance, might offer guarantees or lower-interest loans, while private companies contribute technical know-how. This approach encourages private investors to enter markets they might otherwise view as too risky.

7.5 *Market Development and Innovation*

By bringing diverse groups together, collaboration fuels innovation in the sustainable finance space.

Accelerating Progress: When entire industries commit to higher sustainability standards, it prompts faster overall change. One example is The Climate Pledge, with 375 signatory companies pledging to hit net-zero emissions by 2040 (Gomez, 2023). This joint pledge pushes suppliers, partners, and even competitors to adapt to greener practices.

New Financial Instruments: Partnerships among financial institutions, corporations, and NGOs often lead to novel financial products, like green bonds or sustainability-linked loans. These offerings draw in investors who want solid financial returns along with environmental or social benefits.

7.6 *Knowledge Sharing and Capacity Building*

Another benefit of teamwork is the open exchange of ideas. This collaborative mindset helps spread successful methods across different sectors and regions.

Information Exchange: Rival banks, for example, may come together under umbrella organizations like the Net-Zero Banking Alliance to share data and solutions for tracking carbon emissions or setting reduction targets (Gomez, 2023). By learning from each other, they can move more quickly toward net-zero goals.

Skill Development: Joint projects often include training sessions or workshops. Businesses refine their methods for ESG reporting, while nonprofits gain insights into finance, and government agencies improve their policy-making tools. Over time, this boosts the overall capacity for sustainable finance across the globe.

7.7 Public–Private Partnerships

Public–private partnerships (PPPs) are a prime model of how government support can combine with private sector efficiency to make big changes happen. When a government aligns its policies and resources with a company's operational expertise, they can tackle large-scale tasks, such as renewable energy plants, big infrastructure upgrades, or new transit systems.

Blended Finance: As seen with the Zarafshan wind project in Uzbekistan, funding came from both commercial banks like First Abu Dhabi Bank and development institutions. By joining forces, these entities mobilized significant capital to install large wind farms, advancing the nation's clean energy goals.

Policy Implementation: Even with robust sustainability policies, execution can be tricky. Government agencies often lack the on-the-ground experience and networks that private businesses have. By teaming up, they can transform policies from well-intended regulations into concrete projects that benefit people and the planet (United Nations Environment, Finance Initiative, 2025).

7.8 Conclusion

Collaboration among governments, businesses, and nonprofit organizations is no longer just a nice idea, it has become a necessity for building a finance system that genuinely supports sustainability. Each partner contributes distinct abilities: governments set the stage with regulations and incentives, companies drive innovation and market solutions, and nonprofits ensure environmental and social responsibility remain at the heart of every plan.

Recent initiatives demonstrate that when these groups come together, sharing risks, knowledge, and resources, they can move the needle on big issues like climate change, renewable energy, and social equity. The continued success of sustainable finance will depend on these partnerships becoming stronger and more widespread. By forging common goals and working in unison, we can create a financial ecosystem that delivers lasting economic growth and a healthier, more equitable world for everyone.

8 Challenges in Financing Sustainability

The world of sustainable finance is growing quickly, but it also faces real difficulties that stand in the way of aligning financial goals with social and environmental needs. One major problem is the variety of regulations and standards that differ from one country to another. According to Simone Ruiz-Vergote (2025), this global patchwork of rules makes it hard for financial institutions that work across borders to follow guidelines in each location. The European Union has been a leader in creating new regulations, such as the Corporate Sustainability Reporting Directive (CSRD) and the Sustainable Finance Disclosure Regulation (SFDR). However, even in the EU, many firms are struggling to meet these detailed reporting rules (Lorenzo Saa et al., 2025). Such uneven regulations can raise compliance costs and discourage companies from embracing sustainability.

Alongside regulatory confusion, the risk of "greenwashing," where an organization's claims of environmental responsibility turn out to be misleading, creates deep concerns about integrity and transparency (Fiestas, 2023). Investors and regulators are demanding stronger evidence of genuine sustainability. BlackRock's withdrawal from the Net Zero Asset Managers initiative (NZAM), highlighted by Lorenzo Saa et al. (2025), shows just how delicate these commitments can be. Although BlackRock manages assets worth over $10 trillion, it left NZAM due to the complex push-and-pull between client expectations and strict environmental targets. This move caused the NZAM group to temporarily halt its activities and review its goals, reflecting broader uncertainty across the financial sector.

Another significant challenge lies in assessing and measuring risks linked to environmental, social, and governance (ESG) factors. Many institutions struggle with inadequate or inconsistent data, making it tough to gauge long-term climate threats or measure the social impact of investments (S&P Global, 2025). Climate risk remains particularly tricky because of the uncertainty surrounding future regulations and the possibility that weather patterns will become even more erratic. In addition, sustainable finance often involves balancing immediate returns with long-term transformation. Some organizations funnel funds into high-emission industries on the condition that they shift toward cleaner operations, but such efforts can clash with short-term environmental targets (Deloitte UK, 2025).

Political events and policy changes add yet another layer of instability. Elections in 2024, for instance, have raised new questions about the future direction of sustainable investments, including concerns about protectionist trade measures that might disrupt the flow of goods in global supply chains (S&P Global, 2025). All of these factors, regulatory inconsistencies, fears of greenwashing, weak data, and policy uncertainty, must be managed by everyone involved in sustainable finance. Addressing them is crucial if the industry hopes to stay credible and continue expanding.

9 Recommendations for Improvement

A range of suggestions have emerged from both academic research and industry practice to tackle these hurdles. One vital step involves strengthening the public's knowledge of financial matters so that more people understand how their spending and investing decisions affect the environment. Research has shown that individuals with higher financial literacy are often more willing to adopt practices that help slow climate change, such as installing clean energy solutions in their homes

(Senaya, 2024). To build this capacity, schools can introduce sustainability topics into finance courses. At the same time, targeted training for finance professionals can explore how ESG issues shape market risks and returns. Easy-to-use online tools can also guide everyday investors through the maze of sustainable finance options.

Standardizing the way environmental impact is measured will also boost confidence in sustainable investments. Studies show that leadership teams with a stronger background in finance are more likely to issue green bonds or sustainability-linked loans, yet many small- and medium-sized enterprises remain uncertain about how to track and report their environmental performance (Senaya, 2024). Developing universal ESG reporting standards and requiring third-party checks could address this gap. Reliable methods to calculate carbon footprints and other key indicators would allow investors to compare different opportunities fairly, preventing exaggerated claims and helping direct money toward projects that deliver real results.

Another area of focus is raising public support for greener solutions. People who understand financial concepts in depth are more likely to spend and invest in ways that align with sustainable goals (McAlpine, 2025). By contrast, those with minimal knowledge often lack the means to handle unforeseen expenses like climate-related disasters, according to the OECD (McAlpine, 2025). Governments and advocacy groups can hold campaigns to showcase successful sustainability stories, offer meaningful incentives for choosing energy-efficient upgrades, and highlight the personal and communal benefits of investing in green ventures.

Encouraging innovation in sustainable finance is also a priority. The reports predict that sustainable bonds will grow by about 25% each year in 2023, and the combined forces of financial technology (FinTech) and ESG analytics are opening new doors for investors. Tools powered by artificial intelligence can sift through massive data sets to spot firms that meet strict sustainability criteria. Governments and institutions can back the creation of innovative instruments, such as sustainability-linked bonds, that reward organizations for meeting environmental targets. By blending technological progress and new financial products, the market can give investors more ways to support both returns and responsibility (Ahmad et al., 2023; Russu, 2024; Ozili, 2021; Ryszawska, 2016; Tribelhorn, 2023).

Overall, these improvements, promoting financial literacy, creating global standards, increasing public backing, and fostering invention, can help the sustainable finance field meet its goals. If these ideas are put into action, then current obstacles, from data gaps to credibility worries, could be substantially overcome, and the sector can play a more effective part in global sustainable development.

10 Conclusion: The Urgent Imperative for Sustainable Finance

As we near key deadlines in global sustainability efforts, particularly the 2030 goals, financing environmental preservation has become essential rather than optional. The final message of this chapter is that every stakeholder needs to recognize the strong connection between economic stability and the health of our planet. Recent findings highlight the growing surge of private investment in nature projects, rising from $9.4 billion to over $102 billion in a span of 4 years, along with an increase in biodiversity-related bonds from 5% of issuances in 2020 to 16% in 2023 (Fisher & Matteini, 2025). These figures show that many investors now see sustainability as a real financial opportunity, not just an ethical side project. Yet the stakes remain high, since large sectors of the economy stand to lose trillions of dollars if they fail to adapt to changing ecological conditions.

Global commitments like the Global Biodiversity Framework (GBF), agreed upon in December 2022, outline a clear goal of halting and reversing biodiversity loss by 2030, including protecting 30% of the planet's land and water (UN Environment Programme, 2025b). Achieving these aims calls for a fundamental rethinking of how we direct and manage capital. It means giving nature the same priority we give to infrastructure or human-made assets. It also means creating new types of finance that can channel funds quickly and effectively to the places where they are most needed, particularly areas that hold critical biodiversity.

The year 2025 marks a turning point for nature financing. Efforts such as green loans that tie interest rates to eco-friendly performance, along with stricter rules in the European Union for labeling green bonds, show that sustainable finance can solve environmental problems and unlock large economic gains (Arbor.eco, 2025; Bioy, 2025). While an annual injection of $2.7 trillion may be necessary to transform the global economic system, the potential rewards, estimated at $10.1 trillion every year, are too large to ignore.

It is clear that everyone, from individual consumers to international corporations, must step up. Shifting away from chasing only short-term profit in favor of long-term resilience will require policies that integrate ESG considerations, investments in research for new solutions, and the alignment of national laws with global climate and biodiversity targets. Financial institutions can become leaders of this movement by steering investments toward projects that restore or safeguard ecosystems, rather than funding those that endanger them. Governments and civil society must also play their part, whether that means providing incentives, tightening regulations, or holding businesses accountable for their impact on natural systems.

In the end, sustainable finance is about more than just preserving coral reefs or forests, it is about preserving our collective future. Our economies rely on the stable functioning of the natural world, from clean water to pollinated crops. As we look beyond 2030, the decisions we make right now will shape the stability of our financial systems, the vitality of our environment, and the security of generations to come. The path toward net-zero emissions, nature-positive growth, and resilient communities is not always easy, but the opportunities are enormous. By acting decisively today, we can ensure a world where economic progress and a healthy planet go hand in hand, creating a better future for everyone.

References

Ahmad, I., Alni, R., Arni, S., & Mamduh, M. H. (2023). Financial literacy to improve sustainability: A bibliometric analysis. *Studies in Business and Economics, 18*(3), 24–43. https://doi.org/10.2478/SBE-2023-0043

Arbor.eco. (2025). *A roadmap to green financing and carbon management in 2025*. https://www.arbor.eco/blog/a-roadmap-to-green-financing-and-carbon-management-in-2024

Bioy, H. (2025). *6 sustainable-investing trends to watch in 2025*. https://www.morningstar.ca/ca/news/258561/6-sustainable-investing-trends-to-watch-in-2025.aspx

Bonnefon, J.-F., Landier, A., Sastry, P., Thesmar, D., Tse, Hec, I., Mit, P., Mit, S., & Cepr, S. (2019). *Do investors care about corporate externalities? Experimental evidence*.

Brown, J. L. (2025). *Environmental economics | Benefits, costs & solutions | Britannica Money*. https://www.britannica.com/money/environmental-economics/Permit-markets

Brundtland Commission. (1987). Our common future: Report of the world commission on environment and development.

Change Oracle. (2022). *Types of economies and their implications for sustainability*. https://changeoracle.com/2022/06/28/primer-on-four-economic-systems-and/

Climate Bonds Team. (2018). Green bonds—A bridge to the SDGs.

Davidson, C. (2000). *Economic growth and the environment: Alternatives to the limits paradigm*. https://academic.oup.com/bioscience/article-abstract/50/5/433/264211?redirectedFrom=fulltext

de Souza Cunha, F. A. F., Meira, E., & Orsato, R. J. (2021). Sustainable finance and investment: Review and research agenda. *Business Strategy and the Environment, 30*(8), 3821–3838. https://doi.org/10.1002/BSE.2842

Deloitte. (2024). *SEC commissioner's remarks on ESG challenges at the annual US-Central and Eastern European Connection weekend*. https://www.iasplus.com/en-ca/news/esg/2024/sec-commissioners-remarks-on-esg-challenges-at-the-annual-us-central-and-eastern-european-connection-weekend

Deloitte UK. (2025). *Sustainable finance*. https://www.deloitte.com/uk/en/Industries/financial-services/analysis/sustainable-finance.html

Department of Economic and Social Affairs Sustainable Development. (2025). *THE 17 GOALS | Sustainable development*. https://sdgs.un.org/goals

Derrick, M. (2024). *Google aims for carbon-free data centres by 2030 | Energy Magazine*. https://energydigital.com/technology-and-ai/googles-journey-towards-carbon-free-data-centres

Ditlev-Simonsen, C. D., & Midttun, A. (2011). What motivates managers to pursue corporate responsibility? A survey among key stakeholders. *Corporate Social Responsibility and Environmental Management, 18*(1), 25–38. https://doi.org/10.1002/CSR.237

Eccles, G. R., & Klimenko, S. (2019). The investor revolution. *Harvard Business Review, 97*, 106–116. https://hbr.org/2019/05/the-investor-revolution?fbclid=IwAR32Akwz5vQ1MW4fLUZWB2hVwCSrovo9TO7aPgu-7T7GVolUioqNXSKkH8Y

EPA. (2024). *Learn about sustainability | US EPA*. https://www.epa.gov/sustainability/learn-about-sustainability

EricssonSCR. (2024). *Ericsson among top global 100 sustainable companies*. https://www.ericsson.com/en/news/2024/1/ericsson-among-top-100-sustainable-companies
ESGDIVE. (2024). *Conflicting views on SEC's authority set the tone for climate disclosure rule vote | ESG Dive*. https://www.esgdive.com/news/sec-climate-disclosure-rule-vote-recap-commissioners-disagree-on-authority/709643/
Fiestas, H. V. (2023). *Clear regulation for sustainable finance | SDG Action*. https://sdg-action.org/clear-regulation-for-sustainable-finance/
Fisher, L., & Matteini, A. (2025). *3 reasons why 2025 is the year for nature-positive finance*. https://www.weforum.org/stories/2025/01/3-reasons-2025-is-the-year-for-nature-positive-finance/
Gomez, L. (2023). *Why industry collaboration is necessary for a sustainable future*. https://www.weforum.org/stories/2023/03/radical-collaboration-for-a-sustainable-future-the-case-for-sustainability-collaboration/
Granier, C., Rigot, S., Granier, C., & Rigot, S. (2021). *Sustainable finance*. https://doi.org/10.48611/ISBN.978-2-406-12203-6.P.0149
Grantham Research Institute on Climate Change and the Environment. (2014). *Why do economists describe climate change as a market failure?* https://www.lse.ac.uk/granthaminstitute/explainers/why-do-economists-describe-climate-change-as-a-market-failure/
Green Finance Platform. (2025). *Case studies*. https://www.greenfinanceplatform.org/case-studies
Gutterman, A. S. (2024). Sustainable finance. *SSRN Electronic Journal*. https://doi.org/10.2139/SSRN.3804435
Haanaes, K. B., & Chauvet, M. (2022). *Ørsted: On the path to net zero—IMD Business School for Management and Leadership Courses*. https://www.imd.org/research-knowledge/sustainability/case-studies/orsted-on-the-path-to-net-zero/
Haanaes, K. B., & Olynec, N. (2022). *Why all businesses should embrace sustainability & how they can do it*. https://www.imd.org/research-knowledge/strategy/articles/why-all-businesses-should-embrace-sustainability/
HSBC USA. (2020). HSBC sustainable structured investments.
Hester M. Peirce. (2021). Statement on the IFRS Foundation's Proposed Constitutional Amendments Relating to Sustainability Standards. https://www.sec.gov/newsroom/speeches-statements/peirce-ifrs-2021-07-01
ICMA. (2023). Green, social and sustainability bonds: A high-level mapping to the sustainable development goals principles 2 contents.
International Finance Corporation (IFC). (2023). *Green bonds*. https://www.ifc.org/en/about/investor-relations/green-bonds
Jackson, A. (2024). *Google's journey towards carbon-free data centres | Data Centre Magazine*. https://datacentremagazine.com/critical-environments/googles-journey-towards-carbon-free-data-centres
Janssens, K. (2024). *How green finance is driving sustainable development goals (SDGs) and the role of IT technology–Digital Finance Transformation at NTT DATA*. https://www.digitalfinancetransformation.com/how-green-finance-is-driving-sustainable-development-goals-sdgs-and-the-role-of-it-technology/
Kerch, S. (2024). *Sustainable investing in 2025: The 7 top trends the experts say to watch in the coming year*—Equities.com. https://www.equities.com/impact-investing/sustainable-investing-in-2025-the-7-top-trends-the-experts-say-to-watch-in-the-coming-year/
Krüger, P., Krauss, A., Meyer, J. (2016). *Sustainable finance in Switzerland: Where do we stand?* https://www.sfi.ch/en/publications/sustainable-finance-in-switzerland-where-do-we-stand
Kuhn, B. M. (2022). Sustainable finance in Germany: Mapping discourses, stakeholders, and policy initiatives. *Journal of Sustainable Finance & Investment, 12*(2), 497–524. https://doi.org/10.1080/20430795.2020.1783151
Kumar, P. (2017). The economics of ecosystems and biodiversity.
L'équipe Manutan. (2024). *Circular economy limitations: How to overcome them*. https://www.manutan.com/blog/en/circular-economy/circular-economy-limitations-how-to-overcome-them

Lartey, T. A., Amankwah-Amoah, J., Danso, A., Adomako, S., Khan, Z., & Tarba, S. Y. (2021). Environmental sustainability practices and offshoring activities of multinational corporations across emerging and developed markets. *International Business Review, 30*(5), 101789. https://doi.org/10.1016/J.IBUSREV.2020.101789

McAlpine, T. (2025). *How can financial literacy contribute to global climate change goals?* https://www.currencymarketing.ca/blog/global-climate-change-goals

Mccarthy, S. (2024). *The 100 most sustainable companies of 2024 | Corporate Knights.* https://www.corporateknights.com/rankings/global-100-rankings/2024-global-100-rankings/the-20th-annual-global-100/

McCormick & Company. (2024). *McCormick & Company named to Corporate Knights Global 100 sustainability index for 8th consecutive year.* https://mccormickcorporation.gcs-web.com/news-releases/news-release-details/mccormick-company-named-corporate-knights-global-100

McKinsey. (2020). *Ørsted's renewable-energy transformation.* https://www.mckinsey.com/capabilities/sustainability/our-insights/orsteds-renewable-energy-transformation

Migliorelli, M. (2021). What do we mean by sustainable finance? Assessing existing frameworks and policy risks. *Sustainability, 13*(2), 975. https://doi.org/10.3390/SU13020975

Ozili, P. K. (2021). Making sustainable finance sustainable. *Financial Internet Quarterly, 17*(3), 64–70. https://doi.org/10.2478/fiqf-2021-0020

Ozili, P. K. (2022). *Theories of sustainable finance.* https://papers.ssrn.com/abstract=4055371

Park, S. (2018). *Investors as regulators: Green bonds and the governance challenges of the sustainable finance revolution by Stephen Park:: SSRN.* https://papers.ssrn.com/sol3/papers.cfm?abstract_id=3142887

PRI. (2022). *A sustainable finance policy engagement handbook.* https://www.unpri.org/policy/global-policy/policy-engagement-handbook

PRI. (2025). *How policy makers can implement reforms for a sustainable financial system.* https://www.unpri.org/policy/how-policy-makers-can-implement-reforms-for-a-sustainable-financial-system/6917.article

Purvis, B., Mao, Y., & Robinson, D. (2019). Three pillars of sustainability: In search of conceptual origins. *Sustainability Science, 14*(3), 681–695. https://doi.org/10.1007/S11625-018-0627-5/FIGURES/1

Quatrini, S. (2021). Challenges and opportunities to scale up sustainable finance after the COVID-19 crisis: Lessons and promising innovations from science and practice. *Ecosystem Services, 48*, 101240. https://doi.org/10.1016/J.ECOSER.2020.101240

Ruiz-Vergote, S. (2025). *Sustainable-finance regulation—A look ahead for 2025.* https://www.msci.com/www/blog-posts/sustainable-finance-regulation/05149752758

Russu, C. (2024). *Does financial education matter for a sustainable future?* https://www.developmentaid.org/news-stream/post/175118/does-financial-education-matter-for-a-sustainable-future

Ryszawska, B. (2016). Sustainability transition needs sustainable finance. *Copernican Journal of Finance & Accounting, 5*(1), 185–194. https://EconPapers.repec.org/RePEc:cpn:umkcjf:v:5:y:2016:i:1:p:185-194

S&P Global. (2025). *Top 10 sustainability trends to watch in 2025.* https://www.spglobal.com/esg/insights/2025-esg-trends

Saa, L., Willman, T., & Pina, P. (2025). *Sustainable investing trends to watch in 2025.* https://clarity.ai/research-and-insights/ai/sustainable-investing-trends-to-watch-in-2025-key-insights-for-investors/

Santa Fe Institute. (2020). *Events: The complexity of sustainability and investing.* https://www.santafe.edu/events/complexity-sustainability-investing

Schneider Electric Blog. (2023). *Top 5 inspiring examples of corporate sustainability initiatives.* https://blog.se.com/sustainability/2023/03/08/corporate-sustainability-initiatives-5-inspiring-examples/

Science Based Targets. (2025). *Case study—Ørsted—Science based targets initiative.* https://sciencebasedtargets.org/companies-taking-action/case-studies/orsted

Senaya, G. M. (2024). Financial literacy and its role in promoting sustainable investment. *World Journal of Advanced Research and Reviews, 24*(1), 212–232. https://doi.org/10.30574/wjarr.2024.24.1.2986

Setyowati, A. B. (2023). Governing sustainable finance: Insights from Indonesia. *Climate Policy, 23*(1), 108–121. https://doi.org/10.1080/14693062.2020.1858741

Signatory directory | PRI. (2025). *Principles for responsible investment*. https://www.unpri.org/signatories/signatory-resources/signatory-directory

Spiliakos, A. (2018). *What is sustainability in business? | HBS online*. https://online.hbs.edu/blog/post/what-is-sustainability-in-business

Steenmans, K., & Lesniewska, F. (2023). Limitations of the circular economy concept in law and policy. *Frontiers in Sustainability, 4*, 1154059. https://doi.org/10.3389/FRSUS.2023.1154059/BIBTEX

Tribelhorn, S. (2023). *Financial education for a sustainable future*. https://sdg-action.org/financial-education-for-a-sustainable-future/

UN Environment Programme. (2023). *Net-zero banking alliance*.

UN Environment Programme. (2025a). *Green financing | UNEP—UN Environment Programme*. https://www.unep.org/regions/asia-and-pacific/regional-initiatives/supporting-resource-efficiency/green-financing

UN Environment Programme. (2025b). *Trend report nature and finance: Looking ahead to 2025*. https://www.unepfi.org/themes/ecosystems/trend-report-nature-finance-looking-ahead-to-2025/

Unilever. (2010). *Unilever sustainable living plan*.

Unilever. (2020a). *Four ways that sustainability has made us a stronger business | Unilever*. https://www.unilever.com/news/news-search/2020/four-ways-that-sustainability-has-made-us-a-stronger-business/

Unilever. (2020b). *Making sustainable living commonplace for 8 billion people | Unilever*. https://www.unileverusa.com/news/2020/making-sustainable-living-commonplace-for-8-billion-people/

Unilever. (2021). Unilever sustainable living plan 2010 to 2020.

United Nations Environment – Finance Initiative. (2025). *Policy and regulation*. https://www.unepfi.org/policy-and-regulation/

Vernon, R. (1979). The product cycle hypothesis in a new international environment. *Oxford Bulletin of Economics and Statistics, 41*(4), 255–267. https://EconPapers.repec.org/RePEc:bla:obuest:v:41:y:1979:i:4:p:255-67

World Bank Group. (2024). *IFC and HSBC asset management expand partnership to support sustainability in emerging markets*. https://www.ifc.org/en/pressroom/2024/ifc-and-hsbc-asset-management-expand-partnership-to-support-sustainability-in-emerging-markets

World Economic Forum. (2024). What are green bonds and how can they help the climate? https://www.weforum.org/stories/2024/11/what-are-green-bonds-climate-change/

Green Society and Sustainable Development

Jahangeer Ahmed Soomro, Muhammad Suhrab, and Basit Ali Abbasi

1 Introduction

The thought of green society is based on sustainability principle, environmental preservation, equality, and social accountability. A green society emphasizes sustainable behaviors, optimal resource usage, sustainable production methods, responsible consumption, and regulations to reduce environmental damage (Gibson, 2016; Chiang et al., 2024). Newman and Kenworthy (2015) state that a green society is an urban economic system structured in such a way to adopt sustainable practices, minimizing carbon emissions, advocating renewable energy, and implementing green transportation and infrastructure. Green society is a system that integrates circular economy principles, ensuring minimal waste production, optimal resource usage, efficient production methods, and a core focus on renewable energy sources (Geissdoerfer et al., 2017). This viewpoint corresponds with the economic sustainability via reducing acquisition of carbon producing resources and modifying individual behaviors to adopt sustainable practices in daily life (Mondejar et al., 2021).

J. A. Soomro (✉)
Department of Business Administration, ILMA University, Karachi, Pakistan

School of Economics and Management, Yanshan University, Qinhuangdao, China

M. Suhrab
School of Management, Huazhong University of Science and Technology (HUST), Wuhan, China

B. A. Abbasi
Department of Business Administration, ILMA University, Karachi, Pakistan
e-mail: basit.ali@ilmauniversity.edu.pk

A. Chakir et al. (eds.), *Green Society, Environmental Strategies and Sustainable Development*, https://doi.org/10.1007/978-3-031-94086-6_11

Sustainable development encompasses environmental preservation, economic advancement, and social fairness to maintain ecological equilibrium (Barbier, 1998). To guarantee that human development does not compromise the environmental integrity of the earth, a green society combines environmental friendly practices across many spheres, including energy, transportation, agriculture, industry practices, and urban planning (Rockstrom et al., 2009).

The concept is hegemonic and is now part of the national constitution and laws of numerous countries worldwide since the Earth Summit (Luke, 2005). The same notion has been adopted in matters pertain to business, agricultural production, industrial issues, and urban development and has acquired the structural base of theoretical and practical approaches like green economy and circular economy (Amran et al., 2015; Shcherbina et al., 2017; Schroeder et al., 2019).

1.1 The Necessity for Green Society

The present environmental condition of the planet requires the immediate need of creating and implementing the concept of green societies worldwide. Different human activities related to industry, rapid urbanization, and deforestation over the last century have seriously degraded the environment. Pollution, loss of biodiversity, climate change, and resources depletion are some of serious issued need to address immediately (Chung, 2020). Particularly in the most susceptible areas, these difficulties compromise not just the ecosystems of the globe but also the human livelihoods. Encouragement of need-based consumption, equality in resource distribution, and safeguarding of natural environment help a green society to provide a road map to solve these challenges (Wu et al., 2021a, 2021b).

The idea of the green society aims to change society structures and norms from exploitation to environmental care. Communities, business, and government need to work together to build such a framework of reforms that promote ecological, economic, social well-being, and sustainable development.

2 Literature Review

Within the realms of academia and public policy, notion of sustainable development and green societies has profound roots. When it comes to the concept of creating a sustainable future, scholars and practitioners adopted a number of approaches. A survey of the most important literature in the disciplines of environmental stewardship, economic viability, social equality, and cultural changes is presented in the following sub-sections.

2.1 Sustainable Environment Stewardship

Sustainability literature revolves mostly on environmental stewardship. According to the idea, individuals have obligations to use their natural resources in a manner that guarantees their long-term survival and output. Environmental stewardship is a useful strategy for sustainable development as much as a philosophical idea (Österblom et al., 2022). The United Nations define sustainability as meeting current generation demands without being losing capacity to meet demands of coming generations. SDGs (Sustainable Development Goals) of UN draw attention on the need of preserving the Earth while advancing wealth (Chapin et al., 2022).

Environmental stewardship is based on broad-spectrum activities that support sustainable living, including cutting waste by means of recycling and waste reduction strategies, funding clean energy projects, and safeguarding of natural ecosystem. Aiming to complete the loop in production and consumption, the circular economy tries to lower waste, lengthen the product life cycle, and make use of renewable resources (Geissdoerfer et al., 2017).

Inclusive understanding of sustainability has emerged as a result of increased awareness of interdependence of economic, social, and environmental systems among themselves. Promoting sustainability requires to rethink the current design of goods, processes, and consumption pattern in order to guarantee that they are sustainable and environmentally friendly (McDonough & Braungart, 2002). In this approach, they highlight the need of such sustainable systems that are designed to imitate the natural processes of regeneration and promote closed loop systems that are raw material efficient and minimize wastage.

2.2 Green Societies and Economic Viability

The creation of a green society is dependent on the existence of economic viability and taking responsibility behavior by individuals, industries, and governments. Redefining economic structures such that they support long-term sustainability is necessary in order to create a green society (Adamowicz, 2022). One of the most important idea in this transition is green growth, which is a notion that was articulated by the OECD (2017). It involves working for economic progress along with minimizing the environmental damage and distributing growth benefit equally to all and especially to people in economic distress. Promoting clean energy projects, raising efficiency of available resources, and lowering carbon emissions are all the components of green growth (Chien et al., 2021).

One of the most important aspects of green economic model is the real cost accounting concept which incorporates the environmental costs of economic activities into the decision-making process (Fernandes et al., 2021). Traditional models of economics sometimes failed in accounting the social and environmental costs involved in production. These costs include things like pollution, resource depletion, and effects on public health. Spash (2017) states that true cost accounting is an approach that seeks to address this error by assigning a particular value to ecosystem services and environmental consequences to promote sustainability.

For the purpose of facilitating green growth, many models have been presented. In order to solve environmental concerns while also supporting sustainable development, the green economy strategy places an emphasis on the role that the market plays. It places an emphasis on investments in environmentally friendly technology, infrastructure, and sectors that reduce environmental impact (Nosheen et al., 2021). According to UN Environment Programme (2011), sustainable livelihoods are an essential component of the green economy. This guarantees the growth benefit is distributed to all society components, especially those in developing nations (Belmonte-Ureña et al., 2021).

2.3 *Social Equity and Cultural Norms*

Social equity is complementary for sustainable development. Without making it certain the advantages of sustainability cannot be evenly distributed among all members of society, and eventually green society cannot be realized (Hariram et al., 2023). Promoting sustainable development requires an inclusive strategy that works to alleviate poverty and inequality from the society completely while providing access to educational and medical facilities, job opportunities, and empowering socially marginalized communities (Pritchett et al., 2022).

Society-based development and social inclusion are essential building blocks for construction of eco-friendly communities (Leach & Scoones, 2007). Movements at the grassroots level in which communities at the local level take action on environmental concerns are crucial to ensure that sustainability efforts are not only founded in local needs and beliefs but also come from the top down (Zheng et al., 2021). Moreover, participatory governance, local knowledge, and collaborative decision-making are all elements that are often included in these movements and they all contribute to the formation of green societies.

Cultural norms are significant in leading individual behaviors toward sustainability. Sustainability is incorporated into people's daily lives and into society as a whole depending on how cultures treat environmental challenges and assume personal or group obligations (Emon & Khan, 2023). Stern (2000) states that understanding social and psychological factors is helpful in shaping pro-environmental behavior. Cultural attitudes toward demand for green products (Wojnarowskais et al., 2022), balanced consumption, waste management, and efficient resource management are deeply embedded in societal norms and must be shifted to achieve a green society.

2.4 *Industry 5.0 and Green Society*

Industry 5.0 is a human-centric design solution that aims to solve the difficulties that have been brought to light by Industry 4.0. It is a solution that involves people and cobots working together (Mourtzis, 2021). This shared working environment designed by I5.0 is gaining serious attention over the last few years. Furthermore, in these years, a similar concept Society 5.0, a more realistic form of green society, has come into view as a solution to issues that are plaguing the current society (Fukuyama, 2018). This super-smart society concept enables everyone to live quality and comfortable lives via combination of physical space cyberspace by efficient use of information and communication technology (ICT).

The Japanese government has developed this plan as a reaction to socio-technological programs like I4.0 and Made in China 2025, which are being implemented in Europe and China, respectively. In a similar vein, I5.0 is connected to S5.0 (Society 5.0) when artificial intelligence is taught to reflect and guide organizational processes on its own, which is anticipated to take place in the year 2030 (Fukuyama, 2018). Within the context of Society 5.0, it is envisaged that the CPS that was built in I4.0 will continue to facilitate the evolution of human and technical connections, reduce environmental concerns, and support SDGs (Shiroishi et al., 2018; Narvaez Rojas et al., 2021) (Fig. 1).

Positive engagement between people and technology is vital to ensure sustained progress in all facets of society, including education, economy, health, and democracy (Ferreira & Serpa, 2018). The focus on sustainability in Industry 5.0 is consistent with the tenets of a Green Society, which promote social justice and ecologically beneficial behavior. Industry 5.0 uses sustainable manufacturing practices, ethical behavior, and carbon footprint reduction to help build a society that values ecological balance and social well-being (Dhayal et al., 2023) (Fig. 2).

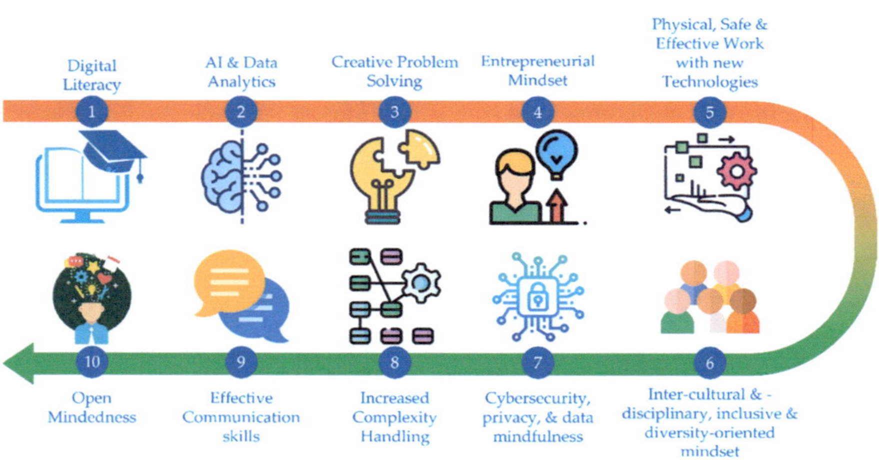

Fig. 1 Potential Skills facilitating Society 5.0. (Source: World Maufacturing Forum Report, 2019)

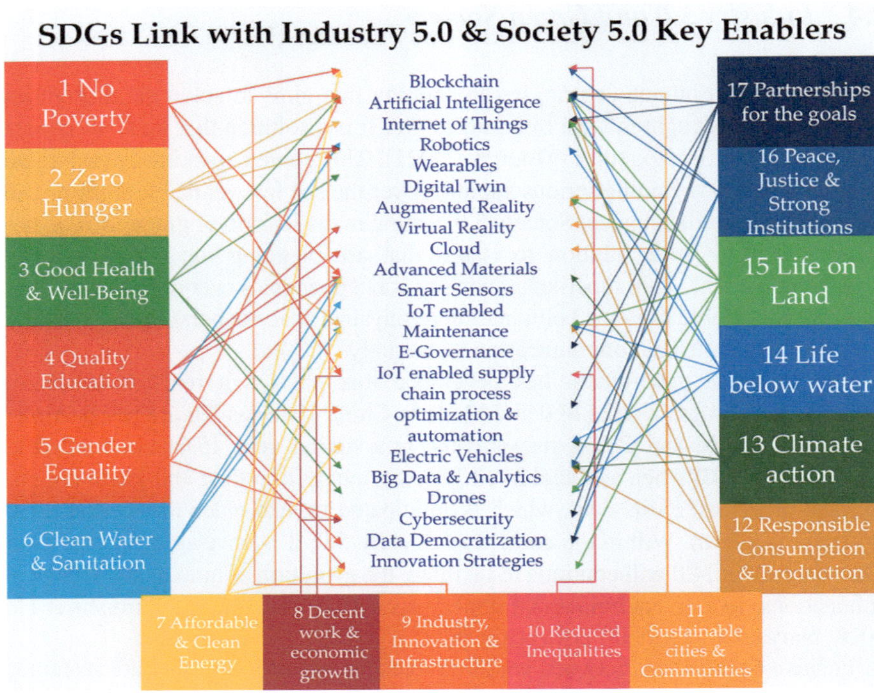

Fig. 2 Society 5.0 for Sustainable Development Goals. (Source: Mourtzis et al., 2022)

3 Methodology

This chapter adopts qualitative research design and is based on the discussion related to key factors promoting the green society and sustainable development, analyzing several worldwide case studies related to green societies and sustainable development. This chapter also assesses the significance of awareness and educational programs in promoting green society and sustainable development, along with investigating the critical role played by different stakeholders. This chapter further highlights the future opportunities and obstacles related to green society and sustainable development.

4 Key Factors Promoting Green Society and Sustainability

4.1 Social Entrepreneurship

A significant facet of rural entrepreneurship is aimed at fostering a secure, inclusive, and sustainable future. This encompasses several types of entrepreneurship that include women entrepreneurship, women empowerment, youth entrepreneurship, and entrepreneurship among indigenous groups (Cahn, 2008; Bhandari, 2017; Mishra, 2021; Tabares et al., 2022; Anand & Josse, 2002). Social entrepreneurship may provide poverty alleviation techniques rooted on farmers' initiatives and flexibility while fostering solutions to sustainable livelihood via collective organizations (Chowdhury et al., 2017; Adeyonu et al., 2022). Scholarly research indicates that such entrepreneurial initiatives occur to assist society during crises and to surmount obstacles to accessing education and training, resulting in diminished prospects (Shaw, 2017; Mishra, 2021). Given that education is vital in all job opportunities (Igwe et al., 2020), various forms of entrepreneurship need governmental assistance particularly during the first phases (Chipfupa & Wale, 2018). Moreover, societal entrepreneurship promotes the concepts of equality, environmental promotion, and fair distribution of wealth, which are all key elements of green society (Figueroa-Domecq et al., 2020).

4.2 Rural Entrepreneurship

This entrepreneurship type views the entrepreneur as an agent affected by various socio-ecological situations (Chong et al., 2019); hence, it is essential to determine the critical aspects contributing to success or failure of rural enterprises (Pato & Teixeira, 2016). Concerning to climate issues, majority of women engaged in crop farming exhibit a heightened awareness of climatic alterations, acknowledging that climate change has significantly extended dry season and impacted soil fertility (Akinbami et al., 2019). Consequently, the output of rural entrepreneurship has to be sustainable if longevity of seasonal crops in various rural regions worldwide is anticipated, particularly as old-style rural livelihoods reduce due to the depletion of natural resources, climate challenges, and modernization (Samal et al., 2016; Kimbu et al., 2022).

Consequently, rural entrepreneurship ecological dimension must not be overlooked if it facilitates sustainable development. Research pertain to environmental preservation provides insights for long-term poverty alleviation and conserves endangered natural resources (Shahraki & Heydari, 2019).

5 Case Studies

Several nations have adopted green community models and green urban deigns to promote sustainability and improve the lives of the public without compromising the environment. Some of the case studies are discussed below.

5.1 *Sustainable Rural Livelihood*

The Development Alternatives Group, an Indian social corporation, has implemented solutions that are sustainable and low in carbon emissions in order to enhance the lives and earning of rural population. One notable program aims to develop knowledge and capacities in rural communities via the application of sustainable practices that support economic development and environmental preservation (Majid, 2020). These efforts have resulted in adoption of environmental friendly tactics and technologies in daily life practices which have contributed to the poverty reduction and preservation of environment for long-term sustainability (Lal et al., 2022).

5.2 *Transitioning to Clean and Green Energy Sources in Denmark*

Denmark is pioneer in clean energy field with the goal to completely eliminate fossil fuel usage for production capacity by 2050. The country has made major investments in clean energy projects like wind energy and solar energy making it world's principal producer of green energy (Hvelplund, 2013). Denmark was able to fulfill almost 50% electricity demand from wind turbines by 2020 (Lund et al., 2022). Policies like energy tariffs and incentives for energy efficient buildings have helped to promote sustainable practices. These rules together with collaboration between the government and private firms have accelerated green innovation and technology adoption at faster rate. As a result, Denmark has expanded the number of green jobs and the economic benefits of exporting renewable energy sources while also lowering down the CO_2 emissions by 40% since 1990 (Pinson et al., 2017; Wang et al., 2021).

5.3 *Curitiba, Brazil: Green Urban Design*

Curitiba city which serves as the capital in the state of Parana (Southern Brazil) is well known for green and sustainable urban development projects. Green urbanization model has been developed by the city to keep balance between natural environment and the expansion of the metropolis (Macedo, 2013; Follador et al., 2022). The implementation of a BRT system (Bus Rapid Transit) that is environment friendly and cost-effective has proved to be a major step to reduce the traffic congestion and pollution of the city. By now more than 70% of the population relies on public transportation. A recycling scheme that encourages the inhabitants to trade recyclable garbage for food and tickets to public transportation has been implemented in Curitiba which has also made environmental sustainability a priority by including more than 50 square meters of green space per person. Carbon emissions have decreased, air quality has improved, and the urban environment has become more sustainable as a result of these activities (Rabinovitch & Leitman, 2015; Gustafsson & Kelly, 2016; Zingoni de Baro, 2022).

5.4 Circular Economy Model of Sweden

Sweden has successfully developed and implemented the circular economy model with the objective of reducing waste and increasing resource efficiency (Heshmati & Rashidghalam, 2021). The producers are liable to dispose and recycle products in accordance with policies like Extended Producer Responsibility (EPR) (Brydges, 2021). According to Swedish Garbage Management, the nation has also established a sophisticated Waste to Electricity Program which has resulted in 99% of domestic garbage being recycled for the creation of electricity. Moreover, stringent laws guarantee the preservation of biodiversity and sustainable forestry techniques. Higher recycle rate, less dependence on landfills, and effective integration of waste to energy operations into district heating systems are all outcomes that have resulted from Sweden's efforts to eliminate waste and reduce environmental concerns (Milios, 2021).

5.5 Toward Low Carbon Society: Germany

A large-scale attempt to shift the society from fossil fuels toward clean energy sources with aim to lower carbon emission is being undertaken by Germany as part of its Energiewende (Energy Transition) program. This clean energy program provided up to 50% of Germany's total power supply as of 2021 (Rechsteiner, 2021; Schischke et al., 2023). Germany took considerable step in escalating its solar and wind energy capacity since the program's inception. Through the use of subsidies, feed-in tariffs, and governmental investments, businesses and people have been pushed to embrace environmentally friendly energy solutions. Germany has achieved a 40% decrease in carbon emissions in comparison to the levels that were existed in 1990, despite the fact that there are hurdles such as grid stability and the progressive phase out of coal and nuclear power (Chen, 2024). As a result of the strategy, high employment opportunities have been created in the clean energy industry and also energy efficiency has been improved.

5.6 *Green Urban Development of Singapore*

According to Tan et al. (2013), Singapore has emerged as a model for ecologically responsible urban development because it has included sustainability into the infrastructure of its cities. The government via Green Mark Certification System has mandated environmentally friendly designs which are part of the broad green construction programs (Zhang et al., 2022). Moreover, the city has also developed a complex water management system that integrates rainwater collection, water recycling, and desalination plants in order to ensure the availability of water throughout year and saves from urban flooding. Singapore has added more urban green spaces such as vertical forests and rooftop gardens to minimize urban heat island effect besides boosting biodiversity (Shamsuzzoha et al., 2021). These projects have helped Singapore become one of the most sustainable cities in the world by lowering the average temperature coupled with maintaining high living standards and thus decreasing the energy used per person (Dell'Anna et al., 2022) (Fig. 3).

Fig. 3 Rooftop gardens—Singapore. (Source: Pandit et al., 2018)

5.7 *Ecological Civilization Strategy of China*

China has introduced the Ecological Civilization Strategy into its national strategy in order to support green development and ecologically sustainable growth (Wu et al., 2021b). The strategy consists on great scale afforestation initiatives, strict environmental rules, and large investment in clean energy projects (Yan et al., 2021). With more than 30% of the total power output of the country coming from renewable sources, the country has become the top nation in the world in the manufacturing of solar and wind technology. China has also started ecosystem restoration projects, aiming at lower pollution levels and more green space by means of strict emission reduction targets (Song et al., 2023). These combined efforts have helped China to become a global leader in sustainability.

5.8 *Bhutan's Gross National Happiness Model for Sustainable Development*

Bhutan has used Gross National Happiness (GNH) in unique way and incorporated environmental sustainability into its development model (Ura et al., 2012). Simultaneously, the country is striving for carbon neutrality and its constitution mandates that at least 60% of its land be kept as forest (Rai, 2024). Hydropower is the key electricity generation source in Bhutan and sells any excess energy to nations that are located in its immediate vicinity. Additionally, government has imposed stringent rules on the tourism industry as well as industrial activity to reduce individuals' ecological footprints (Brooks, 2013). With a high level of biodiversity protection and sustainable economic development, Bhutan is one of the few nations in the world that has achieved carbon neutrality (Lepeley, 2017) (Fig. 4).

5.9 *Toward Green Republic: Costa Rica*

The fact that Costa Rica has been producing almost all of its energy from renewable sources for a number of years in a row is a sign of its strong engagement in sustainability (Milla Quesada, 2023). The country utilizes an extensive range of energy sources, including wind, geothermal, solar, and hydropower in order to fulfill its electric power requirements (Teske et al., 2020). Moreover, Costa Rica made noteworthy investment in protection of its forests which has led to a rate of reforestation that resulted in increase in land that is covered by forest (Evans, 2010; Morán-Ordóñez et al., 2022). Providing financial incentives to communities and farms is one of the ways that the country promotes biodiversity and reduces carbon emission. The implementation of these measures has decreased emission of greenhouse

Fig. 4 Gross national happiness dimensions. (Source: Ura et al., 2012)

gases, increased tourism revenue, and improved the resilience against climate change impacts (Hernández-Chaverri & Buenrostro-Figueroa, 2021).

5.10 Social Entrepreneurship for a Green Society

EcoPost program is a social venture of Kenya that offers a sustainable solution to tackling the issue of plastic waste, while also simultaneously promoting the green employment and economic sustainability. Plastic trash is recycled by EcoPost into long-lasting fence posts which offers an alternative to lumber and helps to reduce the amount of deforestation that occurs (Samantarai & Dutta, 2024). Thousands of employment has been created by this initiative in terms of collecting garbage, manufacturing, and distribution. The initiative has improved people's earnings while also lowering the amount of environmental damage (Heuër et al., 2015). EcoPost highlights the potential of circular economy models in developing nations by recycling waste plastic into useful items. This helps to promote both the sustainability of environment and growth of the socio-economic system (Koech et al., 2023).

5.11 Sustainable Urban Development: Masdar City

Masdar city of UAE is planned in such an eco-friendly manner that incorporates green building design, smart water management, and renewable energy (Alzaabi et al., 2018; Pandita et al., 2024). With support of solar energy and buildings that are energy efficient, the city intends to achieve its goal of having zero waste and zero carbon emissions. The city is run by a solar farm that generates 10 MW of power (Ismael, 2023). Masdar city has a managed water consumption system that is 40% lower compared to traditional cities. Additionally, the city uses autonomous electric transportation systems to further reduce the emissions (Rietmann, 2021).

6 Opportunities for Green Society Development

A green society is a concept that places an emphasis on sustainable living, the conservation of resources, and behaviors that are favorable to the environment. There are several potential for economic progress, environmental conservation, and social well-being that may be found in the creation of green societies. This chapter further discusses the most important potentials for the creation of a green society, focusing on the roles that technology, legislation, and community engagement play.

6.1 Clean Energy Integration

Solar, wind, and hydroelectric power are common examples of renewable and clean energy sources that must be included into green society. This is one of the most promising options for green civilizations. These sources lessen reliance on fossil fuels, bring about a reduction in greenhouse gases, and contribute to promote energy sustainability. The infrastructure for renewable energy is receiving a growing amount of investment from both public and private sectors which is making clean energy more accessible and cheap (Østergaard et al., 2020).

6.2 Sustainable Urban Planning Projects

Buildings and infrastructure that are environmentally friendly, public transit that is efficient, public parks, community halls, and smart city projects are all examples of sustainable urban design that are beneficial to green societies. According to Jones and Miller's research from 2020, urban centers that implement energy efficient buildings, green roofs, and rainwater harvesting systems are able to reduce the carbon footprints and improve the urban resilience simultaneously. For the purpose of enhancing air quality and fostering biodiversity, green spaces are an integral component of intelligent urban development (Naess, 2001; Anthony, 2024).

6.3 Circular Economy Model and Waste Disposal System

Implementing circular economy model provides an opportunity to reduce waste and improves efficiency of resource utilization. Recycling, composting, and sustainable manufacturing methods contribute to reduce pollution and create green employment opportunities (Nikolaou & Tsagarakis, 2021). Extended producer responsibility (EPR) policies are being promoted by governments and organizations at an increasing rate to support sustainable production and consumption. Further reduction in reliance on landfills is made possible by the implementation of cutting-edge waste to energy technologies which also make it possible to transform waste materials into energy sources that may be used (Schöggl et al., 2024).

6.4 Green Transportation Projects

There is a huge influence that transportation has on green society in terms of the economy, society itself, environment, and governance. Effective management of travel demand, automobile growth trends, and proficient land-use contribute to promote sustainability goals in urban settings (Amatuni et al., 2020). Smart transportation strategies help to achieve these goals. Green transportation is a kind of transportation system, having less effect on both environment and human health (Björklund, 2011). Green transportation provides a more environmentally friendly mode of transportation. Reduced risk, decreased traffic mobbing, enhanced energy usage, decreased pollution, accident prevention, security assurance, and maximized traveling speed are all potential outcomes that may be achieved with the implementation of an efficient GT system (Lee et al., 2017; Shah et al., 2021).

6.5 Sustainable Agriculture and Food Systems

A significant potential for a green society is presented by sustainable agriculture. Agricultural methods such as organic farming, permaculture, and precision and technological advanced agriculture methods contribute in reducing environmental damage and improving food security (Muhie, 2022). According to El Bilali et al. (2021), vertical farming and urban agricultural projects make it possible to produce food inside cities, therefore minimizing the carbon footprint that is linked with the transportation of food. In addition, developments in alternative sources of protein such as meat generated in a laboratory and diets based on plant-based foods also contribute to the reduction of the environmental effect that is caused by conventional cattle production (Fenibo et al., 2021).

6.6 Green Technologies and Innovation

A significant number of firms and environmental responsible government activities are looking for environmentally friendly solutions which is driving up the need for environmentally friendly technology. At the vanguard of this transformation are environmentally friendly technologies like wind turbines, solar panels, electric cars, and smart grid systems (Guo et al., 2020). There has been a marked increase in number of businesses that are focusing on energy efficiency, renewable energy, and sustainable practices. There is a high probability that the clean technology sector will continue to expand which will result in the development of considerable economic possibilities and jobs (Walz et al., 2017). Other options for environmentally responsible development are provided by technical development in areas like sustainable agriculture, water management, and carbon capture. In order to assist nations in achieving their climate objectives and decreasing their dependency on fossil fuels, these inventions play a very important role (Shahzad et al., 2022).

7 Significance of Education and Awareness in Fostering Green Society

Education and awareness have long-lasting impact on public attitudes, actions, and acceptability of sustainable practices (Hnatyuk et al., 2024). Individuals and groups face difficulties in realizing the implication of their activities on surrounding environment in the lack of proper knowledge. Sustainable growth and ecological balance require proper environmental awareness and ensure that education on sustainability is included into all levels of society (Debrah et al., 2021).

7.1 Role of Environmental Awareness

Environmental awareness refers to understanding the issues pertain to environmental as well as keep oneself motivated to adopt action in order to solve such problems. It serves as the basis of informed decisions related to responsible consumption and environmentally friendly initiatives. It is more probable that people will engage in sustainable activities like reduced garbage, energy conservation, and promoting green laws when they have awareness regarding negative impacts of environmental deterioration and climate change (Kollmuss & Agyeman, 2002).

The objective of boosting environmental awareness may be accomplished via a variety of channels like educational seminars, green community activities, and media campaigns. Awareness campaigns are frequently organized by governments and NGOs (non-governmental organizations) in order to educate the general public about noteworthy environmental issues like pollution, deforestation, and climate change (Soomro et al., 2025). Coping with climate issues and endorsing sustainable development, United Nations' SDGs put a strong emphasis to promote the need of environmental education in general public (Bexell & Jönsson, 2017). The action's related to consuming an excessive amount of plastic and improperly disposing of garbage are being brought to the notice of the public via awareness campaigns. By gaining an understanding of this knowledge, the general public is able to take responsibility and motivate themselves to make changes in their behaviors and ways of living that are more environmentally friendly (Hajj-Hassan et al., 2024).

7.2 Education as a Tool for Green Society

Environmental education is a potent instrument for developing a green society because it equips people with the information and skills essential to participate in sustainable behaviors. This makes environmental education an incredibly effective strategy. The author Palmer (2002) asserts that environmental education nurtures critical thinking and problem-solving abilities which in turn help people to successfully solve environmental concerns. It is the responsibility of religious institutions along with educational institutions like schools to modify pro-environmental behavior among children, universities to modify adult mentality, and other educational establishments to ensure that students gain a knowledge of environmental concerns and contribute to the incorporation of sustainability into the curriculum (Meng & Hao, 2024).

It is the need of time to cultivate the young minds with formal education related to green society and sustainability in order to prepare them with the mindset necessary to address environmental issues. The incorporation of environmental science courses into the curriculum may assist pupils in comprehending the significance of biodiversity, renewable energy, and the preservation of resources. Furthermore, experiential learning approaches like eco-clubs and outdoor education programs provide students the opportunity to get practical experience that reinforces the ideals of sustainability (Tilbury & Ryan, 2011).

Not only does education play a role in the formation of a green workforce but also vocational training and professional development programs do so as well. In order to respond growing demand of eco-friendly firms and environmentally responsible technology, it is imperative that workers get training in energy efficiency, sustainable working practices, renewable energy systems, and sustainable farming (Asprey & Nash, 2006). The European Commission's 2020 report, "Green Skills Agenda of the European Union," highlights the importance of equipping businesses and workers with the skills required for the shift to a low-carbon economy. This highlights education importance in fostering economic sustainability and preparing society for a future that is environmentally friendly.

7.3 Encouraging Sustainable Behavior

Change in individual behavior is necessary for establishment of a green society and several studies have shown that awareness and education proved to be an active and effective driver of such changes (Van Trijp, 2013). Theory of planned behavior states that individuals actively participate in environmentally responsible activities if having information about environmental issues along with a positive attitude toward sustainability and a feeling of personal control over their actions (Bosnjak et al., 2020). People have a greater tendency to adopt the sustainable behaviors such as utilizing public transportation, lowering their energy use, and supporting sustainable products when they are taught about adverse effect of several daily actions on environment (White et al., 2019).

Study by Stern (2000) highlighted that environmental education along with awareness campaigns dramatically increases the number of people who engage in environmentally conscious behavior. There have been observed reduction in the amount of water usage in a number of societies as a result of projects to teach people about water conservation. In a similar vein, communities have been motivated to reduce their usage of plastic bags and embrace alternatives such as reusable bags and biodegradable packaging as a result of awareness programs highlighting the hazards of plastic pollution (Sloot et al., 2018).

7.4 Community Programs Related to Sustainability

Participation in community activities is regarded as an essential tool for encouraging people to participate in sustainable behaviors. When individuals work together on environmental agendas, they are able to strengthen each other's commitment toward sustainability. There are a number of noteworthy programs that contribute to the development of an environmentally responsible culture and behavior (Harvey et al., 2014). Some examples of these programs are go green behavior, community cleanup drives, planting activities, and proper garbage dumping seminars. Moreover, grassroots level initiatives play an important role in the process of supporting legislative changes, such as prohibition of the use of plastic bags and the provision of incentives for using renewable energy (O'Brien & Wolf, 2010).

8 Green Societies and Sustainable Development: Lensing Out the Role of Stakeholders

The shift from traditional societies toward green societies requires the strong collaborative engagement of stakeholders, including governmental entities, corporations, non-governmental organizations (NGOs), communities, and international organizations. Every stakeholder plays an essential role in forming and executing policies that promote innovation, enhance awareness, and promote sustainability (Steurer et al., 2005).

8.1 Government and Policymakers

Regulations pertain to forestation, CO_2 emission, renewable energy, waste management, and land usage are the areas in which governments have a role in providing a legal framework that promotes sustainable activities (OECD, 2017). Welcoming eco-friendly initiatives is required by policies like carbon pricing, subsidies for renewable energy, and pollution restrictions (Stern, 2008; Usman et al., 2024).

8.2 Businesses and Corporate Sector

Industry and businesses have a meaningful influence on use of resources and degradation of the environment (Suhrab et al., 2023). Therefore, corporate sector is an essential component of the green societies. A considerable number of businesses are shifting to more sustainable business models with the implementation of CSR initiatives related to reduction of carbon footprints and investments in technology advancement (Sheehy & Farneti, 2021). Environmental impact may be mitigated by the use of sustainable supply chain and production methods, using energy efficiency technology and waste reduction (Le, 2022).

8.3 *Civil Society and Non-governmental Organizations*

Non-governmental organizations and civil societies are essential to promote green societies, increasing public awareness and ensuring that governments, businesses, and communities comply with their environmental commitments (Doh & Guay, 2006). Greenpeace, the World Wildlife Fund (WWF), and Friends of the Earth are a few of the world's leading environmental organizations that actively advocate for laws protecting biodiversity, reducing pollution, and addressing climate change issues (Doh & Guay, 2006).

8.4 *Role of Communities and Individuals*

It seems impossible to put seeds of green societies without the active participation of community and individuals. For the reason, sustainable development is mainly dependent on changes in individual behavior as well as active engagement from local communities. By adopting sustainable lives, individuals may make a contribution to the environment. These lifestyles include lowering energy usage, limiting waste, utilizing public transportation, and supporting ecologically friendly products (Steg & Vlek, 2009; Steg, 2016).

8.5 *Significant Role of Academia and Research Institutions*

Academic and research institutes make a significant contribution in the development of green societies in a number of different ways (Abad-Segura & González-Zamar, 2021). These ways include conducting significant scientific investigations, developing innovative solutions for sustainability, and formulating effective policies. According to Berchin et al. (2021), educational institutions and research institutes have consistent focus on the development and enhancement of renewable energy technology, climate modeling, and sustainable resource management techniques. Moreover, they are also providing crucial information to businesses and politicians to form and implement sustainable practices.

9 Obstacles in Achieving a Green Society

A number of factors hinder the way to establish green societies. These factors depend upon the economic, social, and legislative power of nations.

9.1 Economic and Financial Barriers

The main hurdles to paving the way for a green society are the economic and financial conditions of the nation. Considerable initial investments is necessary to acquire advanced technologies, renewable energy infrastructure, electric vehicles, environmentally friendly production methods, and treatment plants of environmental deterioration substances (OECD, 2017; Ghisetti et al., 2017). Despite the long-term advantages and increase in revenues, still numerous developing countries and modest businesses are unable to afford these initial expenses associated with implementing the sustainable alternatives (Abid et al., 2022).

The transition to a green economy involves substantial structural modifications such as financial incentives for businesses that implement sustainable practices, investments in clean technologies, and subsidies for renewable energy (Akermi & Triki, 2017; Blincoe, 2022).

9.2 Lack of Awareness and Education

Lack of awareness and education regarding environmental concerns is a critical obstacle to sustainability. Many individuals are unaware of the advantages of sustainable living and issues pertaining to the environment (Priya & Thenmozhi, 2021). Individuals are less inclined to implement eco-friendly behaviors, including support for green businesses, water conservation, and pollution reduction in the absence of adequate environmental education (Obiagu et al., 2024).

9.3 Resistance to Change and Societal Norms

A green society depends on behavioral changes; however, these are sometimes opposed. Particularly with regard to convenience-driven behaviors including single use plastics, automobile reliance, and quick fashion consumption (Verplanken & Whitmarsh, 2021), people often object to changes that throw off their routines. Furthermore, societal and cultural standards also greatly influence conduct. Many countries have strong established consumerism which makes it challenging to move toward a minimalist and sustainable living.

Corporate interests support this opposition as well. Profiting from unsustainable activities, big businesses can advocate against environmental rules and support customer behavior that puts economic development above environmental preservation (Brulle, 2020). Strong government action, business accountability, and grassroots organizations advocating cultural changes toward sustainability can help one overcome this difficulty.

9.4 Lack of Enforcement and Weak Environmental Policies

Increased awareness of environmental concerns on a global basis has been observed, but still many governments, particularly in underdeveloped countries, are having trouble putting effective environmental laws into practice and enforcing them. According to Akella and Cannon (2017), considerable obstacles to sustainability are created by insufficient enforcement mechanisms, weak legislation, and political hostility to environmentally friendly economic strategies. For instance, despite the fact that international accords like the Paris Agreement have the intention of lowering carbon emissions, many nations are unable to reach their commitments because of the pressures that they face from both the political and economic spheres.

9.5 Technological Constraints and Infrastructure Issues

In order to make the shift to a green society, technical advancements are very necessary; however, limits in both technology and infrastructure advancement come with significant obstacles (Pandey, 2021). Despite the fact that renewable energy sources are increasingly readily accessible, but high initial cost, ineffective infrastructure, and the lack of efficient energy storage systems make it difficult to adopt it (Al-Emran & Griffy-Brown, 2023). In a similar vein, environmentally friendly transportation modes like public transportation and electric vehicles need significant monetary investments in infrastructure which some areas do not have (Shah et al., 2021).

9.6 Differences in Social and Economic Status

A further obstacle to the realization of a green society is the unequal distribution of wealth among communities at different geographical locations. Lower income populations often lack the financial means to participate in a variety of sustainable behaviors and activities (Martinez-Alier, 2021). Moreover, climate change has a more negative impact on groups who are already vulnerable and living in low income nations. These people have made the least contribution to global emissions but they are experiencing severe consequences of climate change, such as extreme weather conditions, food insecurity, physical displacement, and less job opportunities (Jaiswal et al., 2022).

10 Discussion

The transition toward a green society from traditional one and sustainable development is absolutely necessary in order to address global environmental problems like pollution, resource depletion, and climate change. Within the context of social frameworks, a green society places an emphasis on environmental sustainability via inclusion of clean energy, sustainable urban development, responsible consumption, and technological innovation (Glavič, 2021). According to Alamoush et al. (2021), in order to achieve sustainable development, one must have broad strategy that incorporates economic growth, social fairness, and environmental preservation.

It is necessary for the development of green societies to make use of clean energy sources like solar, hydro, wind, and geothermal power. This is because using these sources reduces emissions of greenhouse gases and reduces dependency on fossil fuels (Energy & IEA, 2020). Transition to circular economy is essential for sustainable growth. Technological improvements enable enterprises to transition from the conventional linear economic model to closed-loop solutions that reduce environmental effect (Raihan et al., 2023).

Sustainable urban planning is essential for development of resilient and eco-supporting cities. Green infrastructure, smart urban technology, and sustainable urban transportation systems mitigate urban carbon emissions and enhance air quality in cities which is a major issue in several cities of developed and under-developing countries (Rabinovitch & Leitman, 2015). Conversely, the rapid and effective deployment of these environmentally friendly alternatives are hampered by things like costly initial expenditures, inadequate infrastructure, and regulatory framework weaknesses (OECD, 2017). Further, in order to realize a green society, ecologically sound agricultural and food systems are essential. As stated by Alzaabi et al. (2018), the environmental effect of food production may be mitigated by the use of precision farming, agro-ecological techniques, and alternative food sources. These include lab-grown meat and plant-based diets. According to Muhie (2022), these solutions not only prioritize the preservation of biodiversity but also address the issue of food security.

Despite advancements in sustainable development, some obstacles still persist. Economic and political obstacles, insufficient public knowledge, and opposition to change hinder the transition to green societies. Numerous developing countries have financial and technical limitations hindering their capacity to adopt sustainable solutions efficiently (Martinez-Alier, 2021). Strong and effective environmental legislative framework, significant global collaboration, awareness seminars, seed funding, and investing in sustainable technology are crucial measures to address these challenges.

11 Conclusion

The notion of a green society and sustainable development embodies a comprehensive strategy for attaining environmental, economic, and social prosperity. Integrating renewable energy, circular economy concepts, sustainable urban design, and innovative technology enables communities to substantially diminish their environmental impact and foster long-term sustainability (Hoang & Nguyen, 2021). Achieving this change requires robust regulatory frameworks, financial investments, and coordinated efforts among governments, enterprises, and community engagement (Xie et al., 2024).

The future of green society relies on ongoing technological developments, increasing awareness programs, and the enactment of policies that emphasize sustainability and sustainable practices. The use of clean energy, smart city projects, waste management advances, sustainable infrastructure, and sustainable farming methods are essential in alleviating environmental deterioration and fostering a sustainable future (Allam & Cheshmehzangi, 2024). Despite ongoing problems, the worldwide dedication to attaining sustainability via multilateral accords like the Paris Agreement and the UN SDGs demonstrates an increasing acknowledgment of the need for environmental action (Carlsen & Bruggemann, 2022).

Ultimately, cultivating green societies requires a transformation in societal values, norms, and behaviors, whereby environmental preservation and conscientious consumerism serve as the foundation of economic and social policy (Voulvoulis et al., 2022). By adopting sustainability as a fundamental concept, societies may secure a healthy world for future generations while promoting economic success and social equality (Hariram et al., 2023). Continuous efforts in education, policy reform, and technological innovation are crucial for attaining a genuinely sustainable world.

References

Abad-Segura, E., & González-Zamar, M. D. (2021). Sustainable economic development in higher education institutions: A global analysis within the SDGs framework. *Journal of Cleaner Production, 294*, 126133.

Abid, N., Ceci, F., Ahmad, F., & Aftab, J. (2022). Financial development and green innovation, the ultimate solutions to an environmentally sustainable society: Evidence from leading economies. *Journal of Cleaner Production, 369*, 133223.

Adamowicz, M. (2022). Green deal, green growth and green economy as a means of support for attaining the sustainable development goals. *Sustainability, 14*(10), 5901.

Adeyonu, A. G., Balogun, O. L., Amao, I. O., & Agboola, T. O. (2022). Does farmers' entrepreneurial competencies explain their household poverty status? Evidence from rural areas of Kwara State, Nigeria. *Cogent Economics & Finance, 10*(1), 2071384.

Akella, A. S., & Cannon, J. B. (2017). Strengthening the weakest links: Strategies for improving the enforcement of environmental laws globally. In *Transnational environmental crime* (pp. 459–492). Routledge.

Akermi, R., & Triki, A. (2017). The green energy transition and civil society in Tunisia: Actions, motivations and barriers. *Energy Procedia, 136*, 79–84.
Akinbami, C. A. O., Olawoye, J. E., Adesina, F. A., & Nelson, V. (2019). Exploring potential climate-related entrepreneurship opportunities and challenges for rural Nigerian women. *Journal of Global Entrepreneurship Research, 9*, 1–28.
Alamoush, A. S., Ballini, F., & Ölçer, A. I. (2021). Revisiting port sustainability as a foundation for the implementation of the United Nations Sustainable Development Goals (UN SDGs). *Journal of Shipping and Trade, 6*, 1–40.
Al-Emran, M., & Griffy-Brown, C. (2023). The role of technology adoption in sustainable development: Overview, opportunities, challenges, and future research agendas. *Technology in Society, 73*, 102240.
Allam, Z., & Cheshmehzangi, A. (2024). Technological innovation and sustainable transitions. In *Sustainable futures and green new deals* (pp. 55–79). Cham.
Alzaabi, M., Rizk, Z., & Mezher, T. (2018). Linking smart cities concept to energy-water-food nexus: The case of Masdar City in Abu Dhabi, UAE. In *Smart cities in the Gulf: Current state, opportunities, and challenges* (pp. 85–106). Singapore.
Amatuni, L., Ottelin, J., Steubing, B., & Mogollón, J. M. (2020). Does car sharing reduce greenhouse gas emissions? Assessing the modal shift and lifetime shift rebound effects from a life cycle perspective. *Journal of Cleaner Production, 266*, 121869.
Amran, A., Ooi, S. K., Mydin, R. T., & Devi, S. S. (2015). The impact of business strategies on online sustainability disclosures. *Business Strategy and the Environment, 24*, 551–564. https://doi.org/10.1002/bse.1837
Anand, A., & Josse, O. (2002). Celebrating mountain women: Moving mountains, moving women. *Mountain Research and Development, 22*(3), 233–235.
Anthony, B., Jr. (2024). The role of community engagement in urban innovation towards the co-creation of smart sustainable cities. *Journal of the Knowledge Economy, 15*(1), 1592–1624.
Asprey, A., & Nash, T. (2006). The importance of awareness and communication for the inclusion of young people with life-limiting and life-threatening conditions in mainstream schools and colleges. *British Journal of Special Education, 33*(1), 10–18.
Barbier, E. (1998). The concept of sustainable economic development. In *The economics of environment and development* (pp. 1–21). Edward Elgar Publishing.
Belmonte-Ureña, L. J., Plaza-Úbeda, J. A., Vazquez-Brust, D., & Yakovleva, N. (2021). Circular economy, degrowth and green growth as pathways for research on sustainable development goals: A global analysis and future agenda. *Ecological Economics, 185*, 107050.
Berchin, I. I., de Aguiar Dutra, A. R., & Guerra, J. B. S. O. D. A. (2021). How do higher education institutions promote sustainable development? A literature review. *Sustainable Development, 29*(6), 1204–1222.
Bexell, M., & Jönsson, K. (2017). Responsibility and the United Nations' sustainable development goals. In *Forum for development studies* (Vol. 44(1), pp. 13–29). Routledge.
Bhandari, V. (2017). New directions for social enterprises: The role of design in empowerment. *Scope: Contemporary Research Topics (Art & Design), 2017*(15), 24–28.
Björklund, M. (2011). Influence from the business environment on environmental purchasing—Drivers and hinders of purchasing green transportation services. *Journal of Purchasing and Supply Management, 17*(1), 11–22.
Blincoe, K. (2022). Barriers hindering transition to sustainability. In *Achieving sustainability: The ultimate human challenge: Critical barriers and future perspectives* (pp. 121–137). Springer International Publishing.
Bosnjak, M., Ajzen, I., & Schmidt, P. (2020). The theory of planned behavior: Selected recent advances and applications. *Europe's Journal of Psychology, 16*(3), 352.
Brooks, J. S. (2013). Avoiding the limits to growth: Gross National Happiness in Bhutan as a model for sustainable development. *Sustainability, 5*(9), 3640–3664.
Brulle, R. J. (2020). Denialism: Organized opposition to climate change action in the United States. In *Handbook of US environmental policy* (pp. 328–341). Edward Elgar Publishing.

Brydges, T. (2021). Closing the loop on take, make, waste: Investigating circular economy practices in the Swedish fashion industry. *Journal of Cleaner Production, 293*, 126245.
Cahn, M. (2008). Indigenous entrepreneurship, culture and micro-enterprise in the Pacific Islands: Case studies from Samoa. *Entrepreneurship and Regional Development, 20*(1), 1–18.
Carlsen, L., & Bruggemann, R. (2022). The 17 United Nations' sustainable development goals: A status by 2020. *International Journal of Sustainable Development & World Ecology, 29*(3), 219–229.
Chapin, F. S., III, Weber, E. U., Bennett, E. M., Biggs, R., Van Den Bergh, J., Adger, W. N., et al. (2022). Earth stewardship: Shaping a sustainable future through interacting policy and norm shifts. *Ambio, 51*(9), 1907–1920.
Chen, C. C. (2024). Comparative impacts of energy sources on environmental quality: A five-decade analysis of Germany's Energiewende. *Energy Reports, 11*, 3550–3561.
Chiang, P. C., Ma, H. W., Wen, L., & Lin, C. H. (2024). Sustainable Development. In *Introduction to green science and technology for green economy: Principles and applications* (pp. 9–45). Singapore.
Chien, F., Ananzeh, M., Mirza, F., Bakar, A., Vu, H. M., & Ngo, T. Q. (2021). The effects of green growth, environmental-related tax, and eco-innovation towards carbon neutrality target in the US economy. *Journal of Environmental Management, 299*, 113633.
Chipfupa, U., & Wale, E. (2018). Farmer typology formulation accounting for psychological capital: Implications for on-farm entrepreneurial development. *Development in Practice, 28*(5), 600–614.
Chong, M. D., Chong, E. D., & Díaz, D. G. (2019). Effect of rural entrepreneurship and innovation adoption in the reduction of inequalities. *Journal of Science and Research: Revista Ciencia e Investigación, 4*(1), 268–290.
Chowdhury, A., Naz, A., & Maiti, S. K. (2017). Health risk assessment of 'tiger prawn seed'collectors exposed to heavy metal pollution in the conserved mangrove forest of Indian Sundarbans: A socio-environmental perspective. *Human and Ecological Risk Assessment: An International Journal, 23*(2), 203–224.
Chung, K. C. (2020). Green marketing orientation: Achieving sustainable development in green hotel management. *Journal of Hospitality Marketing & Management, 29*(6), 722–738.
Debrah, J. K., Vidal, D. G., & Dinis, M. A. P. (2021). Raising awareness on solid waste management through formal education for sustainability: A developing countries evidence review. *Recycling, 6*(1), 6.
Dell'Anna, F., Bravi, M., & Bottero, M. (2022). Urban green infrastructures: How much did they affect property prices in Singapore? *Urban Forestry & Urban Greening, 68*, 127475.
Dhayal, K. S., Giri, A. K., Kumar, A., Samadhiya, A., Agrawal, S., & Agrawal, R. (2023). Can green finance facilitate industry 5.0 transition to achieve sustainability? A systematic review with future research directions. *Environmental Science and Pollution Research, 30*(46), 102158–102180.
Doh, J. P., & Guay, T. R. (2006). Corporate social responsibility, public policy, and NGO activism in Europe and the United States: An institutional-stakeholder perspective. *Journal of Management Studies, 43*(1), 47–73.
El Bilali, H., Strassner, C., & Ben Hassen, T. (2021). Sustainable agri-food systems: Environment, economy, society, and policy. *Sustainability, 13*(11), 6260.
Emon, M. M. H., & Khan, T. (2023). The impact of cultural norms on sustainable entrepreneurship practices in SMEs of Bangladesh. *Indonesian Journal of Innovation and Applied Sciences (IJIAS), 3*(3), 201–209.
Energy, I. W., & IEA, A. (2020). IEA. Renewable Energy Market Update.
Evans, S. (2010). *The green republic: A conservation history of Costa Rica*. University of Texas Press.
Fenibo, E. O., Ijoma, G. N., & Matambo, T. (2021). Biopesticides in sustainable agriculture: A critical sustainable development driver governed by green chemistry principles. *Frontiers in Sustainable Food Systems, 5*, 619058.

Fernandes, C. I., Veiga, P. M., Ferreira, J. J., & Hughes, M. (2021). Green growth versus economic growth: Do sustainable technology transfer and innovations lead to an imperfect choice? *Business Strategy and the Environment, 30*(4), 2021–2037.
Ferreira, C. M., & Serpa, S. (2018). Society 5.0 and social development. *Management and Organizational Studies, 5*(4), 26–31.
Figueroa-Domecq, C., De Jong, A., & Williams, A. M. (2020). Gender, tourism & entrepreneurship: A critical review. *Annals of Tourism Research, 84*, 102980.
Follador, D., Duarte, F., & Carrier, M. (2022). Informal institutions and path dependence in urban planning: The case of Curitiba, Brazil. *Journal of Urban Affairs, 44*(1), 2–21.
Fukuyama, M. (2018). Society 5.0: Aiming for a new human-centered society. *Japan Spotlight, 27*(5), 47–50.
Geissdoerfer, M., Savaget, P., Bocken, N. M., & Hultink, E. J. (2017). The circular economy—A new sustainability paradigm? *Journal of Cleaner Production, 143*, 757–768.
Ghisetti, C., Mancinelli, S., Mazzanti, M., & Zoli, M. (2017). Financial barriers and environmental innovations: Evidence from EU manufacturing firms. *Climate Policy, 17*(suppl 1), S131–S147.
Gibson, R. (Ed.). (2016). *Sustainability assessment: Applications and opportunities*. Taylor & Francis.
Glavič, P. (2021). Evolution and current challenges of sustainable consumption and production. *Sustainability, 13*(16), 9379.
Guo, R., Lv, S., Liao, T., Xi, F., Zhang, J., Zuo, X., et al. (2020). Classifying green technologies for sustainable innovation and investment. *Resources, Conservation and Recycling, 153*, 104580.
Gustafsson, H. R., & Kelly, E. A. (2016). Developing the sustainable city: Curitiba, Brazil, as a case study. In *How cities will save the world* (pp. 81–96). Routledge.
Hajj-Hassan, M., Chaker, R., Karameh Shayya, J., & Lo Presti, V. (2024). Environmental citizenship in Lebanon: A study in the higher education context. *International Journal of Adolescence and Youth, 29*(1), 2387079.
Hariram, N. P., Mekha, K. B., Suganthan, V., & Sudhakar, K. (2023). Sustainalism: An integrated socio-economic-environmental model to address sustainable development and sustainability. *Sustainability, 15*(13), 10682.
Harvey, J., Heidrich, O., & Cairns, K. (2014). Psychological factors to motivate sustainable behaviours. *Proceedings of the Institution of Civil Engineers-Urban Design and Planning, 167*(4), 165–174.
Hernández-Chaverri, R. A., & Buenrostro-Figueroa, J. J. (2021). Biomass: Biorefinery as a model to boost the bioeconomy in Costa Rica, a review. *Agronomía Mesoamericana, 32*(3), 1047–1070.
Heshmati, A., & Rashidghalam, M. (2021). Assessment of the urban circular economy in Sweden. *Journal of Cleaner Production, 310*, 127475.
Heuër, A., Kloibhofer, M., & Marquard, H. (2015). *EcoPost: Preventing deforestation, combatting pollution and creating youth employment by transforming waste into plastic posts in Kenya*. SEED.
Hnatyuk, V., Pshenychna, N., Kara, S., Kolodii, V., & Yaroshchuk, L. (2024). Education's role in fostering environmental awareness and advancing sustainable development within a holistic framework. *Multidisciplinary Reviews, 7*, 2024spe012.
Hoang, A. T., & Nguyen, X. P. (2021). Integrating renewable sources into energy system for smart city as a sagacious strategy towards clean and sustainable process. *Journal of Cleaner Production, 305*, 127161.
Hvelplund, F. (2013). Innovative democracy, political economy, and the transition to renewable energy. A full-scale experiment in Denmark 1976-2013. *Environmental Research, Engineering and Management, 66*(4), 5–21.
Igwe, P. A., Rahman, M., Odunukan, K., Ochinanwata, N., Egbo, O. P., & Ochinanwata, C. (2020). Drivers of diversification and pluriactivity among smallholder farmers—Evidence from Nigeria. *Green Finance, 2*(3), 263.

Ismael, N. I. (2023). Sustainability criteria used in designing energy-efficient smart cities A study of Masdar City as a model for one of the smart cities that realize the idea of sustainable development. *International Journal of Advanced Engineering and Business Sciences, 4*(1), 42–71.

Jaiswal, K. K., Chowdhury, C. R., Yadav, D., Verma, R., Dutta, S., Jaiswal, K. S., & Karuppasamy, K. S. K. (2022). Renewable and sustainable clean energy development and impact on social, economic, and environmental health. *Energy Nexus, 7*, 100118.

Kimbu, A. N., Booyens, I., & Winchenbach, A. (2022). Livelihood diversification through tourism: Identity, well-being, and potential in rural coastal communities. *Tourism Review International, 26*(1), 25–40.

Koech, M. K., Munene, K. J., Kinoti, M. K., & Situma, D. K. (2023). Circular economy policies and innovations in Africa: Pillars for achieving sustainable development. In *Circular economy adoption: Catalysing decarbonisation through policy instruments* (pp. 99–130). Singapore.

Kollmuss, A., & Agyeman, J. (2002). Mind the gap: Why do people act environmentally and what are the barriers to pro-environmental behavior? *Environmental Education Research, 8*(3), 239–260.

Lal, S. R. S., Herbert, G. M. J., Arjunan, P., & Suryan, A. (2022). Advancements in renewable energy transition in India: A review. In *Energy Sources, part A: Recovery, utilization, and environmental effects* (pp. 1–31). Taylor & Francis.

Le, T. T. (2022). How do corporate social responsibility and green innovation transform corporate green strategy into sustainable firm performance? *Journal of Cleaner Production, 362*, 132228.

Leach, M., & Scoones, I. (2007). Mobilising citizens: Social movements and the politics of knowledge.

Lee, C. T., Hashim, H., Ho, C. S., Van Fan, Y., & Klemeš, J. J. (2017). Sustaining the low-carbon emission development in Asia and beyond: Sustainable energy, water, transportation and low-carbon emission technology. *Journal of Cleaner Production, 146*, 1–13.

Lepeley, M. T. (2017). Bhutan's gross national happiness: An approach to human centred sustainable development. *South Asian Journal of Human Resources Management, 4*(2), 174–184.

Luke, T. W. (2005). Neither sustainable nor development: Reconsidering sustainability in development. *Sustainable Development, 13*, 228–238. https://doi.org/10.1002/sd.284

Lund, H., Thellufsen, J. Z., Sorknæs, P., Mathiesen, B. V., Chang, M., Madsen, P. T., et al. (2022). Smart energy Denmark. A consistent and detailed strategy for a fully decarbonized society. *Renewable and Sustainable Energy Reviews, 168*, 112777.

McDonough, W., & Braungart, M. (2002). Design for the triple top line: new tools for sustainable commerce. *Corporate Environmental Strategy, 9*(3), 251–258.

Macedo, J. (2013). Planning a sustainable city: The making of Curitiba, Brazil. *Journal of Planning History, 12*(4), 334–353.

Majid, M. (2020). Renewable energy for sustainable development in India: Current status, future prospects, challenges, employment, and investment opportunities. *Energy, Sustainability and Society, 10*(1), 1–36.

Martinez-Alier, J. (2021). Mapping ecological distribution conflicts: The EJAtlas. *The Extractive Industries and Society, 8*(4), 100883.

Meng, B., & Hao, Z. (2024). Role of green finance and higher education in fostering the sustainability and energy transition practices. *Humanities and Social Sciences Communications, 11*(1), 1–8.

Milios, L. (2021). Towards a circular economy taxation framework: Expectations and challenges of implementation. *Circular Economy and Sustainability, 1*(2), 477–498.

Milla Quesada, V. (2023). *Unveiling the role of state intervention in green transformations within Costa Rica's forestry and agriculture land use sectors*.

Mishra, O. (2021). Principles of frugal innovation and its application by social entrepreneurs in times of adversity: An inductive single-case approach. *Journal of Entrepreneurship in Emerging Economies, 13*(4), 547–574.

Mondejar, M. E., Avtar, R., Diaz, H. L. B., Dubey, R. K., Esteban, J., Gómez-Morales, A., et al. (2021). Digitalization to achieve sustainable development goals: Steps towards a Smart Green Planet. *Science of the Total Environment, 794*, 148539.

Morán-Ordóñez, A., Hermoso, V., & Martínez-Salinas, A. (2022). Multi-objective forest restoration planning in Costa Rica: Balancing landscape connectivity and ecosystem service provisioning with sustainable development. *Journal of Environmental Management, 310*, 114717.

Mourtzis, D. (2021). Towards the 5th industrial revolution: A literature review and a framework for process optimization based on big data analytics and semantics. *Journal of Machine Engineering, 21*(3).

Mourtzis, D., Angelopoulos, J., & Panopoulos, N. (2022). A literature review of the challenges and opportunities of the transition from Industry 4.0 to Society 5.0. *Energies, 15*(17), 6276.

Muhie, S. H. (2022). Novel approaches and practices to sustainable agriculture. *Journal of Agriculture and Food Research, 10*, 100446.

Naess, P. (2001). Urban planning and sustainable development. *European Planning Studies, 9*(4), 503–524.

Narvaez Rojas, C., Alomia Peñafiel, G. A., Loaiza Buitrago, D. F., & Tavera Romero, C. A. (2021). Society 5.0: A Japanese concept for a superintelligent society. *Sustainability, 13*(12), 6567.

Newman, P., & Kenworthy, J. (2015). *The end of automobile dependence* (pp. 105–140). Island Press/Center for Resource Economics.

Nikolaou, I. E., & Tsagarakis, K. P. (2021). An introduction to circular economy and sustainability: Some existing lessons and future directions. *Sustainable Production and Consumption, 28*, 600–609.

Nosheen, M., Iqbal, J., & Abbasi, M. A. (2021). Do technological innovations promote green growth in the European Union? *Environmental Science and Pollution Research, 28*, 21717–21729.

O'Brien, K. L., & Wolf, J. (2010). A values-based approach to vulnerability and adaptation to climate change. *Wiley Interdisciplinary Reviews: Climate Change, 1*(2), 232–242.

Obiagu, A., Ocheje, J., Ofodum, I., & Eze, E. (2024). Fostering environmental personal norms: The role of environmental education in Nigerian pre-service teachers' environmental knowledge, pro-environmental beliefs and behaviours. *Environmental Education Research, 30*(8), 1231–1246.

OECD. (2017, June). Measuring distance to the SDG targets: An assessment of where OECD countries stand.

Österblom, H., Bebbington, J., Blasiak, R., Sobkowiak, M., & Folke, C. (2022). Transnational corporations, biosphere stewardship, and sustainable futures. *Annual Review of Environment and Resources, 47*(1), 609–635.

Østergaard, P. A., Duic, N., Noorollahi, Y., Mikulcic, H., & Kalogirou, S. (2020). Sustainable development using renewable energy technology. *Renewable Energy, 146*, 2430–2437.

Palmer, J. (2002). *Environmental education in the 21st century: Theory, practice, progress and promise*. Routledge.

Pandit, R., Parrota, J., Anker, Y., Coudel, E., Diaz Morejón, C. F., Harris, J., et al. (2018). *Responses to halt land degradation and to restore degraded land*. IPBES.

Pandita, A., Pradeep, A., Anjukandan, R., & Vignesh, T. (2024, June). Exploring Masdar City: Smart innovations and future eco-city model in urban sustainability. In *2024 Advances in Science and Engineering Technology International Conferences (ASET)* (pp. 1–6). IEEE.

Pandey, V. (2021). Energy infrastructure for sustainable development. *In Affordable and clean energy* (pp. 427–440). Cham: Springer International Publishing.

Pato, M. L., & Teixeira, A. A. (2016). Twenty years of rural entrepreneurship: A bibliometric survey. *Sociologia Ruralis, 56*(1), 3–28.

Pinson, P., Mitridati, L., Ordoudis, C., & Østergaard, J. (2017). Towards fully renewable energy systems: Experience and trends in Denmark. *CSEE Journal of Power and Energy Systems, 3*(1), 26–35.

Pritchett, M., Ala'i-Rosales, S., Cruz, A. R., & Cihon, T. M. (2022). Social justice is the spirit and aim of an applied science of human behavior: Moving from colonial to participatory research practices. *Behavior Analysis in Practice, 15*(4), 1074–1092.

Priya, B. K., & Thenmozhi, S. (2021). Study on self-efficacy and pro-environmental behavior among school students. *International Journal of Indian Psychology, 9*(2), 1851–1858.

Rabinovitch, J., & Leitman, J. (2015). Urban planning in Curitiba. In *The City Reader* (pp. 548–554). Routledge.

Rai, K. (2024). *The contribution of the United Nations Sustainable Development Goals (SDGs) to enhancing Gross National Happiness (GNH) in Bhutan: A case study of Bhutan* [Doctoral dissertation, Murdoch University].

Raihan, A., Pavel, M. I., Muhtasim, D. A., Farhana, S., Faruk, O., & Paul, A. (2023). The role of renewable energy use, technological innovation, and forest cover toward green development: Evidence from Indonesia. *Innovation and Green Development, 2*(1), 100035.

Rechsteiner, R. (2021). German energy transition (Energiewende) and what politicians can learn for environmental and climate policy. *Clean Technologies and Environmental Policy, 23*, 305–342.

Rietmann, H. (2021). *Ecological modernization in The United Arab Emirates?: The case of Masdar Eco-City* (Vol. 119). Walter de Gruyter GmbH & Co KG.

Rockstrom, J., Steffen, W., Noone, K., Persson, A., Chapin, S., Lambin, E. F., Lenton, T. M., et al. (2009). A safe operating space for humanity. *Nature, 461*(7263), 472–475.

Samal, P. K., Lodhi, E. M. S., Arya, S. C., Sundriyal, R. C., & Dhyani, P. P. (2016). Eco-technologies for agricultural and rural livelihoods in north East India. *Current Science, 111*, 1929–1935.

Samantarai, M., & Dutta, S. (2024). Lorna Rutto's EcoPost: Creating a sustainable solution to Kenya's plastic waste problem. *The CASE Journal.*

Schischke, A., Papenfuß, P., Brem, M., Kurz, P., & Rathgeber, A. W. (2023). Sustainable energy transition and its demand for scarce resources: Insights into the German Energiewende through a new risk assessment framework. *Renewable and Sustainable Energy Reviews, 176*, 113190.

Schöggl, J. P., Stumpf, L., & Baumgartner, R. J. (2024). The role of interorganizational collaboration and digital technologies in the implementation of circular economy practices—Empirical evidence from manufacturing firms. *Business Strategy and the Environment, 33*(3), 2225–2249.

Schroeder, P., Anggraeni, K., & Weber, U. (2019). The relevance of circular economy practices to the sustainable development goals. *Journal of Industrial Ecology, 23*, 77–95. https://doi.org/10.1111/jiec.12732

Shah, K. J., Pan, S. Y., Lee, I., Kim, H., You, Z., Zheng, J. M., & Chiang, P. C. (2021). Green transportation for sustainability: Review of current barriers, strategies, and innovative technologies. *Journal of Cleaner Production, 326*, 129392.

Shahraki, H., & Heydari, E. (2019). Rethinking rural entrepreneurship in the era of globalization: Some observations from Iran. *Journal of Global Entrepreneurship Research, 9*(1), 42.

Shahzad, M., Qu, Y., Rehman, S. U., & Zafar, A. U. (2022). Adoption of green innovation technology to accelerate sustainable development among manufacturing industry. *Journal of Innovation & Knowledge, 7*(4), 100231.

Shamsuzzoha, A., Nieminen, J., Piya, S., & Rutledge, K. (2021). Smart city for sustainable environment: A comparison of participatory strategies from Helsinki, Singapore and London. *Cities, 114*, 103194.

Shaw, B. (2017). The whole village project Saxon village restoration in rural Romania. In *The science and practice of landscape stewardship* (pp. 164–166). Cambridge University Press.

Shcherbina, E., Gorbenkova, E., & Slepnev, M. (2017). Urban-planning sustainability problems in a city natural framework. In *MATEC Web of Conferences* (p. 106). https://doi.org/10.1051/matecconf/201710601032

Sheehy, B., & Farneti, F. (2021). Corporate social responsibility, sustainability, sustainable development and corporate sustainability: What is the difference, and does it matter? *Sustainability, 13*(11), 5965.

Shiroishi, Y., Uchiyama, K., & Suzuki, N. (2018). Society 5.0: For human security and well-being. *Computer, 51*(7), 91–95.

Sloot, D., Jans, L., & Steg, L. (2018). Can community energy initiatives motivate sustainable energy behaviours? The role of initiative involvement and personal pro-environmental motivation. *Journal of Environmental Psychology, 57*, 99–106.

Song, M., Hu, C., Yuan, J., Zhang, A., & Liu, X. (2023). Toward an ecological civilization: Exploring changes in China's land use policy over the past 35 years using text mining. *Journal of Cleaner Production, 427*, 139265.

Soomro, J. A., Suhrab, M., & Abbasi, B. A. (2025). Navigating green regulations in the business ecosystem: Government support, compliance, and obstacles in the context of Pakistan. In *Government influences on eco-friendly practices in business* (pp. 55–86). IGI Global.

Spash, C. L. (2017). *Routledge handbook of ecological economics*. Routledge.

Steg, L. (2016). Values, norms, and intrinsic motivation to act proenvironmentally. *Annual Review of Environment and Resources, 41*(1), 277–292.

Steg, L., & Vlek, C. (2009). Social science and environmental behaviour. In *Principles of environmental sciences* (pp. 97–141). Springer Netherlands.

Stern, P. C. (2000). New environmental theories: Toward a coherent theory of environmentally significant behavior. *Journal of Social Issues, 56*(3), 407–424.

Stern, N. (2008). The economics of climate change. *American Economic Review, 98*(2), 1–37.

Steurer, R., Langer, M. E., Konrad, A., & Martinuzzi, A. (2005). Corporations, stakeholders and sustainable development I: A theoretical exploration of business–society relations. *Journal of Business Ethics, 61*, 263–281.

Suhrab, M., Soomro, J. A., Ullah, S., & Chavara, J. (2023). The effect of gross domestic product, urbanization, trade openness, financial development, and renewable energy on CO2 emission. *Environmental Science and Pollution Research, 30*(9), 22985–22991.

Tabares, A., Londoño-Pineda, A., Cano, J. A., & Gómez-Montoya, R. (2022). Rural entrepreneurship: An analysis of current and emerging issues from the sustainable livelihood framework. *Economies, 10*(6), 142.

Tan, P. Y., Wang, J., & Sia, A. (2013). Perspectives on five decades of the urban greening of Singapore. *Cities, 32*, 24–32.

Teske, S., Morris, T., & Nagrath, K. (2020). 100% renewable energy for Costa Rica. Report prepared by ISF for the World Future Council/Germany and the One Earth Foundation, USA.

Tilbury, D., & Ryan, A. (2011). Today becomes tomorrow: Re-thinking business practice, education and learning in the context of sustainability. *Journal of Global Responsibility, 2*(2), 137–150.

United Nations Environment Programme. Division of Early Warning, & Assessment. (2011). UNEP year book 2011: emerging issues in our global environment. UNEP/Earthprint.

Ura, K., Alkire, S., & Zangmo, T. (2012). *GNH and GNH index* (pp. 1–60). The Centre for Bhutan Studies.

Usman, M., Khan, N., & Omri, A. (2024). Environmental policy stringency, ICT, and technological innovation for achieving sustainable development: Assessing the importance of governance and infrastructure. *Journal of Environmental Management, 365*, 121581.

Van Trijp, H. C. (2013). *Encouraging sustainable behavior: Psychology and the environment*. Psychology Press.

Verplanken, B., & Whitmarsh, L. (2021). Habit and climate change. *Current Opinion in Behavioral Sciences, 42*, 42–46.

Voulvoulis, N., Giakoumis, T., Hunt, C., Kioupi, V., Petrou, N., Souliotis, I., & Vaghela, C. J. G. E. C. (2022). Systems thinking as a paradigm shift for sustainability transformation. *Global Environmental Change, 75*, 102544.

Walz, R., Pfaff, M., Marscheider-Weidemann, F., & Glöser-Chahoud, S. (2017). Innovations for reaching the green sustainable development goals—Where will they come from? *International Economics and Economic Policy, 14*, 449–480.

Wang, W. H., Moreno-Casas, V., & Huerta de Soto, J. (2021). A free-market environmentalist transition toward renewable energy: The cases of Germany, Denmark, and the United Kingdom. *Energies, 14*(15), 4659.
White, K., Habib, R., & Hardisty, D. J. (2019). How to SHIFT consumer behaviors to be more sustainable: A literature review and guiding framework. *Journal of Marketing, 83*(3), 22–49.
Wojnarowskais, M., Ćwiklicki, M., & Ingrao, C. (2022). Sustainable products in the circular economy.
World Maufacturing Forum. (2019). *Report 2019: Skills for the future of manufacturing*. https://worldmanufacturing.org/report/report-2019/
Wu, C. H., Tsai, S. B., Liu, W., Shao, X. F., Xia, Y. K., & Wacławek, M. (2021a). Green environment and sustainable development: Methods and applications. *Ecological Chemistry and Engineering S, 28*(4), 467–470.
Wu, M., Liu, Y., Xu, Z., Yan, G., Ma, M., Zhou, S., & Qian, Y. (2021b). Spatio-temporal dynamics of China's ecological civilization progress after implementing national conservation strategy. *Journal of Cleaner Production, 285*, 124886.
Xie, H., Ahmad, T., Zhang, D., Goh, H. H., & Wu, T. (2024). Community-based virtual power plants' technology and circular economy models in the energy sector: A techno-economy study. *Renewable and Sustainable Energy Reviews, 192*, 114189.
Yan, L., Zhang, X., Pan, H., Wu, J., Lin, L., Zhang, Y., et al. (2021). Progress of Chinese ecological civilization construction and obstacles during 2003–2020: Implications from one set of emergy-based indicator system. *Ecological Indicators, 130*, 108112.
Zhang, X., Han, L., Wei, H., Tan, X., Zhou, W., Li, W., & Qian, Y. (2022). Linking urbanization and air quality together: A review and a perspective on the future sustainable urban development. *Journal of Cleaner Production, 346*, 130988.
Zheng, X., Wang, R., Hoekstra, A. Y., Krol, M. S., Zhang, Y., Guo, K., et al. (2021). Consideration of culture is vital if we are to achieve the Sustainable Development Goals. *One Earth, 4*(2), 307–319.
Zingoni de Baro, M. E. (2022). Curitiba case study. In *Regenerating cities: Reviving places and planet* (pp. 117–162). Springer International Publishing.

Investing in Sustainability: Economic Models and Sustainable Finance Products

Raja Rehan, Auwal Adam Sa'ad, Hina Fatima, and Liaqat Ali

1 Introduction

The world is facing several mounting challenges of climate alteration, resource depletion, social inequalities, biodiversity losses, etc. that enhance the importance of sustainability and become a core concern for businesses and investors (Tanchangya et al., 2024). Thus, it forced policymakers and investigators to rethink the existing financial and economic systems and come up with an alternative sustainable finance which offers sustainable growth with considerations of environmental and ethical issues. Sustainable finance explains a transformative change in how financial products and services support an equitable, greener, and sustainable future (Adeoye et al., 2024). Technically, sustainable finance mentions the procedure of considering environmental, social, and governance (ESG) measures when finalizing long-term investment-related decisions and returns of financial activities of businesses (Boiral

R. Rehan (✉)
IIUM Institute of Islamic Banking and Finance (IIiBF), International Islamic University Malaysia (IIUM), Kuala Lumpur, Malaysia

Khadim Ali Shah Bukhari Institute of Technology (KASBIT), Karachi, Pakistan
e-mail: rajarehan@iium.edu.my; dr.rajarehan@kasbit.edu.pk

A. A. Sa'ad
IIUM Institute of Islamic Banking and Finance (IIiBF), International Islamic University Malaysia (IIUM), Kuala Lumpur, Malaysia
e-mail: auwal@iium.edu.my

H. Fatima
Department of Management Sciences, Muhammad Ali Jinnah University, Karachi, Pakistan
e-mail: hina.fatima@jinnah.edu

L. Ali
Department of Management Studies, Bahria University, Karachi, Pakistan
e-mail: liaquatali.bukc@bahria.edu.pk

A. Chakir et al. (eds.), *Green Society, Environmental Strategies and Sustainable Development*, https://doi.org/10.1007/978-3-031-94086-6_12

et al., 2024). Importantly, the traditional growth-based economic models that offer guidelines for investment decisions generally ignore the capital costs of businesses that they face because of executing numerous obligatory environmental and social measures (Newell et al., 2024). For instance, the core components of sustainable finance such as ESG measures are required to be implemented by investors but the costs associated with executing these measures are unnoticed or hardly discussed (Hampl & Vágnerová Linnertová, 2024). This is because the traditional economic models focus only on investment outcomes and overlook the concept of sustainable investments. An investment that produces positive, assessable environmental and social impact together with a financial return is called a sustainable investment (Marti et al., 2024). Remarkably, sustainable investments are measured as a core component that helps attain the United Nations' Sustainability Development Goals (SDGs) (Singhania et al., 2024a). However, the scarcity of finance and investments for sustainability-connected plans is the main reason that has forced numerous nations to take off track from the SDGs agenda and devastating climate change (United Nations, 2024). Noticeably, global financial institutes are facing a lot of pressure to arrange funds for preserving SDG-connected activities through their investment choices and allocation of credits (Ozili & Iorember, 2024). The recent estimates of SDGs maintaining and implementing costs over the last 10 years vary between 20% and 80% of global gross domestic products (Cingolani, 2024). Hence, attracting investments for financing sustainability is now no longer purely an ethical consideration but it has also become a critical financial and strategic imperative. Visibly, there is mounting recognition of the need to develop more systematic ways that produce sustainable revenues and investments to accomplish the sustainability-connected project financing challenges (Joshipura et al., 2024).

In the quest to develop more investment opportunities for financing sustainability, it is unavoidable to ignore the vital role of current economic models and their offered sustainability-connected products for attracting more investments. Interestingly, current economic models such as the green finance model, impact investing model, green bond financing, socially responsible investing financing, blended finance, circular economy, climate finance model, ESG integration-based model, blended finance model, etc. of sustainable finance are raised to attract more investments and generate funds for financing sustainability-connected projects by introducing innovative financial products (Myronchuk et al., 2024). Fundamentally, these topical economic models are produced by considering sustainable growth for nominees in all periods. In comparison with the traditional economic models (see Table 1), these recent models offered products such as green bonds, green venture capital, and impact investing and present numerous new opportunities that generate investments for financing sustainability-connected plans (Hossain et al., 2024). Therefore, they are rapidly accepted by investors as they help in accomplishing long-term financial, environmental, and social goals. Hence, the upsurge of recent economic models signals a paradigm shift, which explains how capital flows move toward projects and businesses that prioritize financial resilience, environmental health, and social equity. Visibly, the investment landscape has evolved, with growing recognition that the concept of sustainability is not only an ethical obligation but it is also a key source of investment opportunity (Nartey, 2025). Moreover, investors are now more aware of the dissimilar risks linked with maintaining unsustainable practices such as executing traditional economic models and the potential return from investing in sustainable businesses. Remarkably, the traditional growth-based economic models focused more on profit maximization and achieving short-term financial goals (see Table 1). As a result, sustainability-based current economic models and their introduced financial products are evolving because of the mounting demand for responsible and impactful investments (Cui et al., 2020). In addition, the products and services offered by the firms are also useful in capturing more investments to finance sustainability-related projects.

Table 1 displays the key differences between traditional and current sustainable finance economic models. Remarkably, the traditional economic models ignore ecological effects and involve activities that are not measured as sustainable. In

Table 1 Comparison between traditional and sustainable finance models

S#	Features	Traditional finance models	Sustainable finance models
1	Primary focus	Profit maximization, financial returns	Long-term value creation, balancing profit and sustainability
2	Objectives	Short-term financial goals	Long-term economic, environmental, and social outcomes
3	Risk assessment	Primarily financial risks (market, credit, operational)	Include environmental, social, and governance, i.e., ESG risks
4	Capital allocation	Allocate capital based on financial metrics (ROI, NPV)	Allocate capital with consideration of ESG impacts, alongside financial returns
5	Decision-making criteria	Financial metrics such as ROI and profit margins	ESG criteria alongside financial returns and risks
6	Environmental impacts	Limited considerations of environmental issues	Significant focus on minimizing environmental harm and fostering sustainability
7	Social impact	Minimal focus on social consequences	Strong emphasis on social welfare, equity, and inclusion
8	Stakeholder focus	Primarily shareholders and investors	Broad focus including society, environment, employees, and communities
9	Investment products	Traditional stocks, bonds, derivatives	Green bonds, social impact funds, ESG funds, sustainability-linked loans
10	Regulatory environment	Less emphasis on regulations for social and environmental goals	Stronger regulations encouraging sustainability and responsible practices
11	Time horizon	Short-term (quarterly or annual)	Long-term (years or decades)
12	Return expectations	Focus more on financial return and market performance	Focus on financial return but with added emphasis on ESG performance
13	Profit model	Financial profit, short-term focus	Profit integrated with sustainable development and societal well-being

Source: Authors' Own Elaboration

contrast, the sustainable finance economic models encourage investments that align financial return with social and environmental goals.

Considering the above discussion, this chapter is developed to inspect the various current economic models that are measured essential and offer guidelines that may help in funding sustainability-connected initiatives. It also delves into how various innovative economic models and their offered sustainability-connected products can align financial growth with sustainable development. Thus, this chapter begins with deliberating economic models and their offered investment tools that underpin the concept of sustainability. Additionally, this chapter also deliberates on the difficulties and challenges of discussed models in generating funds and investments for financing sustainability-connected projects. Subsequently, the following section explores the various economic models that facilitate dissimilar sustainability initiatives.

2 Current Key Economic Models for Funding Sustainability

Traditional economic models often prioritize short-term revenues and mostly do not consider the cost of long-term social and environmental well-being (see Table 1). Typically, these models rely on existing market mechanisms that mostly undervalue natural resources and environmental costs (Ari & Koc, 2021). Therefore, the limitations of these models in encouraging sustainability are becoming evident and demanding the development of more integrated approaches that can generate revenue and bring more investments for financing sustainability (Zhang & Zhao, 2024). The topical core economic models that support sustainability and their connected products produce finance by capturing investments for sustainable projects, including green finance sustainable financing model, impact investing financing model, green bond financing model, socially responsible investing financing model, blended finance model, circular economy model, sustainable development goals financing model, etc. Figure 1 mentions the key types of current economic models developed to finance sustainability-linked plans and projects.

Figure 1 explains the core types of sustainability-connected economic models. Some other sustainability-linked economic models include climate finance financing models, the regenerative finance financing model, the well-being economy financing model, the pay-for-success economic model, the carbon finance financing model, the natural capital financing model, the decentralized finance (DeFI) financing model, etc. A brief overview of current sustainability-supportive economic models and how their offered products generate investments for financing sustainability are discussed below in the following subsections.

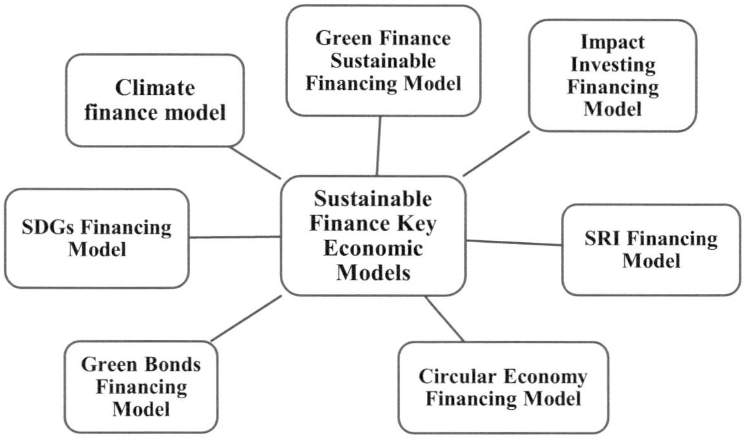

Fig. 1 Sustainability-connected key economic models. (Source: Authors' Own Elaboration)

2.1 Green Finance Sustainable Financing Model

This model refers to the plans created to support those initiatives that sustain businesses' financial sustainability while considering the environmental benefits (Kumar et al., 2024). Besides, this model offers products that are used to produce more capital and attract investments to finance sustainability link plans. Typically, it comprises energy efficiency improvement projects, renewal energy projects, and sustainable agriculture that generate revenue to finance more sustainable projects. Hence, the green finance investment-based model and its connected products are built to promote financial and environmental sustainability (Chen et al., 2024). Also, this model pursues the transition from a traditional economy to a low-carbon and sustainable economy by integrating environmental considerations into financial decision-making. Moreover, it highlights investments that not only offer financial returns but also produce significant environmental outcomes. It functions on the principles that are aligned with financial returns and environmental sustainability to attain a resilient and sustainable economy. Notably, the three core principles on which the green finance investment model depends are environmental impact, sustainable goals, and risk management of the project.

The environmental impact explains that investments must possess a direct or indirect significant impact on the environment. Similarly, the sustainability goals support long-term environmental goals that are developed to mitigate risks that are connected with natural resource depletion and climate change (Zhang et al., 2021). Then, risk management is a process that assesses dissimilar types of environmental risks. These risks comprise physical, reputational, financial, and regulatory risks of sustainable finance projects that may harm the environment and reduce the earning capacity of the project (Mumtaz & Smith, 2019). The key green finance investment model financial products used for financing sustainability and contributing to building a green economy are green bonds, sustainable investing, and eco-lending

Table 2 Green finance model products and their sustainable financing goals

S#	Green finance model products	Description	Sustainable financing-related projects
1	Green bonds	Debt-based financial instruments issued by governments, corporations, and institutions to produce finance for sustainable projects	Green bonds are introduced to finance environmental-related sustainable projects, such as energy efficiency, carbon emission, renewable energy, sustainable water management, etc.
2	Sustainable investing	Sustainable investing refers to the mixing of ESG measures into investment decisions to finance sustainability connect projects	These products are produced to generate funds for sustainability-sustaining plans that encourage the flow of capital into businesses that meet sustainability standards
3	Green venture capital	This financial product refers to sustainable investments in early-stage businesses that generate evolving solutions for environmental problems	These products generate finance for those businesses that generate sustainable technologies, lessen carbon emissions, enhance energy efficiency, and produce sustainable plans, such as solar energy startups
4	Impact investing	Impact investing products produce finance for sustainable, measurable, social, and environmental impacts with financial returns	Impact investing is introduced to attain finance for sustainable projects, such as affordable solar power, and sustainable urban development that generate returns with environmental considerations

Source: Authors' Own Elaboration

(Beebeejaun & Maharoo, 2024). Table 2 clarifies the key products of the green finance model that are used to generate funds for financing sustainability.

Table 2 clarifies the green finance model's main products and their description. Principally, these products are offered to create funds to finance sustainability-linked projects. Afterward, this table also elucidates the financing model goals that businesses attain by considering environmental impacts to obtain finance for sustainability-connected projects.

2.2 *Impact Investing Financing Model*

Impact investment is a recent investment strategy that is used to generate financial returns alongside with measurable environmental and social impact of financing. Technically, this model aligns with the up-surging demand for sustainable finance with the recognition that invested reserves to finance sustainability can bring a positive change (Agrawal & Hockerts, 2021). Remarkably, this model focuses on several sectors such as sustainable agriculture, education, health, sustainable infrastructure, renewable energy, affordable housing, and social enterprises. This model offers sustainable products that attract investments to finance sustainable projects, including green financing and social impact bonds (SIBs). Technically, the SIBs are an outcome-based supporting approach that permits funders to finance sustainability-linked social and environmental plans with the potential to generate returns while attaining measurable results (Phillips & Johnson, 2021). In SIBs, the capital of the private sector is used to finance sustainability-grounded social programs. Later, the government will repay to investor after achieving the predefined outcomes. Interestingly, the impact investing financing model encourages investors to invest their capital in sustainability-linked projects that offer financial return and measurable social and environmental impact (La Torre et al., 2019). Furthermore, impact investors measure the environmental and social impact of their investments by using defined metrics and frameworks to ensure the achievement of the desired positive outcomes. Remarkably, the impact investors accept dissimilar sorts of risk, but their focus is aligning the risk–return profile with the investor's goals. Thus, the key principles of the impact investing model are investors' intention, financial return, risk, and return trade-offs (Brandstetter & Lehner, 2014).

The main types of impact investment models that attract investors and produce funds for more sustainability-linked plans include private equity, debt instruments, real assets, public equity, social impact bonds, etc. Also, impact investing adopts several strategies to finance sustainability plans and to achieve the target of the investments. These strategies include socially responsible investing (SRI), ESG investing, thematic investing, community investing, and program-related investments, i.e., in short known as PRIs (Chiappini et al., 2023). The SRI emphasizes excluding all those sorts of investments that have negative social or environmental impacts, such as firearms and tobacco. ESG investing incorporates environmental, social, and governance measures into the all-investment-related decisions. In the same way, thematic investing focuses more on particular sectors, such as sustainable agriculture, renewable energy, and affordable housing schemes. Then, community investing is the investments in all those sorts of projects that are beneficial for underserved societies, such as affordable houses and microfinance programs. Likewise, PRIs are projects that are developed by any foundation or charities to support them typically in the shape of equity-based investments or loans with lower interest rates (Cojoianu et al., 2022).

Table 3 displays the dissimilar types of impact investing models, such as private equity, debt instruments, real assets, public equity, and social impact bonds. Principally, private equity investors are required to invest in private firms or new startups that address social or environmental connected problems, such as healthcare, clean energy, and education. Similarly, debt instruments are loans given to firms to be used to finance sustainable and impactful projects like community and microfinance institution-related projects (Brest & Born, 2013). In the same vein, real assets are sustainable financing in the property or infrastructure sectors to contribute positively to society or the environment, for instance, sustainable agricultural projects, renewal energy projects, etc. After that, public equity is financing sustainability in publicly traded firms that maintain strong ESG-based performance and also contribute significantly to attaining societal goals. Last but not least, social impact bonds are a form of agreement where private financiers finance

Table 3 Impact investing types, strategies, and sustainable projects

S#	Types of impact investing models	Impact investing strategies	Sustainable financing-related projects
1	Private equity	Socially responsible investing (SRI), investing in private firms, creating long-term value, equity financing utilization, generating sustainable return	Clean energy startups, sustainable agriculture, water and sanitation projects, healthcare innovations, education and skill development projects
2	Debt instruments	Investing in debt securities, ESG investing, debt investments, social development projects, impact-linked loans	Renewable energy, affordable housing, sustainable agriculture, water, and sanitation, workforce projects
3	Real assets	Thematic investing, investment in tangible assets, focusing on long-term value projects, developing assets that generate social and environmental outcomes	Renewable energy infrastructure, sustainable real estate, sustainable agriculture, climate resilience infrastructure, forest and natural resource management projects
4	Public equity	Focusing on generating both financial return and measurable social/environmental impact	Renewable energy, clean water projects, affordable housing projects
5	Social impact bonds (SIBs)	Program-related investments (PRI), outcome-based payments, risk-sharing	Mental health addiction recovery, childhood education program, job training, financial literacy projects

Source: Authors' Own Elaboration

sustainability-connected social programs (Ambrose-Oji et al., 2024). Later, the government repays the social impact on bond investors if the program achieves the targeted goals. Additionally, Table 2 also specifies the strategies that investors use to finance sustainability plans and to attain the goals of their impact investing projects. Also, some examples of sustainable financing projects such as microfinance, affordable housing, renewable energy, and healthcare are revealed in the above table.

2.3 *Green Bonds Financing Model*

A green bond is a type of financing instrument formed particularly to finance sustainability-connected environmental beneficial plans and projects. Remarkably, governments, businesses, financial institutions, etc. utilize these bonds to attract investments and generate finance for sustainability-connected projects such as pollution control, sustainable agriculture, renewable energy, clean water, and energy efficiency projects (Ahmed et al., 2024). Visibly, in the last few years, the market of green bonds has expanded rapidly due to increasing demand for financial products that help to attain environmental and sustainability-connected goals. Typically, the green bond is a type of fixed-income sustainable asset that is issued to attract

Table 4 Green bond types, descriptions, and sustainable projects

S#	Types of green bonds	Descriptions	Sustainable financing-related projects
1	Standard green proceed bonds	Green bonds that are issued by firms or governments, particularly for environmentally friendly projects	Renewable energy projects such as solar, wind, energy efficiency improvement
2	Green securitization bonds	Bonds backed by a pool of green assets, such as green loans or lease	Green mortgages for energy-efficient homes, financing electric vehicle infrastructure, or low-carbon buildings
3	Green revenue bonds	Green bonds secured by revenue generated from green projects	Clean energy projects such as hydropower plants, waste-to-energy facilities, sustainable transportation projects
4	Green project bonds	Bonds issued to finance a specific green project	Financing for wind power plants, large solar farms, eco-friendly infrastructure like green buildings
5	Green hybrid bonds	Green hybrid bonds contain features of both debt and equity, where proceeds are used for green initiatives	Financing for a combination of renewable energy and energy efficiency projects, such as hybrid solar-storage systems

Source: Authors' Own Elaboration

investments and generate an amount for environmentally favorable initiatives (Cheng et al., 2024). In comparison, these bonds are similar to traditional bonds, though, they are nominated mainly to fund sustainable green initiatives. The green bond is available to finance renewable energy projects that contain solar, wind, and hydroelectric power energy schemes, as well as energy efficiency improvements in developments, such as buildings, industrial facilities, clean water, waste management, sustainable agricultural and forestry, and pollution prevention and control. The main characteristic of green bonds is that the money raised from them is only used for financing sustainability plans and capable environmental initiatives (Makpotche et al., 2024). Also, nominees are usually given transparency about the usage of their funds which they invested in green bonds. Several types of green bonds are introduced and offered in the markets (see Table 4).

Table 4 demonstrates the key types of green sustainable finance bonds, their descriptions, and their sustainable financing projects. The core green bond types are standard green proceed bonds, green securitization bonds, green revenue bonds, green project bonds, and green hybrid bonds (Dutordoir et al., 2024). Standard green proceeds bonds are the most common type of green bonds. They are primarily used to attract investment and raise funds for financing projects that promote environmental sustainability. The issuer pledges to use the funds raised toward a particular group of green initiatives. Importantly, the bondholder provides a guarantee that the funds obtained from the issuance of these bonds will go toward sustainable finance and environment-friendly activities. These bonds are financed by the profits made from a specific green project or set of connected initiatives, such as a renewable energy power plant that generates electricity and retails it to the utilities (Onabowale, 2024). Likewise, green securitization bonds are also one of the main types of green bond financing models. This bond is the procedure of financing sustainability by merging environment-friendly assets into a security that depositors can purchase. These assets could contain leases or loans for energy-efficient businesses or setting up renewable energy projects. Next, sustainability-linked green bonds are also considered sustainable finance products of green bonds that are used to finance sustainability-linked plans. These bonds incorporate sustainability performance objectives with green finance. Interestingly, in this type of bond, issuers receive a lower coupon rate if they meet particular sustainability-connected standards. On the other side, if targeted goals are not obtained, then the bondholder may obtain better interest.

Subsequently, this model's another innovative product is the green revenue bonds. These bonds are financed by the income that is generated by a particular green project or set of projects such as renewable energy projects that generate electricity and then sell it to the grid (Ji & Zhang, 2024). Also, a firm offered green bonds to fund a wind farm project, where the profit that will be attained from selling electricity is used to reimburse the bondholders. The green project bond is another important type of green bond that is typically a type of loan instrument that is used to fund environment-friendly schemes such as conservation energy, renewable energy, and energy efficiency projects. These bonds give depositors a chance to earn profit from their money by supporting green projects. Generally, the funds collected from these green project bonds are used to support initiatives that fulfill certain green standards and are consistent with environmental goals (Mertzanis & Tebourbi, 2025).

2.4 Socially Responsible Investing Financing Model

Socially responsible investing (SRI) is a sort of investment tactic that is developed to attain societal and environmental goals while concurrently generating financial gains to support more sustainable finance products. Remarkably, it blends ethical considerations with several conventional-based investment approaches, thus, highlighting not only financial achievements but also how companies and initiatives fit with specific social and environmental standards (Ferrat et al., 2025). Notably, selecting sustainable finance products according to ESG-defined standards is a common exercise in SRI. The supporting initiatives and businesses that share the investor's values which may contain ethical governance, labor standards, human rights, sustainable practices, and environmental preservation are the key objectives of the SRI financing model. Typically, SRI maintains value-based screening which means it excludes all those businesses that are involved in operations that are harmful to society or the environment, such as fossil fuels, arms manufacturing, tobacco, and child labor. Thus, SRI investing actively considers and allows those businesses that maintain ethical values, such as businesses that are involved in healthcare, renewable energy, clean technologies, and education. Besides, SRI investments also focused on those businesses that maintain ESG criteria. Thus, SRI procedures assess how a firm manages environmental risk, social impacts, financial risks, and maintains corporate governance practices (Ebekozien et al., 2025).

Moreover, SRI-based investments offer active ownership and engagement of investors. Therefore, SRI fund-based investors are allowed to regularly interact with businesses to promote enhanced ESG implementations. Essentially, this could be discussing dissimilar types of resolutions with firms' officials, voting rights for shareholders, and putting pressure on firms to boost their sustainability strategies (Al-Daihani et al., 2025). Besides, SRI is normally connected to a long-term investment horizon because socially conscious funders believe those companies that prioritize ESG measures will eventually experience sustainable financial growth. Importantly, with focusing on financial return, SRI is also closely related to impact-based investing which specifies its goals of producing measurable environmental and social impact. In short, SRI's ultimate objective is to produce productive change while generating competitive financial revenues.

Table 5 clarifies SRI-based financing models, their descriptions, and their offered sustainable financing projects. Essentially, SRI offers several financial instruments to generate funds for social and environmental projects, such as SRI mutual funds, SRI exchange-traded funds, green bonds, social impact bonds, and ESG funds.

Importantly, SRI possesses positive social and environmental impact, enabling it to manage dissimilar sorts of risk, deliver long-term financial performance, and

Table 5 SRI financing model types, descriptions, and sustainable projects

S#	Types of SRI investments	Descriptions	Sustainable financing-related projects
1	SRI mutual funds	These are mutual funds that invest in firms, bonds, or sustainable projects that meet ethical, social, and environmental criteria. The goal is to attain financial return	Environmental projects, social projects, governance projects, renewable energy projects, etc.
2	SRI exchange-traded funds (SRI ETFs)	SRI ETFs offer diversified exposure to firms that attain ESG criteria	Green building and infrastructure, renewable energy projects, clean water projects, sustainable agriculture, etc.
3	SRI social impact bonds	A unique type of financial product that is used to finance plans with measurable environmental and social impacts	Education improvement projects, youth employment programs, climate justice reforms, poverty alleviation, etc.
4	SRI ESG funds	SRI ESG funds integrate two different concepts that are SRI and ESG factors	Renewable energy projects, green building and infrastructure, healthcare initiatives, electric transportation, etc.

Source: Authors' Own Elaboration

align with personal values. On the other side, several challenges also exist for SRI-based investments such as it holds limited investment opportunities, SRI measurement, and reporting standards that are still not clear; likewise, its performance is also uncertain. Interestingly, greenwashing is also a noteworthy challenge that is faced by SRI-based investments (Nartey, 2025). Technically, greenwashing occurs when a firm misinforms and claims about its environmental activities that in actuality do not exist (Santos et al., 2025).

2.5 Blended Finance Model

The blended finance model indicates the deliberate use of development financing together with private investments to support social and environmental initiatives or to achieve sustainable financing-connected goals (Flammer et al., 2025). It combines private investment with public or charitable finance to boost investments in those projects that are considered too risky or unprofitable for private investors. Thus, to address global matters, including poverty, health, and climate change, a blended finance model is executed to raise private sector capital while creating long-term financial gains. Remarkably, the blended finance model is beneficial for investors in numerous ways, such as mobilizing private capital, addressing global changes, de-risk investments, risk–return imbalance, and providing sustainable development. For instance, the Global Infrastructure Facility (GIF), the Green Climate Fund (GCF), and the Development Impact Bond (DIB) are some core examples of blended finance sustainable financing products that are measured as beneficial for investors and also generate funds for sustainable projects (Jung, 2020).

The GIF is a forum for collaboration designed to raise private investment in high-quality sustainable finance infrastructure-connected projects of deprived nations. Likewise, the GCF is a worldwide initiative recognized to support the efforts of emerging nations to handle climate change. Also, a DIB is a financial product that is developed to address environmental and social challenges via result-based financing. Remarkably, the key components of the blended finance model include concessional finance, private capital, risk mitigation, and impact investing.

Usually, concessional or development finance involves financing from charitable or public sources that offer better terms that are lower interest rates, guarantees, or grants attained from international organizations, development banks, or governments. Thus, these attributes of the blended finance model offer sustainable investments in emerging or high-risk markets that are more attractive for private investors (Anshori et al., 2023). Next, private capital is one of the main products of the blended finance model. The majority of the money comes from private sector investors, including corporate organizations, investment funds, and commercial banks. Although they might need the de-risking offered by concessional financing, they are frequently driven by the possibility of financial gains. Moreover, risk mitigation is another product of the blended finance model. These arrangements are frequently

Table 6 Blended finance model types, descriptions, and sustainable project

S#	Types of blended finance products	Descriptions	Sustainable financing-related projects
1	Concessional finance	These are investments with below-market terms mostly provided by development finance institutions or charitable firms to eliminate risk and attract private capital	Renewable energy (solar and wind) projects, affordable housing in low-income areas, clean drinking water accessibility, etc.
2	Private capital/ equity capital	Private investors invest in a project intending to generate returns while addressing SDGs	Sustainable agriculture, green technology startups, ECO-tourism developments, etc.
3	Risk mitigation instruments	These are tools that mitigate the risk of private-sector investments	Climate-resilient infrastructure, green bonds with government-backed guarantees, agricultural ventures in emerging capital markets, etc.
4	Impact investments	Investments designed for producing both financial return and positive, measurable, social or environmental impacts	Clean energy projects, financial inclusion initiatives, affordable housing schemes, etc.

Source: Authors' Own Elaboration

used in blended financing to increase a project's appeal to private investors. This could include first-loss capital where public funds absorb the initial losses, safeguarding private investors, insurance, or guarantees. Hence, impact funding is also a significant product of the blended finance model (Khan & Badjie, 2022). Therefore, all those projects that generate quantifiable social, environmental, or development outcomes are usually the focus of blended finance. Also, all those initiatives in these fields including affordable housing, healthcare, education, and renewable energy fall under this category (see Table 6).

Table 6 explains the key blended finance model products, their descriptions, and their sustainable finance projects. Notably, the blended finance investment models are facing several challenges such as it is measured as a complex model, it is difficult to measure blended finance project impacts, and investors do not have much awareness about these financing models and imbalances of risk and return (Sa'ad et al., 2024).

2.6 Circular Economy Financing Model

A circular economy financing model is developed to support the transition from a traditional linear economy to a circular economy. Notably, in a circular economy, the emphasis is on enhancing the product life cycle, preserving the product value, less wastage, materials, and resource availability in the economy for as long as possible (Aranda-Usón et al., 2019). To adopt circular economy practices, investments play an important role in enabling businesses and projects to adopt circular model practices. The circular economy model is built on key principles such as sustainable supply chains, recycling, refusing and reducing waste, refurbishment and remanufacturing, regeneration and resource recovery, offering products as services, leveraging data and digital technologies, and designing products for durability and longevity. Likewise, the main components of circular economy financing models contain capital allocation for circular projects, investment instruments, performance-based financing, innovative financing tools, collaboration, and partnerships (Ghisetti & Montresor, 2020).

Technically, the circular economy model directs capital to those sustainable finance projects that accept circular economy principles. Thus, investment allocation, i.e., the main component of the circular economy model is used for circular sustainable finance projects that generate revenues and focus more on upcycling, recycling, waste reduction, resource recovery, etc. (Dewick et al., 2020). This model uses several investment instruments to attract investors and to finance sustainability-linked products, such as green bonds, sustainability bonds, impact investing, blended finance, and equity financing. Moreover, circular economy models are also considered performance-based sustainable financing models as they pay for performance mechanisms where payment is linked with certain sustainability-based outcomes. For instance, reduction in waste, carbon emission, energy consumption, etc. (Saarinen & Aarikka-Stenroos, 2023). Likewise, this model uses innovative financing tools such as circular economy funds and revolving funds to finance sustainability-connected projects. In addition, it also focuses on collaboration and partnerships like public–private partnerships and multi-stakeholder partnerships. The circular economy model offers several sustainable finance products (see Table 7) that attract investors and generate funds for more sustainable financing projects, such as product-as-a-service, extended producer responsibility, take-back schemes, closed-loop supply chains, shared resource models, impact investment funds, and pay-per-use model (Aboulamer et al., 2020).

Table 7 explains the core products of the circular economy model, their descriptions, and the main sustainable financing projects of the stated products. Interestingly, this model contains several benefits for investors and generates funds for financing sustainability. For instance, it is a resource efficiency project that reduces consumption of raw materials, handles waste reductions, offers several business opportunities, mitigates climate change, and delivers long-term revenues. Evidently, the circular economy faces several challenges such as high upfront costs that occur due to the transition of circular models that need significant initial investments in research and development (R&D), infrastructure, new technologies, market education, circular economy model acceptance, measurement and metrics for this model, and regulatory uncertainty (Patra et al., 2024). Also, limited market availability for recycled materials is another main challenge for the circular economy model. Technically, if there are limited resources or markets, then in this model, it is very hard to close the ongoing loop.

Table 7 Circular economy model types, descriptions, and sustainable project

S#	Types of circular economy products	Descriptions	Sustainable financing-related projects
1	Product-as-a-service (PaaS)	In PaaS, instead of selling items outright, industries offer them as services, which guarantees extended use and less waste	Electric vehicle charging infrastructure, software as a service, subscription-based agriculture, technology subscription services
2	Extended producer responsibility (EPR)	In EPR, the manufacturer is liable for the full life cycle of a product, counting collection and recycling at the end of its valuable life	Electronic waste recycling program, agricultural packaging, electronic (e-waste)
3	Take-back schemes	Industries offer incentives for returning belongings for reprocessing, recycling, or refurbishment	Clothing brands with take-back programs
4	Closed-loop supply chains	A product in which products are continually recycled or reused, thus minimizing wastage	Recycling and upcycling of materials
5	Shared resource model	Multiple users share access to the resources	Shared electric bikes or scooters
6	Impact investment funds	Financing sustainability projects focused on funding clear environmental or social impact	Sustainable energy projects
7	Pay-per-use model	Customers pay for products and services based on usage rather than ownership	Shared electric bikes or scooters

Source: Authors' Own Elaboration

2.7 *Sustainable Development Goals (SDGs) Financing Model*

A Sustainable Development Goals (SDGs) investment-based model is a funding structure created to finance sustainability-connected initiatives and projects that support the 17 SDGs established by the United Nations (UN). Principally, these objectives are developed by the UN to achieve the goals of mitigating poverty and attaining sustainable growth by 2030. These objectives seek to address global issues like poverty, inequality, environmental degradation, climate change, and peace (Stenberg et al., 2017). Besides, these sustainable finance funding models make sure that money is raised, handled, and distributed effectively to accomplish the SDGs and to finance sustainability over the long run. Importantly, these sustainability models are financed by private-sector investments, public-sector investments, government institution-provided investments, and international financial institutions offered funds. Likewise, they are also financed by those firms and nations that are interested in carbon offsets, by using blended finance, and by availing local and community-level funds (Ziolo et al., 2021).

In SDGs models of sustainable financing, the private sector model focuses more on funding impact investing sustainability-connected projects, such as individuals, firms, and institutional investors. Thus, it involves in investing sustainability financing schemes that are generated to invest in social and environmental benefit projects to earn the utmost sustainable earnings (Barua, 2020). Hence, the capital generated from this model offers financial instruments that significantly contribute to achieving UN-created SDGs. Specifically, environment-friendly initiatives often focus on areas such as affordable and clean energy (SDG 7), climate action (SDG 13), healthcare projects (SDG 3), and educational programs (SDG 4), among others. Likewise, the funds generated from public and government funding are also used to finance several sustainability-linked projects. For instance, investing in sustainable city development (SDG11) and constructing infrastructure for clean water (SDG 6). Also, funds generated from these projects are used in promoting social welfare and reducing poverty (SDG10 and SDG1). Additionally, some other models of SDGs are used to produce funds for sustainability finance projects, for instance, SDG-based blended finance models that combine lower-cost capital with existing projects to generate sustainable development, such as affordable housing projects (SDG11) in urban areas. Importantly, funding for SDGs financing models is also offered by several international financial institutions such as regional development banks, World Bank, and International Monetary Funds (IMF). These international financial institutions finance sustainability-linked projects by investing in several sustainable schemes such as in projects that are built for carbon offsets or reductions that help climate finance actions (SDG 13), investments in projects created to protect ecosystems to tackle climate changes and protect biodiversity (SDG14, SDG15), etc. (see Table 8). Furthermore, by adopting SDGs financing-based models, businesses can integrate corporate social responsibility (CSR) based strategies into their operations to attain sustainable development. Like other sustainable financing economic models, the SDGs financing model also introduces several types of financing models, such as public financing, private sector financing, blended finance, impact investing, sustainable bonds, philanthropic financing, carbon finance, and development finance (Streimikiene et al., 2023).

Table 8 explains the 17 SDGs of the UN and their related projects to finance sustainability, such as green initiatives, renewable energy projects, and clean water projects.

Financing models for the Sustainable Development Goals (SDGs) face several challenges. These include difficulties in aligning with global standards, a significant gap between the financial resources needed to achieve the SDG targets and the current levels of funding, and external pressures such as economic crises, geopolitical tensions, inflation, and pandemics, all of which create further uncertainty that also affect the flow of financing to attains SDGs model goals. Likewise, a lack of transparency and accountability increases inefficient fund allocation that deters donors' institutions and investors from this financing model (Schwerhoff & Sy, 2017). Another important challenge for SDG financing models is their sustainability. It is a challenging task to judge the adopted SDG financing model's sustainability in the long run.

Table 8 UN 17 SDGs and related sustainable finance projects

SDGs	Sustainable financing-related projects	SDGs	Sustainable financing-related projects
SDG 1 (no poverty)	Microfinance programs for small businesses, social safety nets for vulnerable people	SDG 10 (reduced inequalities)	Social programs for marginalized groups, policies to ensure income equality
SDG 2 (zero hunger)	Urban farming initiatives, sustainable food systems	SDG 11 (sustainable cities and communities)	Urban planning projects for sustainable cities, for instance, green initiatives
SDG 3 (good health and well-being)	Access to affordable health care services in rural areas, vaccination campaigns for infectious diseases	SDG 12 (responsible consumption and production)	Circular economy initiatives, for instance, recycling programs
SDG 4 (quality education)	School enrollment programs in rural areas, digital literacy programs	SDG 13 (climate actions)	Renewable energy project to reduce CO_2 emissions
SDG 5 (gender equality)	Women empowerment programs, legislation for equal pay, and opportunities at the workplace	SDG 14 (life below water)	Marine protected areas (MPAs), initiatives to reduce plastic pollution in oceans
SDG 6 (clean water and sanitation)	Clean water projects in underserved communities, waste water treatment facility improvement	SDG 15 (life on land)	Reforestation projects, protection of endangered species and habitats
SDG 7 (affordable clean energy)	Solar energy initiatives in remote areas, wind power projects for sustainable energy	SDG 16 (peace, justice and strong institutions)	Human rights advocacy, legal support for victims of injustice
SDG 8 (decent work and economic growth)	Job training programs, promotion of SMEs, and entrepreneurship	SDG 17 (partnerships for the goal)	International collaboration on climate change, public–private partnerships for SDG funding
SDG 9 (industry, innovation, and infrastructure)	Green building construction projects, technological innovation in renewal energy systems		

Source: Authors' Own Elaboration

2.8 Climate Finance Financing Model

The sustainability model of climate finance is developed to attract investments and finance long-term sustainability goals, mitigation of climate changes, and mobilization of financial resources via using strategic processes and frameworks. From public sector funds and blended finance to private-sector investments, carbon markets, insurance, etc. these models aim to knob the need for sustainable funds to combat climate change (Kalinowski, 2024). Besides, by integrating ESG measures into financial systems and investment-related frameworks, this financing model aims to promote a resilient, low-carbon, and sustainable future. The investments to handle climate-related sustainable finance initiatives are gathered from private finance, public finance, and philanthropic finance. In private finance, banks, pension funds, and insurance firms play an important role in providing funds, specifically, through injecting direct capital or by procuring green bonds. Likewise, in public finance, governments, international financing institutes such as the World Bank, and development banks provide finance in the shape of concessional loans, guarantees, and grants (Briera & Lefèvre, 2024). In philanthropic finance, charitable foundations and non-profit organizations contribute their investments to support early-stage projects. Likewise, the climate finance model offers some other types such as blended finance, impact investing, sustainable bonds, philanthropic financing, carbon financing, and development impact bonds (see Table 9).

Climate risk sustainable investments refer to financial mechanisms, insurance, and sustainable financing projects that are designed to support businesses, countries, and individuals who are coping with the negative impacts of climate change (Billio et al., 2024). Typically, this model is financed to recover and adapt efforts after facing the natural disasters or to shield vulnerable sectors from forthcoming climate risks. The core products of the climate finance sustainable funding model are catastrophe bonds, climate insurance, philanthropy, non-government organizations (NGOs), and sustainable development goals financing. The catastrophe bonds support businesses to transfer the climate-connected financial risk or climate-linked disasters to capital markets. Then, climate insurance offers coverage for losses produced by worst weather actions or dissimilar climate-linked disruptions (Flögel et al., 2024). Philanthropy and NGOs play a key role in implementing climate finance by offering grants and financing innovative projects. Particularly, they focus on small-scale projects and community-linked initiatives. The SDGs financing model of climate finance model is linked to attain broader sustainable goals, especially, SDG 12 (climate actions). The financing for climate action aligned with the SDGs to guarantee that funds contribute not only to reducing emissions and adapting to climate action but also to refining quality of life and economic development (Diezmartínez & Short Gianotti, 2024).

Table 9 explains the main types of climate finance models, their descriptions, and offered sustainable finance-related projects. Importantly, like other sustainable finance models, the climate financing model also faces several challenges. For example, in its early stages, this model depends heavily on grants and aid. It is often influenced by political decisions, and measuring the impact of climate finance investments can be challenging. Additionally, the model carries high risk due to the constant evolution of markets and sustainable technologies, making implementation difficult (Rawat et al., 2024).

Having discussed the details about the main sustainable finance economic models, the next section discusses the other significant sustainable finance economic models.

Table 9 Climate finance model types, descriptions, and sustainable project

S#	Types of climate finance models	Descriptions	Sustainable financing-related projects
1	Public financing	Government funding and public resources are used to fund sustainable development projects	Solar power infrastructures, health care and educational programs, etc.
2	Private sector financing	Private sector funding attained from businesses impacts investors and financial institutions	Eco-friendly manufacturing procedures, corporate sector investments in green technologies, developing sustainable products, etc.
3	Blended finance	Combine development finance with private sector capital to fund high-risk projects	Clean water and sanitation management projects in low-income areas
4	Impact investing	Investments that are made to create positive social or environmental impacts alongside financial returns	Venture funding that supports affordable health care, clean energy startups, etc. in low-income areas
5	Sustainable bonds (green, SDGs, and social bonds)	Debt instruments that are used to raise capital precisely for SDG-connected projects	Issuing green bonds to generate funds for renewable energy projects, social bonds for housing projects
6	Philanthropic financing	Donations or grants received from charities, foundations, or individuals to support SDG-connected initiatives	Program targeting global health issues, philanthropic grants for sustainable agricultural research
7	Carbon finance	Funding through the sale of carbon credits, i.e., generated by emission reduction procedures, for instance, reforestation or renewable energy projects	Reforestation projects, carbon-offset programs, and clean energy projects in the corporate supply chain
8	Development impact bonds (DIBs)	DIBs are based on investments where investors are reimbursed based on measurable outcomes attained by a development project	Healthcare initiatives, vocational training programs, etc.

Source: Authors' Own Elaboration

3 Other Significant Sustainable Finance Economic Models

Interestingly, numerous other financing models have been developed and are operational to attract investments and generate funds for financing sustainability in the long run, such as the regenerative finance model, well-being economy model, pay-for-success financing model, carbon finance financing model, natural capital financing model, decentralized finance financing model, and microfinance for sustainability model.

Figure 2 clarifies some other types of sustainability-linked economic models that are used to generate finance for funding sustainability. A summary of these

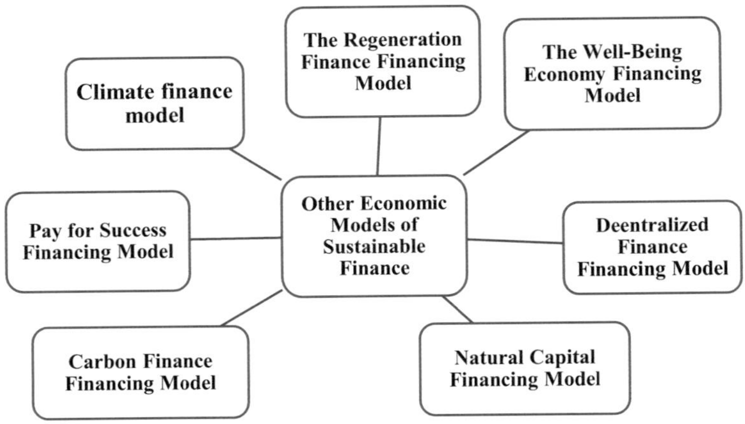

Fig. 2 Sustainability finance connected other economic models. (Source: Authors' Own Elaboration)

sustainability-supportive models and how their presented products create funds for financing sustainability are deliberated below in this section's next subsections.

3.1 The Regenerative Finance Financing Model

The Regenerative Finance Model (ReFi) is a significant sustainable finance system that generates profit by considering the welfare of the environment and community. Technically, the ReFi model is an evolving approach that develops an innovative financial system that is sustainable. This means it is socially and environmentally responsible and also restores the related ecosystems and communities (Čegar et al., 2024). Thus, the ReFi model extends beyond the impression of decreasing harm to essentially redeveloping and enhancing the health of social and natural systems. Therefore, ReFi is considered good for attracting investments and boosts sustainable financing-connected initiatives that help mitigate the effects of climate change, for instance, projects related to carbon offset, renewable energy, etc. Also, this model builds resilience in ecosystems and societies that are vulnerable to its effects such as developing disaster preparedness, climate adaptation strategies, etc. Similarly, the ReFi model-based green bonds are used to fund sustainable finance projects that address environmental challenges and climate change, such as sustainable agricultural, water conservation efforts, and energy projects. Thus, this model channels investments into all those initiatives that decrease carbon emissions and enhance adaptive capacity. Interestingly, this model follows the circular economic concepts that focus on reducing waste and reusing resources. Besides, ReFi investments are directed toward those initiatives, businesses, and projects that generate returns and develop positive social and environmental impacts. This includes regenerative agriculture, community-centered enterprises, and renewal energy projects (Pickerill, 2023).

This model also involves decentralized finance systems that permit cooperative models and peer-to-peer exchange. Hence, this model avoids the centralized, profit-driven incentives of conventional finance. It fosters social equity, encourages a circular economy, creates long-term value, attracts impact-driven investors, encourages transparency and accountability, catalyzes innovations, etc. (Sonetti et al., 2019). Essentially, Regenerative Finance (ReFi) aims to build sustainable wealth while restoring and renewing ecosystems. Examples include investing in regenerative agriculture projects and funding carbon offset initiatives—both of which promote long-term environmental resilience and well-being through financial investment. The core products of the ReFi model are green bonds, impact investing, community investment, sustainable agricultural financing (SIBs), green real estate development, circular economy finance, decentralized finance (DeFi) for sustainability, climate adaptation finance, and natural capital financing. The funds generated from ReFis' offered products are used to finance sustainable projects, such as renewable energy projects, clean water access, sustainable agriculture, social enterprises, recycling innovations, upcycling businesses, closed-loop supply chain systems, zero waste manufacturing, etc. (Das & Bocken, 2024).

This model also faces several challenges such as lack of standardization and metrics as there is no universally accepted framework or set of metrics that measure the success of this model. This model also needs a long-term horizon for revenues, and this may not align with the expectations of traditional investors who are looking for immediate returns. The ReFi model is newly introduced; thus, it is difficult to understand for investors. The ReFi projects are focused on natural resources and newly introduced technologies, thus, carry inherent risks, such as failure of technology and fluctuating market conditions. This model also requires a stable economic and political environment to work (Fullerton, 2020).

3.2 *The Well-Being Economy Sustainable Financing Model*

The well-being financing model is a recent sustainable financing approach that first considers the welfare of people and the environment and then goes for the conventional profit-maximizing objectives (Ari & Koc, 2018). Thus, in the well-being economy model the investors not only think about sustainable financial growth but also prosperity. Moreover, this model intends to bring existing financial systems into line with the more general goals, such as raising sustainability, financing sustainability-linked plans, lowering inequality, generating sustainable revenues, and enhancing the quality of life. For instance, investing in sustainable green technologies, social firms, and initiatives that enhance education and health care is a project where the returns are measured not just in monetary and sustainable terms but also in terms of the social and environmental impact (Diener & Seligman, 2004). The key principles on which this model depends or focuses are human-created development, sustainability, and regeneration, inclusive and equitable growth, collaboration, sustainable financial growth, and shared prosperity. Like other sustainable finance economic models, the well-being economy model also offered several sustainability-connected financial projects and products to generate more investments. For instance, the core products of this model to create more investment for financing sustainability are social impact investing, green bonds, community investments, well-being bonds, impact-driven venture capital, public–private partnerships, fair trade financings, sustainable development funds, local currencies, and mutual credit systems. The products of well-being models help reduce inequalities, promote resilience and sustainability, encourage long-term investments, and enhance social cohesion (Maack & Davidsdottir, 2015).

In a nutshell, this model offers a transformative approach to sustainable finance and its connected investment-based products. By supporting human well-being, social equality, environmental sustainability, and social equity, this model seeks to shift the main focus of economic systems, i.e., financial growth to well-being for all. Therefore, this model holds immense potential for generating a more equitable, inclusive, and sustainable future (Brüggen et al., 2017). Furthermore, this model also faces several challenges such as a lack of established metrics, difficulty in measuring its impact, scaling its impact universally, and difficulties in shifting conventional financial practices.

3.3 *Pay-for-Success Sustainable Financing Model*

The pay-for-success (PFS) financing model is a performance-based sustainable finance investment strategy where investors contribute to upfront investments for social projects (Olson et al., 2024). Notably, the government and other stakeholders only reimburse the funders if the program attains predefined success outcomes. Hence, this financing model combines beneficial social impact with financial incentives. Essentially, this model shifts the focus from paying for actual sustainable outcomes. This model is mostly adopted in educational, healthcare, housing sectors, and climate change programs. Thus, this is a broader sustainable finance approach that provides several sustainable investment avenues and looks for ways to invest in initiatives that contribute to long-term environmental, social, and governance, i.e., ESG goals (Williams, 2024). Typically, this model contains four different key players that are government, service providers, investors, and evaluators. This model is also named an incentivized impact model as its funds are only paid out when real and impactful results are achieved. Moreover, this model allows private investors to invest in sustainable finance projects that reduce the burden on the public budget. Similarly, PFS encourages creative solutions that can attain social goals more effectively. Remarkably, PFS aims for the long-term impact of investments, aligns with sustainable finance principles, and fosters programs that offer lasting sustainable benefits for the environment and communities (Andrikopoulos & Tsekrekos, 2024).

3.4 Carbon Finance Sustainable Financing Model

A carbon finance sustainable financing model is developed to combat climate change by funding initiatives that reduce or eliminate greenhouse gas (GHG) emissions. Remarkably, this notion functions mainly through mechanisms counting carbon credit markets, carbon trading, and funding in carbon-offset projects, trying to improve climate change while producing financial gains (Singhania et al., 2024b). The key elements that attract investors to invest in this sustainable finance model are its sustainability impact. Certainly, the ultimate goal of this model is to facilitate the transition to a low-carbon economy by investing capital into sustainable finance projects that eliminate emissions. Thus, by enabling emission reductions through financial mechanisms, carbon finance supports the battle against climate change and thus aligns with broader sustainability goals. The impact and success of this model-based sustainable finance project are typically ensured and always measured by third-party certified bodies, such as Verified Carbon Standard, i.e., VCS. This sustainable finance model is beneficial in several ways such as encouraging green investments, lowering the cost of emission reductions, supporting global climate goals, fostering innovation-based sustainable finance products for further investments, etc. (Wu, 2024).

Visibly, the carbon finance model plays an important role in the sustainable finance landscape and helps to move investments toward those projects that diminish carbon emissions and contribute to global climate goals. Despite facing several challenges, such as market volatility and verification concerns, this model provides several important sustainable finance tools and products for driving investments in environmental sustainability and easing the transition toward a low-carbon economy.

3.5 Natural Capital Financing Model

A natural capital financing model of sustainable finance is a strategy or structure that is developed to allocate financial resources and investments to the protection, rebuilding, and sustainable management of natural ecosystems to produce sustainable returns (Dai et al., 2024). It promotes long-term environmental sustainability by offering several opportunities to investors to incorporate the genuine worth of nature into exciting financial systems. This model emphasizes identifying and valuing natural assets and ecosystem services within financial systems and decision-making procedures. Moreover, the key goal of this model is to encourage the sustainable use of natural resources while integrating the value of nature into investment-related economic and financial decisions. Thus, the key elements of the natural capital sustainable financing model are valuing nature, integration of natural capital into decision-making, investment in sustainable finance nature base projects, and biodiversity finance. Likewise, this model's core product and mechanism include green bonds and impact investing, ecosystem service markets, ecosystem service markets, public–private partnerships (PPPs), payments for ecosystem services, and sustainable supply chains (Liang et al., 2024). Interestingly, this model offers several benefits, including reducing long-term environmental risks and increasing resilience to climate change. It also plays a key role in preserving ecosystem services and boosting biodiversity. As investor demand for sustainable solutions grows, this model attracts more investment and drives innovation in sustainability-focused financial products, especially with the rising emphasis on natural capital. This fosters new openings for sustainable economic and financial growth.

3.6 Decentralized Finance (DeFI) Sustainable Financing Model

Decentralized Finance (DeFi) for sustainability uses blockchain, smart contracts, and technology to develop transparent, simply manageable financial systems that support environmental and social objectives (Auer et al., 2024). Thus, through this financing model, investors and businesses can directly invest in sustainable projects, such as renewable energy, green enterprises, and carbon-offset initiatives. The key components of DeFI contain blockchain platforms that support smart contracts and sustainable projects for secure and transparent transactions and decentralized applications, i.e., dApps that assist investments in green products and sustainable assets (Aquilina et al., 2024).

Hence, the sustainable products of this model include green bonds, sustainable investment funds, tokenized environmental assets, carbon credit trading, etc. Likewise, the types of DeFi sustainable finance services range from decentralized green lending platforms to tokenized renewable energy-related products. The challenges faced by the DeFI model include the environmental sustainability of blockchain networks themselves, lack of regulatory transparency in green finance, difficulty in validating the authenticity of sustainable projects, and the possibilities of greenwashing. Thus, these challenges slow down the adoption and effectiveness of DeFi in advancing sustainable financing goals (Green et al., 2024).

3.7 Microfinance for Sustainability Development Financing Model

This microfinance for sustainable development provides financial services to low-income people or societies like savings accounts, small loans, and insurance, especially, to people who are living in developing nations. The main goal of this model is to boost economic-related activities that support and enhance social and environmental sustainability (Morse, 2024). Typically, the microfinance financing model can solve environmental-related problems and encourage comprehensive growth by supporting micro-entrepreneurs and small- and medium-sized enterprises by engaging them in sustainable practices, such as renewable energy, organic farming, waste management, or eco-friendly products. The key components of this model include financial institutions, micro-lenders, local community engagement, tailored financial products, etc. Notably, these products address social and environmental challenges, such as sustainable agricultural financing and practices, financing renewable energy products, microloans for clean energy, saving accounts that are tied with eco-friendly investments, and low-interest loans for green projects. Similarly, the types of services offered by this sustainable finance model are green micro-credit, micro-insurance, and climate-resilient agricultural financing (Moiko & Makau, 2024). Moreover, the challenges faced by this model include high administrative costs in offering small debts, capital for micro-lenders, lack of financial literacy in investors, difficulties in ensuring that financed projects meet genuine sustainability measures,

4 Difficulties and Challenges to Fund Sustainability and Solutions

Difficulties to finance sustainability include limited access to capital, high investment risks, a clear and consistent regulatory framework, and insufficient awareness or education about sustainable finance options. Clearly, several investors are discouraged by the long-term horizon and uncertainty of sustainable finance projects which mostly need higher costs and their funds' generation is also uncertain (Barua, 2020). Moreover, the absence of standardization in sustainability metrics also leads to greenwashing. This makes it difficult to distinguish real impactful projects from those that only claim to be sustainable. To address all these sorts of issues, solutions could involve increasing access to financing through green bonds or blended finance sustainability-linked models that combine public and private funds. Likewise, government and other regulatory bodies can play a significant role by introducing clear regulations and incentives for sustainable investments. In addition, the development of universally accepted sustainability standards can build the trust of investors (Park, 2022). Likewise, enhancing education and increasing awareness of the financial benefits of sustainability could encourage more widespread investments. Also, technological advancements, such as artificial intelligence, blockchain, and metaverse could improve accountability and transparency, reduce risks, and help attract more investors to sustainable projects (Turkcu & Tura, 2023). Table 10 elucidates the difficulties that are faced by sustainable financing models and the solution to handle all these problems.

Table 10 displays the key difficulties and challenges, their explanations and solutions to handle the sustainability finance-connected project-related issues. Thus, by addressing these difficulties and challenges with targeted solutions, the global financing ecosystem can unlock the full potential of sustainability-connected projects.

Table 10 Difficulties and challenges to fund sustainability and their solutions

S#	Difficulties and challenges	Descriptions	Solution
1	Limited access to capital	Sustainability finance models are facing difficulties in accessing funds due to high risk and low return on investments	To handle this issue, a blended finance model can be used where public and philanthropic funds can be combined with private capital to reduce risk and attract investors
2	High investment risk and uncertainty	Sustainability products and projects often involve uncertainties related to changing regulations, market volatility, and unproven technologies	Risk mitigation strategies such as insurance of products can reduce the perception of risk for investors. Likewise, impact investing funds that focus on long-term value rather than immediate return can make these projects attractive for investors
3	Lack of clear and consistent regulatory framework	The lack of clear regulatory frameworks can make it difficult for both investors and project developers to navigate the landscape	Governments, regulatory bodies, and international organizations can work together to develop standardized frameworks for sustainability
4	Greenwashing and lack of transparency	Greenwashing occurs when firms misrepresent their operations' environmental and social benefits	Developing third-party certification or audits to ensure that firms present correct information about their environmental and social benefits
5	Insufficient awareness and financial literacy	The project developers and investors both have insufficient awareness and financial literacy about sustainability	A financial literacy program should be developed to educate both investors and project developers about the benefits and opportunities of sustainable finance projects
6	Long payback periods	Sustainable finance projects also have long payback periods due to high initial costs, which can deter those investors who are looking for immediate returns	Innovative financial instruments such as green bonds and sustainability-linked loans can be used to inspire investors who are willing to accept longer payback periods to achieve sustainable outcomes
7	Scalability and small-scale projects	Several sustainability-connected projects, mainly in developing countries are small-scale projects and facing a lack of resources to achieve profitability targets	This challenge can be handled by encouraging regional and local funds that are particularly focused on small-scale community-based projects

Source: Authors' Own Elaboration

5 Conclusion

The world is facing numerous escalating challenges of climate alteration, resource depletion, social inequalities, biodiversity losses, etc. that boost the importance of sustainability and has become a main concern for industries and investors. Thus, it forced policymakers and investigators to rethink the existing financial and economic systems and come up with an alternative sustainable finance which offers sustainable growth with considerations of environmental and ethical issues. However, a major reason many nations move away from sustainability-related plans and climate action is the lack of adequate funding and investment needed to support these programs. Remarkably, global financial institutes are facing a lot of pressure to arrange funds for maintaining sustainability-connected activities. Also, the traditional growth-based economic models typically ignore the expenses of long-term social and environmental well-being in favor of short-term profits. As a result, the limitations of these traditional models in boosting sustainability are becoming evident and demanding the expansion of more combined tactics that can produce revenue for financing sustainability-linked plans. In comparison with the traditional economic models, the recent economic models that fund sustainability finance and their offered products create numerous new opportunities that generate funds for financing sustainability-connected plans. The core recent economic models that support sustainability and produce finance by offering several products for sustainable projects are impact investing financing model, green finance sustainable financing model, sustainable development goals financing model, socially responsible investing financing model, green bond financing model, blended finance model, circular economy model, well-being economy model, pay-for-success financing model, natural capital financing model, the regenerative finance model, carbon finance financing model, decentralized finance financing model and microfinance for sustainability model. Moreover, the offered products by these models contain blended finance, green finance, impact investing, incorporation of ESG measures, SRI ESG funds, impact investment funds, product-as-a-service, philanthropic financing, sustainable bonds, etc. Notably, offered products of sustainable finance models are not only used as key investment vehicles but are also supportive in generating more funds to fulfill the investment needs of financing sustainability-connected projects. Likewise, sustainable finance-connected economic models also face numerous challenges that include limited access to capital, high investment risk and uncertainty, greenwashing and lack of transparency, insufficient awareness and financial literacy, lack of clear and consistent regulatory framework, scalability, long payback periods, and small-scale projects. Thus, this chapter discussed almost all recent economic models and their offered sustainable products which deliver a comprehensive understanding of how recent models of sustainable finance can be used as a tool for achieving positive environmental and social targets while achieving economic growth. Besides, the discussed challenges and their solutions for sustainable finance models are also suggested. This chapter offers policymakers and regulatory bodies an understanding of how recent economic models and their offered products can generate funds for financing sustainability and driving sustainable economic growth while addressing environmental and social challenges.

References

Aboulamer, A., Soufani, K., & Esposito, M. (2020). Financing the circular economic model. *Thunderbird International Business Review, 62*(6), 641–646.

Adeoye, O. B., Chigozie, A. E., Nwakamma, N. E., Danny, J. M., Usman, F. O., & Olu-Lawal, K. A. (2024). A conceptual framework for data-driven sustainable finance in green energy transition. *World Journal of Advanced Research and Reviews, 21*(2), 1791–1801.

Agrawal, A., & Hockerts, K. (2021). Impact investing: Review and research agenda. *Journal of Small Business & Entrepreneurship, 33*(2), 153–181.

Ahmed, R., Yusuf, F., & Ishaque, M. (2024). Green bonds as a bridge to the UN sustainable development goals on environment: A climate change empirical investigation. *International Journal of Finance & Economics, 29*(2), 2428–2451.

Al-Daihani, M., Dirie, K. A., Alam, M. M., & Abdullah, A. S. (2025). Business process model for "crowdfunding cash waqf model". *Journal of Islamic Accounting and Business Research, 16*(1), 147–169.

Ambrose-Oji, B., Pearson, M., Cross, E., Shallcross, J., Polschneider, S., Hattersley, R., & FitzGerald, O. (2024). Community tree nurseries in the UK: Promise, potential and pitfalls. *Forests, 15*(3), 560.

Andrikopoulos, A., & Tsekrekos, A. E. (2024). The pay-for-success contract: A valuation note. *Journal of Futures Markets, 44*(9), 1465–1473.

Anshori, F. A., Salomo, V., & Kusumastuti, R. (2023). Understand blended finance: How different definitions of blended finance result in different inputs and outputs and what to expect. *Technium Social Sciences Journal, 44*, 15.

Aquilina, M., Frost, J., & Schrimpf, A. (2024). Decentralized finance (DeFi): A functional approach. *Journal of Financial Regulation, 10*(1), 1–27.

Aranda-Usón, A., Portillo-Tarragona, P., Marín-Vinuesa, L. M., & Scarpellini, S. (2019). Financial resources for the circular economy: A perspective from businesses. *Sustainability, 11*(3), 888.

Ari, I., & Koc, M. (2018). Sustainable financing for sustainable development: Understanding the interrelations between public investment and sovereign debt. *Sustainability, 10*(11), 3901.

Ari, I., & Koc, M. (2021). Towards sustainable financing models: A proof-of-concept for a waqf-based alternative financing model for renewable energy investments. *Borsa Istanbul Review, 21*, S46–S56.

Auer, R., Haslhofer, B., Kitzler, S., Saggese, P., & Victor, F. (2024). The technology of decentralized finance (DeFi). *Digital Finance, 6*(1), 55–95.

Barua, S. (2020). Financing sustainable development goals: A review of challenges and mitigation strategies. *Business Strategy & Development, 3*(3), 277–293.

Beebeejaun, A., & Maharoo, T. (2024). Green banking laws and regulations in Mauritius: A comparative study with other countries' policies. *International Journal of Law and Management, 66*(4), 518–535.

Billio, M., Murgia, M., & Vismara, S. (2024). Sustainable and climate finance: An integrative framework from corporates to markets and society. *Review of Corporate Finance, 4*(1–2), 1–16.

Boiral, O., Brotherton, M. C., & Talbot, D. (2024). What you see is what you get? Building confidence in ESG disclosures for sustainable finance through external assurance. *Business Ethics, the Environment & Responsibility, 33*(4), 617–632.

Brandstetter, L., & Lehner, O. M. (2014). Impact investment portfolios: Including social risks and returns

Brest, P., & Born, K. (2013). When can impact investing create real impact. *Stanford Social Innovation Review, 11*(4), 22–31.

Briera, T., & Lefèvre, J. (2024). Reducing the cost of capital through international climate finance to accelerate the renewable energy transition in developing countries. *Energy Policy, 188*, 114104.

Brüggen, E. C., Hogreve, J., Holmlund, M., Kabadayi, S., & Löfgren, M. (2017). Financial well-being: A conceptualization and research agenda. *Journal of Business Research, 79*, 228–237.

Čegar, S., Drezgić, S., & Čišić, D. (2024). Exploring the regenerative economy: A comprehensive review of literature. *Journal of Regenerative Economics, 1*(1), 1–40.
Chen, J. M., Umair, M., & Hu, J. (2024). Green finance and renewable energy growth in developing nations: A GMM analysis. *Heliyon, 10*(13), e33879.
Cheng, X., Yan, C., Ye, K., & Chen, K. (2024). Enhancing resource efficiency through the utilization of the green bond market: An empirical analysis of Asian economies. *Resources Policy, 89*, 104623.
Chiappini, H., Marinelli, N., Jalal, R. N. U. D., & Birindelli, G. (2023). Past, present and future of impact investing and closely related financial vehicles: A literature review. *Sustainability Accounting, Management and Policy Journal, 14*(7), 232–257.
Cingolani, M. (2024). Public and private financing of sustainable development goals (SDGs). *Review of Political Economy, 36*(2), 792–826.
Cojoianu, T. F., Hoepner, A. G., & Lin, Y. (2022). Private market impact investing firms: Ownership structure and investment style. *International Review of Financial Analysis, 84*, 102374.
Cui, H., Wang, R., & Wang, H. (2020). An evolutionary analysis of green finance sustainability based on multi-agent game. *Journal of Cleaner Production, 269*, 121799.
Dai, Y., Ding, Y., Fu, S., Zhang, L., Cheng, J., & Zhu, D. (2024). Analyzing the impact of natural capital on socio-economic objectives under the framework of sustainable development goals. *Environmental Impact Assessment Review, 104*, 107322.
Das, A., & Bocken, N. (2024). Regenerative business strategies: A database and typology to inspire business experimentation towards sustainability. *Sustainable Production and Consumption, 49*, 529–544.
Dewick, P., Bengtsson, M., Cohen, M. J., Sarkis, J., & Schröder, P. (2020). Circular economy finance: Clear winner or risky proposition? *Journal of Industrial Ecology, 24*(6), 1192–1200.
Diener, E., & Seligman, M. E. (2004). Beyond money: Toward an economy of well-being. *Psychological Science in the Public Interest, 5*(1), 1–31.
Diezmartínez, C. V., & Short Gianotti, A. G. (2024). Municipal finance shapes urban climate action and justice. *Nature Climate Change, 14*(3), 247–252.
Dutordoir, M., Li, S., & Neto, J. Q. F. (2024). Issuer motivations for corporate green bond offerings. *British Journal of Management, 35*(2), 952–973.
Ebekozien, A., Aigbavboa, C. O., Samsurijan, M. S., Radin Firdaus, R. B., & Rohayati, M. I. (2025). Expanded corporate social responsibility framework: Companies' role in improving higher education institutions infrastructure to Sustainable Development Goal 4. *Journal of Facilities Management, 23*(1), 1–18.
Ferrat, Y., Daty, F., & Burlacu, R. (2025). The role of size effects in moderating the benefits of sustainable investing. *BRQ Business Research Quarterly, 28*(1), 203–219.
Flammer, C., Giroux, T., & Heal, G. M. (2025). Biodiversity finance. *Journal of Financial Economics, 164*, 103987. Retrieved February 10, 2025, from https://www.nber.org/system/files/working_papers/w31022/w31022.pdf
Flögel, F., Schepelmann, P., Zademach, H. M., & Zörner, M. (2024). Injecting climate finance into SME lending in Germany: Opportunities for and limitations of regional savings and cooperative Banks. *ZFW–Advances in Economic Geography, 68*(2), 111–123.
Fullerton, J. (2020). The capital institute presents: An agenda for genuine financial system reform. *Solutions: For a Sustainable & Desirable Future, 11*(4).
Ghisetti, C., & Montresor, S. (2020). On the adoption of circular economy practices by small and medium-size enterprises (SMEs): Does "financing-as-usual" still matter? *Journal of Evolutionary Economics, 30*(2), 559–586.
Green, A. M., Giannattasio, M. P., Erickson, J. S., Seneviratne, O., & Bennett, K. P. (2024). DeFi survival analysis: Insights into the emerging decentralized financial ecosystem. *Distributed Ledger Technologies: Research and Practice, 3*(1), 1–23.
Hampl, F., & Vágnerová Linnertová, D. (2024). How do ESG controversies moderate the nexus between ESG performance and cost of capital? Evidence from European listed companies. *Managerial Finance, 50*(10), 1727–1746.

Hossain, M. R., Rao, A., Sharma, G. D., Dev, D., & Kharbanda, A. (2024). Empowering energy transition: Green innovation, digital finance, and the path to sustainable prosperity through green finance initiatives. *Energy Economics, 136*, 107736.

Ji, M., & Zhang, X. (2024). Assessing the impacts and mechanisms of green bond financing on the enhancement of green management and technological innovation in environmental conservation enterprises. *Journal of the Knowledge Economy, 15*(3), 12709–12750.

Joshipura, M., Mathur, S., & Kedia, N. (2024). Sustainable investing and financing for sustainable development: A hybrid review. *Sustainable Development*.

Jung, H. (2020). Development finance, blended finance and insurance. *International Trade, Politics and Development, 4*(1), 47–60.

Kalinowski, T. (2024). The Green Climate Fund and private sector climate finance in the Global South. *Climate Policy, 24*(3), 281–296.

Khan, T., & Badjie, F. (2022). Islamic blended finance for circular economy impactful SMEs to achieve SDGs. *The Singapore Economic Review, 67*(1), 219–244.

Kumar, B., Kumar, L., Kumar, A., Kumari, R., Tagar, U., & Sassanelli, C. (2024). Green finance in circular economy: A literature review. *Environment, Development and Sustainability, 26*(7), 16419–16459.

La Torre, M., Trotta, A., Chiappini, H., & Rizzello, A. (2019). Business models for sustainable finance: The case study of social impact bonds. *Sustainability, 11*(7), 1887.

Liang, Y., Zhou, H., Zeng, J., & Wang, C. (2024). Do natural resources rent increase green finance in developing countries? The role of education. *Resources Policy, 91*, 104838.

Maack, M., & Davidsdottir, B. (2015). Five capital impact assessment: Appraisal framework based on theory of sustainable well-being. *Renewable and Sustainable Energy Reviews, 50*, 1338–1351.

Makpotche, M., Bouslah, K., & M'Zali, B. (2024). Long-run performance following corporate green bond issuance. *Managerial Finance, 50*(1), 140–178.

Marti, E., Fuchs, M., DesJardine, M. R., Slager, R., & Gond, J. P. (2024). The impact of sustainable investing: A multidisciplinary review. *Journal of Management Studies, 61*(5), 2181–2211.

Mertzanis, C., & Tebourbi, I. (2025). Geopolitical risk and global green bond market growth. *European Financial Management, 31*(1), 26–71.

Moiko, D. O., & Makau, G. K. (2024). Micro-finance accessibility and sustainable growth of entrepreneurial ecosystem among Maasai women owned small holder businesses in Kajiado County. *International Academic Journal of Innovation, Leadership and Entrepreneurship, 2*(4), 128–150.

Morse, S. (2024). Giving and receiving: Faith and the sustainability of institutions providing microfinance services for development. *Sustainability, 16*(5), 1923.

Mumtaz, M. Z., & Smith, Z. A. (2019). Green finance for sustainable development in Pakistan. *IPRI Journal, 19*(2), 1–34.

Myronchuk, V., Yatsenko, O., Riznyk, D., Hurina, O., & Frolov, A. (2024). Financing sustainable development: Analysis of modern approaches and practices in the context of financial and credit activities. *International Journal of Economics and Financial Issues, 14*(5), 317–329.

Nartey, E. K. (2025). Beyond profit: Ethical imperatives and the role of duty of care in sustainable investment strategies. *International Journal of Economics and Finance, 17*(1), 22–47.

Newell, R. G., Pizer, W. A., & Prest, B. C. (2024). The shadow price of capital: Accounting for capital displacement in benefit-cost analysis. *Environmental and Energy Policy and the Economy, 5*(1), 49–69.

Olson, H., Painter, G., Albertson, K., Fox, C., & O'Leary, C. (2024). Are social impact bonds an innovation in finance or do they help finance social innovation? *Journal of Social Policy, 53*(2), 407–431.

Onabowale, O. (2024). Energy policy and sustainable finance: Navigating the future of renewable energy and energy markets. *World Journal of Advanced Research and Reviews, 25*, 2235–2252.

Ozili, P. K., & Iorember, P. T. (2024). Financial stability and sustainable development. *International Journal of Finance & Economics, 29*(3), 2620–2646.

Park, J. (2022). How can we pay for it all? Understanding the global challenge of financing climate change and sustainable development solutions. *Journal of Environmental Studies and Sciences, 12*(1), 91–99.

Patra, S. P., Wankhede, V. A., & Agrawal, R. (2024). Circular economy practices in supply chain finance: A state-of-the-art review. *Benchmarking: An International Journal, 31*(7), 2197–2216.

Phillips, S. D., & Johnson, B. (2021). Inching to impact: The demand side of social impact investing. *Journal of Business Ethics, 168*(3), 615–629.

Pickerill, T. (2023). 10. Designing regenerative investment circles. *Reconnecting the city with nature and history: Towards circular regeneration strategies*, 267. Retrieved February 10, 2024, from https://www.torrossa.com/en/resources/an/5650226#page=269

Rawat, A., Kumar, D., & Khati, B. S. (2024). A review on climate change impacts, models, and its consequences on different sectors: A systematic approach. *Journal of Water and Climate Change, 15*(1), 104–126.

Sa'ad, A. A., Usman, A., & Rehan, R. (2024). Islamic sustainable finance: Gulf experiences. In Bilah et al. (Eds.), *Islamic finance and sustainable development* (pp. 5–14). Routledge. https://doi.org/10.4324/9781003468653

Saarinen, A., & Aarikka-Stenroos, L. (2023). Financing-related drivers and barriers for circular economy business: Developing a conceptual model from a field study. *Circular Economy and Sustainability, 3*(3), 1187–1211.

Santos, C., Coelho, A., & Marques, A. (2025). Greenwashing effects inside organizations: How does it affect organizational citizenship behaviours for the environment (OCBEs)? *Social Responsibility Journal, 21*(1), 35–53.

Schwerhoff, G., & Sy, M. (2017). Financing renewable energy in Africa–Key challenge of the sustainable development goals. *Renewable and Sustainable Energy Reviews, 75*, 393–401.

Singhania, M., Bhan, I., & Chadha, G. (2024a). Sustainable investments: A scientometric review and research agenda. *Managerial Finance, 50*(1), 266–294.

Singhania, M., Chadha, G., & Prasad, R. (2024b). Sustainable finance research: Review and agenda. *International Journal of Finance & Economics, 29*(4), 4010–4045.

Sonetti, G., Brown, M., & Naboni, E. (2019). About the triggering of UN sustainable development goals and regenerative sustainability in higher education. *Sustainability, 11*(1), 254.

Stenberg, K., Hanssen, O., Edejer, T. T. T., Bertram, M., Brindley, C., Meshreky, A., et al. (2017). Financing transformative health systems towards achievement of the health Sustainable Development Goals: A model for projected resource needs in 67 low-income and middle-income countries. *The Lancet Global Health, 5*(9), e875–e887.

Streimikiene, D., Mikalauskiene, A., & Burbaite, G. (2023). The role of sustainable finance in achieving sustainable development goals. *Economics & Sociology, 16*(1), 256–283.

Tanchangya, T., Raihan, A., Chowdhury, S., Islam, M. A., Islam, F., & Chakma, J. (2024). Practices of green marketing and corporate social responsibility (CSR) to promote sustainable business in emerging country. *Frontiers of Finance, 2*(2), 7202.

Turkcu, D., & Tura, N. (2023). The dark side of sustainable packaging: Battling with sustainability tensions. *Sustainable Production and Consumption, 40*, 412–421.

United Nations. (2024). *Inter-agency Task Force on financing for development, financing for sustainable development report 2024: Financing for development at a crossroads*. United Nations, New York. Retrieved January 25, 2025, from https://developmentfinance.un.org/fsdr2024

Williams, J. W. (2024). Finance interrupted: Social impact bonds, spatial politics, and the limits of financial innovation in the social sector. *Environment and Planning A: Economy and Space, 56*(1), 42–58.

Wu, J. (2024). Role of green finance and carbon accounting in achieving sustainability. *Humanities and Social Sciences Communications, 11*(1), 1–7.

Zhang, T., & Zhao, F. (2024). A study on the relationships among green finance, environmental pollution, and economic development. *Energy Strategy Reviews, 51*, 101290.

Zhang, S., Wu, Z., Wang, Y., & Hao, Y. (2021). Fostering green development with green finance: An empirical study on the environmental effect of green credit policy in China. *Journal of Environmental Management, 296*, 113159.

Ziolo, M., Bak, I., & Cheba, K. (2021). The role of sustainable finance in achieving sustainable development goals: Does it work? *Technological and Economic Development of Economy, 27*(1), 45–70.

Role of Green Societies in Sustainable Development: Strategies for a Balanced Future

Suresh Kumar and Amit Kumar

1 Introduction

Human actions such as destruction of habitats, polluting environment, and unsustainable utilization of resources are profoundly diminishing biodiversity. This decline adversely affects ecosystem functions that are vital for human well-being, including clean water, food supply and natural resources (Kurre & Purena, 2024). Numerous human actions also impact agricultural crops in India. These encompass global warming, coral reef demise, reduction in amphibian populations, extinction of animals, land deterioration, deforestation, ocean salinization, harm to ecosystems, loss of biodiversity and exhaustion of natural resources (Rehman et al., 2021). Land use alterations and unsustainable extraction practices are major contributors to the global decline in biodiversity. Due to the rising exports of agri-food products, natural habitats have been transformed into agricultural land, resulting in significant consequences for biodiversity, especially in tropical areas characterized by high species richness (Cabernard et al., 2024). Traditional knowledge systems offer valuable perspectives on resource management, biodiversity conservation and climate adaptation, contributing to improved soil condition and food safeguard (Jakes, 2024). Traditional knowledge is essential for nurturing a green society and attaining sustainable development. It includes the knowledge and customs of indigenous groups developed over many generations to correspond to their surroundings. The main objective of the green society is to establish balance between human endeavors and the environment through sustainable practices, policies and innovations. Green society symbolizes a pioneering structure in Society 5.0 that prioritizes human-centricity, eco-friendliness and enduring well-being (Tornjanski et al., 2024). The concept emphasizes reducing carbon footprints, conserving biodiversity and encouraging renewable energy use. In spite of challenges like diminishing

S. Kumar · A. Kumar (✉)
Department of Biosciences, Himachal Pradesh University, Shimla, H.P., India

A. Chakir et al. (eds.), *Green Society, Environmental Strategies and Sustainable Development*, https://doi.org/10.1007/978-3-031-94086-6_13

resources, contamination and changes in the climate, shift toward a green society has become vital. It promotes environmentally responsible behavior among individuals, industries and governments to ensure a healthy and resilient ecosystem. Sustainable urban development, green technologies, waste reduction and environmental education are key pillars that encourage the transition to an environmentally conscious future. The main goal is to reduce environmental impact and promote practices that support the future generations wellness. The key characteristics of a green society are:

Sustainable resource use: It corresponds to cautious utilization and managing natural assets to satisfy present needs without sacrificing the willingness of subsequent generations. This strategy prioritizes equilibrium between social well-being, revenue generation, and the welfare of the environment. To safeguard the ecosystem and the availability of resources for future generations, natural resource management is essential (Merwe et al., 2024).

Environmental conservation: It entails managing natural ecosystems and resources sustainably, protecting and conserving biodiversity, forests and natural habitats to preserve the health of the Earth for the present as well as future. It seeks to enhance biodiversity and lessen detrimental effects of human activities on the ecosystem. Women frequently take lead in opposing projects that harm the environment and protecting water and land resources from exploitation driven by patriarchal standards (Mathur et al., 2023).

Eco-friendly methods: It refers to actions and lifestyle choices that minimize harm to the environment and promote sustainability. These practices aim to reduce pollution, conserve natural resources and protect ecosystems. According to eco-feminism, women are inherently connected to the natural world because of their role in the management of resources like food and water.

Life quality: Green societies improve quality of life by providing clean air, water and ensure healthy living environment. It emphasizes the importance of well-being and public health.

Endurance and adaptation: A green society builds endurance and adaptability to deal with ecological, societal and financial challenges like natural disasters, climate change and economic traumas. Green society promulgates strategies for community-based endurance, promotion of sustainable livelihoods, effective management of resources to enhance anticipationand recovery with acceptable risk thresholds and preferred steadiness dynamics (Tornjanski et al., 2024).

Public awareness and environmental education: Green society sets up integrated environmental education alongside increased public understanding, emphasizing educational and developmental programs that enhance knowledge regarding environmental concerns and conservation.

Diversity and cultural heritage: Green society recognizes cultural heritage, diversity and indigenous knowledge as important resources that enhance human flourishing, health and resilience. It values cultural privileges and associated practices and activities that enhance responsible management of natural resources, thus promoting distinctiveness and strength of cultural identity.

Ethical utilization and production: Green society promotes schemes for production and consumption that are ethical and emphasize sustainability, responsibility, and equitableness. It promotes consumer preferences, green lifestyles, and ethical corporate governance that reduce environmental impact and enhance social equity.

2 Scope of Green Society

Green society is a community that emphasizes environmental consciousness and adheres to clean, healthy living practices. It entails educating people about sustainable practices like 3R concept, Reduce, Reuse and Recycle, to nurture a culture of environmental stewardship from childhood through adulthood. Education is crucial for the practical, effective use of the tools and mechanisms associated with the green economy and transitioning to green society (Ziankova et al., 2019). Green society promotes holistic and thriving economic progress, dignified employment and viable livelihoods that enhance human progress and contribute to well-being. The conservation of biodiversity through habitat protection and species preservation is a key aspect of maintaining ecological integrity. A green society also fosters a green economy by encouraging industries to embrace sustainable strategies and invest in environment friendly technologies. Public awareness and environmental education play vital role in inspiring communities to adopt sustainable practices while robust environmental policies and governance assure green initiatives. Thus, overall scope of a green society integrates environmental protection, social health and economic expansion to create sustainable and resilient future (Fig. 1).

3 Sustainable Development in the Modern Era

The Millennium Development Goals (MDGs) established by the United Nations in the year 2000 are intended to address global impoverishment and progress the lives of people worldwide. The MDGs sought to reduce poverty in all its forms by focusing on key issues, such as hunger, disease, gender inequality, lack of education, limited access to basic services and environmental degradation. The initiative aimed to create a world where all individuals had access to the necessities of life with a vision for tackling the fundamental reasons for poverty and improving social and economic outcomes globally. The Sustainable Development Goals (SDGs) established by the United Nations in 2015 offer a framework for addressing ecological, social and financial challenges and establishing a more eco-friendly and balanced global environment (Tornjanski et al., 2024). The various SDGs are eradicating poverty and famine, enhancing well-being and educational opportunities and achieving gender parity and inclusivity. Even with the challenges brought about by the COVID-19 pandemic, the war between Ukraine and Russia, and worldwide

Fig. 1 Various scopes of green society

inflation, numerous countries have advanced toward these objectives. However, they may not achieve these SDGs by 2030. The detrimental impacts associated with the lockdowns enforced in response to the COVID-19 pandemic increase violence against women. The ongoing conflict between Ukraine and Russia coupled with rising inflation has led to significant increase in food insecurity (Pandey et al., 2025). The Government of India has adopted Agenda 2030 which includes 17 SDGs along with numerous indicators and targets designed to operate within a sustainable framework (Dwivedi & Sharma, 2025). These goals are vital for managing human necessities, controlling climate change, providing financial stability and maintaining biodiversity (Fig. 2).

Sustainable development can be defined in a broad sense encompassing all aspects of human and environmental life. It is not possible to define it solely. Three pillars of sustainable development are economic sector, social sector and environmental sector (Table 1). In terms of human development and advancement, these three are interconnected. Since humans and society are closely related to the environment, it is impossible to consider human sustainability without a healthy environment (Sharma, 2021). The political viewpoint always differs from that of environmentalists and scientists, prioritizing development and progress over environmental issues (Kishore & Deswal, 2017). It is believed that the three domains are distinct actual spheres of activity. Even though a holistic evaluation must incorporate all three domains. It is essential to identify various environmental sustainability domains and recognize them in practice (Babu & Satya, 2022).

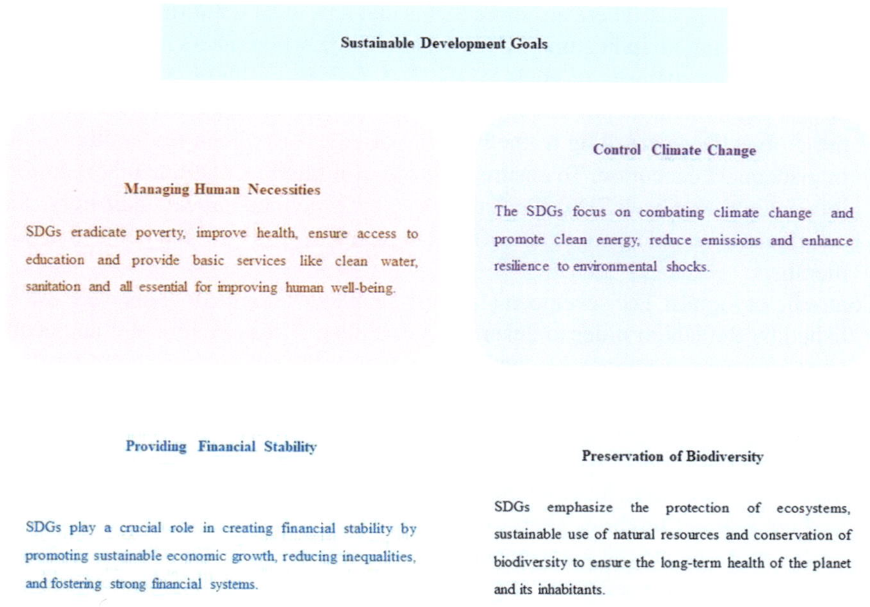

Fig. 2 Importance of sustainable development goals in modern era

Table 1 Principles of three domains of sustainable development

Domain of sustainable development	Principle of domain
Economic domain	Sufficient funding for social development, job creation, fair trade practices, increased income for people, improved living standards, open and collaborative markets, cost reduction and sustainable financing, financial security, and the promotion of a green and circular economy
Social domain	Social fairness and equality, development of social infrastructure, participatory governance, strengthening of social capital, and the preservation of community and cultural values
Environmental domain	Safeguard environmental health, prevent excessive pollution, transition to renewable resources, prioritize intergenerational decision-making, focus on well-being over GDP growth, and emphasize restoration and conservation efforts

4 Challenges for Implementation of Sustainable Development Goals

The execution of SDGs presents number of difficulties for UN member nations. Some of the main hurdles to accomplish the goals of sustainable development are:

Vaguely defined goals: There are some SDGs that lack clear definitions. Some goals have targets but no indicators (SDGs 8 and 14–16), while others lack both (SDGs 5 and 13). The majority of goals (SDGs 1–4, 6–7, 9–11, and 17) have both clear targets and indicators (Biermann et al., 2017). Many nations therefore consider these objectives according to their own interpretations which frequently results in inadequate execution. To ensure its successful implementation and execution, it is obvious that each SDG needs to have clearly defined targets, indicators, and benchmarks in addition to an official agreement with governments that are members.

Contradictory goals: The seventeen (17) sustainable development goals were established by the UN in order to ensure sustainability of the environment and economy. However, there may be conflicts between some goals which could result in some goals being replaced. It is obvious that they cannot be accomplished together particularly once these conflicting goals have been identified (McNeill et al., 2012).

Budgetary limitations: The UN's SDGs cannot be achieved, implemented, or carried out without funding. Therefore, a significant amount of money must be set aside for these kinds of endeavors (Jaiyesimi, 2016).

Lack of requisite competencies: To effectively accomplish the SDGs, majority of UN member countries lack necessary competencies or capabilities such as sufficient skills, suitable tools, competent and relevant education, etc.

Insufficient technology and data: In this age of globalization, nations must move and adjust to the newest technology available because the acquisition or lack of it determines whether their endeavors succeed or fail (Easterly, 2012).

Limiting cultural factors: The people who live in a place, community, state, region, or country are bound by their unique cultural norms, beliefs and ways of living. People in some cultures are not allowed to be receptive to new ideas, development, or contemporary thoughts. And with these deeply held views of maintaining things as they should be for everyone, regardless of global development progress achieving any of the SDGs set forth by the UN would be an enormous undertaking (Ashencaen et al., 2019).

5 Environmental Strategies for Green Society

Some companies develop robust environmental management policies, implement comprehensive environmental management systems and conduct thorough evaluations of health and safety to enhance and effectively control their overall environmental performance. Techniques like recycling and pollution prevention are employed to put policy goals into action (Sarmento et al., 2006). Organizations will have to convince financial institutions, investors, shareholders and most important channels through which the value of green materialism is disseminated as the attitudes of new consumers toward the green society are actually becoming more positive (Roy et al., 2001). Nowadays, the idea of sustainability can be employed to successful commercial endeavors, communities, and ecosystems and to use of natural resources. Business groups are progressively shifting their perspectives on sustainable development and many small businesses are showing governance through innovation. It is now prevalent for environmental sectors, organizations and programs that support the growth of green enterprises. Sustainable development is influenced by several key factors including government support, private sector investment, banking and commercial investment practices, the establishment of international agreements and the adoption of clear community-driven sustainability strategies (Sarmento et al., 2006). Strategies for the environment to promote a green society pay attention to sustainability, conservation and ecological footprints. The following are important environmental tactics for a green society:

Sustainable resource management: Encourage the effective use of natural resources such as water, raw materials, and power. Encourage sustainable agriculture and forestry practices.

Renewable energy adoption: Shift from fossil fuels to solar, wind, hydro and bioenergy. Provide incentives for households and industries to adopt green energy solutions.

Waste reduction and management: Implement the 3Rs: Reduce, Reuse and Recycle. Promote circular economy models to minimize waste.

Green infrastructure and urban planning: Develop eco-friendly buildings using sustainable materials. Integrate green spaces (parks and rooftop gardens) in urban designs.

Water conservation strategies: Implement rainwater harvesting and wastewater recycling. Promote efficient irrigation practices in agriculture.

Environmental education and awareness: Educate communities about sustainable practices. Encourage environmental stewardship through campaigns and workshops.

Sustainable consumption and production: Promote eco-labeling and green products. Encourage responsible consumption to reduce overexploitation.

Policy and governance: Enforce environmental laws and regulations. Incentivize green technologies and penalize polluting industries (Fig. 3).

6 Role of Innovation and Technologies in Green Society

Innovative ideas and advancement of sustainable development require use of digital technologies. These technological developments could affect long-term viability in both positive and negative ways. Gaining a comprehensive understanding of how individuals and organizations are utilizing these technologies is essential for enhancing sustainability outcomes. Small innovations like increased energy and material efficacy in current manufacturing processes are essential for the shift to sustainable economy. But more significant and even drastic progress in technology is also needed. For example, substituting fossil fuels for major technological advancements rather than just incremental efficiency improvements in the transportation

Fig. 3 Environmental strategies for green society

sector and the manufacturing of iron and steel (Ahman et al., 2017). Yet several elements will make revolutionary changes intrinsically challenging. The primary challenge is risk that companies invest in technological advancement such as pilot tests and basic research and development. Insufficient historical data to assess risks may prevent these markets from offering risk management tools for emerging technologies. The international financial markets' liberalization has also raised concerns that private investors now have a shorter-term perspective (Newell, 2013). Furthermore, there might not always be compelling reasons for private investors to contribute to long-term technological advancements. The economics literature has emphasized the risks of underfunding public goods like knowledge gained from research and development initiatives and learning-by-doing. Moreover, incumbent technologies frequently unfairly compete with new green technologies. Due to their ability to expand during periods of laxer environmental regulations, the incumbents will have a relative competitive advantage because their institutions and infrastructures being more or less customized. As a result, they may be close substitutes for their greener competitors. This leads to path-dependencies in which the economy is largely dependent on specific technological avenues (Soderholm, 2020). In the private sector, the aforementioned elements typically impede various forms of sustained technological advancements. However, there is reason for optimism that they may be problematic when it comes to sustainable technologies. Preliminary empirical research indicates that green technologies especially in energy and transportation foster greater knowledge spillovers and innovation impacts than the conventional, less eco-friendly technologies they are replacing (Dechezlepretre et al., 2017). Organizations tend to use patents more frequently to safeguard innovative products rather than novel processes despite the fact that process innovations are essential for the development of sustainable technology in industry. The energy sector represents a significant socio-technical system in the shift toward a green economy. This sector continues to be largely influenced by entrenched technologies such as nuclear power and fossil fuels, exhibiting various traits that contribute to path-dependent behavior.

7 Traditional Knowledge and Sustainable Development

The significance of traditional knowledge and indigenous groups must be acknowledged as a priority in achieving sustainable development goals. A global collaboration consortium composed of individuals from different nations and diverse sectors aims to harness advantages of traditional ethnobotanical wisdom to reduce poverty, eliminate hunger, enhance healthcare services, address climate change, preserve biodiversity and tackle issues related to biodiversity (Kumar et al., 2021). In India, plants are for medicinal purposes from prehistoric period for treatment of various ailments and traditional medicine has deeply ingrained itself into the country culture (Khajuria et al., 2021). Plant-based medicine is the basic source of healthcare for about 88% of world population. Especially in rural regions (Ralte et al., 2024). Agriculture faces a major challenge as a result of increasing demand, wholesome food for growing population, as well as to protect biodiversity and other natural resources. Climate change and the overuse of synthetic fertilizers and pesticides pose a threat to agriculture (Dwivedi et al., 2017). The varied climate zones and cultural legacy of India are reflected in the agricultural landscape historical richness which includes different traditional water management techniques. These methods have been crucial in collecting and preserving water for agricultural uses for centuries because they are firmly anchored in regional knowledge and environmental factors. The most common traditional approaches for water conservation in India are Kuhls, Tanks, Johads, and Bawdi (Rastogi et al., 2024). Unprecedented threats to the world's biodiversity hotspots include habitat loss, climate change, and overuse of natural resources (Shivanna, 2022). These threats endanger not only the existence of innumerable plant and animal species but also the means of subsistence and cultural legacy of indigenous groups that have historically depended on these ecosystems for their health and well-being (Estrada et al., 2022). The primary concern of humanity has been documentation of biodiversity, its rapid decline and the pressing need for its preservation (Shivanna & Sanjappa, 2021). The documentation of indigenous knowledge regarding medicinal use of plants illustrates their potential for new drug discoveries that directly help in conservation of plant diversity (Jena et al., 2025) (Table 2).

Table 2 Traditional knowledge in achieving sustainable development goals

Role of indigenous knowledge	Contribution to sustainable development goals (SDGs)	Description
Traditional medicine	SDG 3: Good HBalth and Well-Being	Indigenous healing methods offer comprehensive medical care, especially in isolated areas with little access to contemporary medical care
Food security	SDG 2: Zero Hunger	Sustainable indigenous agricultural techniques such as terrace farming and mixed cropping enhance food production and soil conservation
Sustainable land use	SDG 15: Life on Land	Indigenous agroforestry, shifting cultivation and biodiversity-friendly farming maintain soil fertility and prevent land degradation
Water conservation	SDG 6: Clean Water and Sanitation	Indigenous practices for water management like step wells, rainwater harvesting and bamboo drip irrigation ensure efficient water usage
Waste reduction	SDG 12: Responsible Consumption and Production	Indigenous practices promote biodegradable materials, composting and circular economy principles to minimize waste
Biodiversity conservation	SDG 13: Climate Action SDG 15: Life on Land	Traditional ecological knowledge preserves ecosystems through sustainable harvesting, sacred groves and seed-saving techniques
Climate change and mitigation	SDG 13: Climate Action	Indigenous practices like controlled burns, afforestation and wetland conservation contribute to carbon sequestration and climate resilience
Sustainable livelihoods	SDG 1: No Poverty SDG 8: Economic Development and Prestigious Work	Indigenous resource management supports local economies, sustainable crafts and eco-tourism, reducing poverty

8 Community Engagement and Social Responsibility for Green Society

Community involvement should be viewed as an integral part of education rather than as an addition. Education is not only a commodity but also a public service if it serves as a catalyst for social change and the realization of social justice rather than merely a way to accumulate personal wealth. Community engagement and social responsibility are fundamental to foster a green society that prioritizes environmental sustainability and ecological balance. Active community participation through awareness programs, participatory projects, and local green initiatives empowers individuals to embrace sustainable practices like reducing water consumption, separating waste, and using renewable energy. Integrating traditional knowledge further enhances resource management by blending indigenous practices with modern sustainability efforts.

9 Policy and Governance for Green Society

A novel approach that fosters the simultaneous advancement of economic and social domains by deeply integrating human intelligence, empathetic consciousness and innovative machine intelligence driven by the principles of open innovation. Thus, it enhances collective intelligence to support the long-term well-being and thriving of diverse ecosystem stakeholders (Tornjanski et al., 2023). A human-centric society actively promotes open collaboration among various stakeholders, scholarly institutions, public administration, business sectors, industrial organizations and nonprofit entities. This takes place within national and international networks through multiple stakeholder partnerships aimed at developing policy improvements, integrated strategies, regulations and incentives that foster eco-friendly sustainability, human-based integration and financial stability. Block chain is a novel innovation that can aid in the development of sustainable solutions due to its capabilities for accountability, clarification, online monitoring and stability as well as enhancing the effectiveness of worldwide cooperation. The enormous escalation in the demand for electricity gives rise to a fresh dilemma. Block chain technology is now recognized as the leading IoT Technology which has gathered attention regarding powerful entities such as enterprises, initiatives, financial banks, legislators and investigators (Senthil et al., 2023).

10 Future Prospects

A green society represents a community that prioritizes sustainability using technologies such as block chain to improve accountability, transparency and efficiency in managing resources. This method seeks to tackle ecological issues while fostering sustainable practices and international cooperation. Green society promotes the administration of eco-friendly patterns and technologies including recyclable energy and sustainable resource management. By prioritizing eco-friendly solutionswe can alleviate environmental deterioration and lessen society eco-related impact. Green society asserts that technological progress focuses on ecological conservation, resource efficacy and resilience in order to reduce environmental disasters and foster sustainable advancement.

11 Conclusion

Green society fosters embraced eco-related education, consciousness, focusing on educational initiatives and sustainable learning endeavors that enhance environmental literacy, sustainability and responsible citizenship. Establishment of green society is not merely an ecological need but also it a crucial avenue for achieving sustainable development. Switching to green society requires sustainable utilization and conservation of natural resources, using solar energy, bioenergy, and neglecting dependence on the fossil fuels. Traditional knowledge and effective policies play a significant role in shifting to green society and achieve sustainable development goal. It also requires collaboration of government with scientists, stakeholders, local organizations, international organizations, industries and communities. Green society promotes sustainable livelihoods, economic opportunitiesand social inclusion contributing to the reduction of poverty and disparity. The establishment of a sustainable society can provide a solid groundwork for economic growth, especially in regions that embrace environmentally friendly enterprises, fair trade initiatives, and community enhancement efforts such as sustainable agriculture, renewable energy sources and eco-tourism. Green society also fosters sustainable farming, thereby ensuring food security and promoting smallholder farmers, agroecological methods and indigenous food systems.

Acknowledgement The financial assistance provided to Mr. Amit Kumar (Award No. NSFDC/E-71131) by the University Grants Commission (UGC) under the Junior Research Fellowship (JRF) scheme is highly appreciated.

References

Ahman, M., Nilsson, L. J., & Johansson, B. (2017). Global climate policy and deep decarbonization of energy-intensive industries. *Climate Policy, 17*(5), 634–649. https://doi.org/10.1080/14693062.2016.1167009

Ashencaen, C. S., Parker, J., Man, Z., Garcia, S. A., & Sylvester, O. (2019). Sustainability, development and devastation: New encounters in indigenous dialogues. *Discover Society, 70*.

Babu, G., & Satya, S. (2022). Understanding the inherent interconnectedness and other salient characteristics of nature crucial for sustainability. *Environment Development and Sustainability, 26*(3), 1–13. https://doi.org/10.1007/s10668-022-02831-4

Biermann, F., Kanie, N., & Kim, R. E. (2017). Global governance by goal-setting: The novel approach of the UN Sustainable Development Goals. *Current Opinion in Environmental Sustainability, 26–27*, 26–31. https://doi.org/10.1016/j.cosust.2017.01.010

Cabernard, L., Pfister, S., & Hellweg, S. (2024). Biodiversity impacts of recent land-use change driven by increases in agri-food imports. *Nature Sustainability, 7*(11), 1512–1524. https://doi.org/10.1038/s41893-024-01433-4

Dechezlepretre, A., Glachant, M., & Mohnen, M. (2017). *Knowledge spillovers from clean and dirty technologies: A patent citation analysis* (p. 135). Grantham Research Institute on Climate Change and the Environment.

Dwivedi, P. P., & Sharma, D. K. (2025). Performance measures of sustainable development goals using SWI MCDM methods: A case of the Indian states. *International Transactions in Operational Research.* https://doi.org/10.1111/itor.13609

Dwivedi, S. L., Lammerts van Bueren, E. T., Ceccarelli, S., Grando, S., Upadhyaya, H. D., & Ortiz, R. (2017). Diversifying food systems in the pursuit of sustainable food production and healthy diets. *Trends in Plant Sciences, 22*(10), 842–856. https://doi.org/10.1016/j.tplants.2017.06.011

Easterly, W. (2012). How the millennium development goals are unfair to Africa. *World Development, 37*(1), 26–35. https://doi.org/10.1016/j.worlddev.2008.02.009

Estrada, A., Garber, P. A., Gouveia, S., Fernandez-Llamazares, A., Ascensao, F., Fuentes, A., Garnett, S. T., Shaffer, C., Bicca-Marques, J., Fa, J. E., Hockings, K., & Shanee, S. (2022). Global importance of Indigenous Peoples, their lands and knowledge systems for saving the world's primates from extinction. *Science Advances, 8*(32), 2927. https://doi.org/10.1126/sciadv.abn2927

Jaiyesimi, R. (2016). The challenge of implementing the sustainable development goals in Africa: The way forward. *African Journal of Reproduction Health, 20*(3), 13–18. https://doi.org/10.29063/ajrh2016/v20i3.1

Jakes, V. (2024). The role of traditional knowledge in sustainable development. *International Journal of Humanity and Social Sciences, 3*(2), 40–55. https://doi.org/10.47941/ijhss.2079

Jena, N., Rout, S., Mishra, S., & Kumar, S. (2025). Evaluation of quantitative ethnobotanical uses in Mayurbhanj district, Odisha, India. *Journal of Biodiversity and Conservation, 9*(1), 1–29. https://doi.org/10.5281/zenodo.14674405

Khajuria, A. K., Manhas, R. K., Kumar, H., & Bisht, N. S. (2021). Ethnobotanical study of traditionally used medicinal plants of Pauri district of Uttarakhand, India. *Journal of Ethnopharmacology, 276*, 114204. https://doi.org/10.1016/j.jep.2021.114204

Kishore, N., & Deswal, S. (2017). Analysis of air pollution in Indian cities: A literature review. *International Journal of Emerging Technology., 8*(1), 191–195.

Kumar, A., Kumar, S., Komal, Ramchiary, N., & Singh, P. (2021). Role of traditional ethnobotanical knowledge and indigenous communities in achieving sustainable development goals. *Sustainability, 13*(6), 3062. https://doi.org/10.3390/su13063062

Kurre, R., & Purena, R. (2024). Human impact on flora and fauna depletion. Futuristic Trends in Biotechnology. IIP series, Volume 3, Book 24, Part 1, Chapter 12.

Mathur, R., Katyal, R., Bhalla, V., Tanwar, L., Mago, P., & Gunwal, I. (2023). Women at the forefront of environmental conservation. *Current World Environment, 18*(2). https://doi.org/10.12944/CWE.18.2.22

McNeill, D., Nesheim, I., & Brouwer, F. (2012). *Land use policies for sustainable development: Exploring integrated assessment approaches*. Edward Elgar Publishing. https://doi.org/10.4337/9781781002476

Merwe, V. J., Wahid, S. M., Cesna, G. P., & Prabowo, D. A. (2024). Improving natural resource management through AI: Quantitative analysis using SmartPLS. *International Transactions on Artificial Intelligence, 2*(2), 135–142.

Newell, P. (2013). *Globalization and the environment: Capitalism, ecology and power*. John Wiley & Sons.

Pandey, S., Paul, P., & Ortiz, G. (2025). Sustainable development goals: A midpoint review of accomplishments and challenges. In *Encyclopedia of Social Work*. National Association of Social Workers. https://doi.org/10.1093/acrefore/9780199975839.013.934.

Ralte, L., Sailo, H., & Singh, Y. T. (2024). Ethnobotanical study of medicinal plants used by the indigenous community of the western region of Mizoram, India. *Journal of Ethnobiology and Ethnomedicine, 20*(1), 2. https://doi.org/10.1186/s13002-023-00642-z

Rastogi, M., Kolur, S. M., Burud, A., Sadineni, T., Sekhar, M., Kumar, R., & Rajput, A. (2024). Advancing water conservation techniques in agriculture for sustainable resource management: A review. *Journal of Geography, Environment and Earth Science International, 28*(3), 41–53. https://doi.org/10.9734/jgeesi/2024/v28i3755

Rehman, G., Hamayun, M., Rahman, A., Haseeb, M., Umar, M., Ali, S., & Pervaiz, R. (2021). Impacts of mining on local fauna of wildlife in District Mardan & District Mohmand Khyber Pakhtunkhwa Pakistan. *Brazilian Journal of Biology, 84*, e251733. https://doi.org/10.1590/1519-6984.251733

Roy, M., Boiral, O., & Lagace, D. (2001). Environmental commitment and manufacturing excellence: A comparative study within Canadian industry. *Journal of Business Strategy and the Environment, 10*(5), 257–268. https://doi.org/10.1002/bse.304

Sarmento, M., Durao, D., & Duarte, M. (2006). Analysis of companies environmental strategies for a green society. *Energy, 31*(13), 2333–2340. https://doi.org/10.1016/j.energy.2006.01.015

Senthil, G. A., Prabha, R., Divya, B., & Sathya, A. (2023). An empirical analysis and challenging era of block chain in green society. In *International Conference on Communication and Intelligent Systems Singapore* (p. 689). Springer Nature. https://doi.org/10.1007/978-981-99-2322-9_15

Sharma, S. K. (2021). Connection of economy and society with environment. *International Journal of Innovative Research in Engineering and Management, 8*(6), 285–289. https://doi.org/10.55524/ijirem.2021.8.6.58

Shivanna, K. R. (2022). Climate change and its impact on biodiversity and human welfare. *Proceeding of Indian National Science Academy, 88*(2), 160–171. https://doi.org/10.1007/s43538-022-00073-6

Shivanna, K. R., & Sanjappa, M. (2021). Conservation of endemic and threatened flowering plants: Challenges and priorities for India. *Journal of Indian Botanical Society, 101*, 269–290.

Soderholm, P. (2020). The green economy transition: The challenges of technological change for sustainability. *Sustainable Earth Reviews, 3*(1), 6. https://doi.org/10.1186/s42055-020-00029-y

Tornjanski, V., Knezevic, S., & Mircetic, V. (2023). Human-Centric Society: A holistic overview of sustainable and vibrant future of humanity and the world. In *Book of Abstracts of the IX International Scientific & Professional Conference MEFkon 2023–Innovation as an Initiator of the Development*, University Business Academy, Faculty of Applied Management, Economics and Finance, Serbia.

Tornjanski, V., Knezevic, S., Mirceti, C., Alzoubi, H. M., Juraev, A. D., & Verma, K. R. (2024). The role of Green Society in Society 5.0: Tango diamond in a collective intelligence (hybrid) ecosystem founded on human-centricity and sustainability. *Journal of Business and Social Sciences, 10*, 1–26. https://doi.org/10.61453/jobss.v2024no10

Ziankova, I., Lisichonak, A., & Yemialyanau, A., (2019). Development of inclusive society with the instruments of "green" economy and education. In *Environment. Technologies. Resources. Proceedings of the International Scientific and Practical Conference*, (1) (pp. 353–358). https://doi.org/10.17770/etr2019vol1.4057.

Index

A. Chakir et al. (eds.), *Green Society, Environmental Strategies and Sustainable Development*, https://doi.org/10.1007/978-3-031-94086-6

MIX
Papier aus verantwortungsvollen Quellen
Paper from responsible sources
FSC® C105338

FSC
www.fsc.org

If you have any concerns about our products,
you can contact us on
ProductSafety@springernature.com

In case Publisher is established outside the EU,
the EU authorized representative is:
Springer Nature Customer Service Center GmbH
Europaplatz 3, 69115 Heidelberg, Germany

Printed by Libri Plureos GmbH
in Hamburg, Germany